ERGONOMICS AND HUMAN FACTORS IN SAFETY MANAGEMENT

INDUSTRIAL AND SYSTEMS ENGINEERING SERIES

Series Editor
Waldemar Karwowski

PUBLISHED TITLES:

Ergonomics and Human Factors in Safety Management
Pedro Miguel Ferreira Martins Arezes and Paulo Victor Rodrigues de Carvalho

Manufacturing Productivity in China
Li Zheng, Simin Huang, and Zhihai Zhang

Supply Chain Management and Logistics: Innovative Strategies and Practical Solutions
Zhe Liang, Wanpracha Art Chaovalitwongse, and Leyuan Shi

Mobile Electronic Commerce: Foundations, Development, and Applications
June Wei

Managing Professional Service Delivery: 9 Rules for Success
Barry Mundt, Francis J. Smith, and Stephen D. Egan Jr.

Laser and Photonic Systems: Design and Integration
Shimon Y. Nof, Andrew M. Weiner, and Gary J. Cheng

Design and Construction of an RFID-enabled Infrastructure:
The Next Avatar of the Internet
Nagabhushana Prabhu

Cultural Factors in Systems Design: Decision Making and Action
Robert W. Proctor, Shimon Y. Nof, and Yuehwern Yih

Handbook of Healthcare Delivery Systems
Yuehwern Yih

ERGONOMICS AND HUMAN FACTORS IN SAFETY MANAGEMENT

EDITED BY

Pedro M. Arezes
Paulo Victor Rodrigues de Carvalho

CRC Press
Taylor & Francis Group
Boca Raton London New York

CRC Press is an imprint of the
Taylor & Francis Group, an **informa** business

CRC Press
Taylor & Francis Group
6000 Broken Sound Parkway NW, Suite 300
Boca Raton, FL 33487-2742

First issued in paperback 2019

© 2016 by Taylor & Francis Group, LLC
CRC Press is an imprint of Taylor & Francis Group, an Informa business

No claim to original U.S. Government works

ISBN-13: 978-1-4987-2756-3 (hbk)
ISBN-13: 978-0-367-87333-2 (pbk)

This book contains information obtained from authentic and highly regarded sources. Reasonable efforts have been made to publish reliable data and information, but the author and publisher cannot assume responsibility for the validity of all materials or the consequences of their use. The authors and publishers have attempted to trace the copyright holders of all material reproduced in this publication and apologize to copyright holders if permission to publish in this form has not been obtained. If any copyright material has not been acknowledged please write and let us know so we may rectify in any future reprint.

Except as permitted under U.S. Copyright Law, no part of this book may be reprinted, reproduced, transmitted, or utilized in any form by any electronic, mechanical, or other means, now known or hereafter invented, including photocopying, microfilming, and recording, or in any information storage or retrieval system, without written permission from the publishers.

For permission to photocopy or use material electronically from this work, please access www.copyright. com (http://www.copyright.com/) or contact the Copyright Clearance Center, Inc. (CCC), 222 Rosewood Drive, Danvers, MA 01923, 978-750-8400. CCC is a not-for-profit organization that provides licenses and registration for a variety of users. For organizations that have been granted a photocopy license by the CCC, a separate system of payment has been arranged.

Trademark Notice: Product or corporate names may be trademarks or registered trademarks, and are used only for identification and explanation without intent to infringe.

Library of Congress Cataloging-in-Publication Data

Names: Arezes, Pedro M., editor. | Carvalho, Paulo Victor Rodrigues de, editor.
Title: Ergonomics and human factors in safety management / [edited by] Pedro Miguel Ferreira Martins Arezes and Paulo Victor Rodrigues de Carvalho.
Description: Boca Raton : CRC Press, 2016. | Series: Industrial and systems engineering series | Includes bibliographical references.
Identifiers: LCCN 2016010604 | ISBN 9781498727563 (hard cover)
Subjects: LCSH: Industrial safety. | Human engineering. | Manufacturing processes--Human factors.
Classification: LCC T55 .E69 2016 | DDC 658.4/08--dc23
LC record available at https://lccn.loc.gov/2016010604

**Visit the Taylor & Francis Web site at
http://www.taylorandfrancis.com**

**and the CRC Press Web site at
http://www.crcpress.com**

Contents

SECTION I Occupation Safety

SECTION II   Safety and Human Factors in Training and Simulation

SECTION III   Models and Other Topics

Preface

This book is a compilation of contributions from invited authors organized in 18 chapters and grouped by three main topics. All of the authors were invited after their participation in the 2nd and 3rd International Conferences on Safety Management and Human Factors, which are affiliated with the International Conference on Applied Human Factors and Ergonomics.

This book has contributions from 60 authors from 11 countries, and it intends to cover specific aspects of safety and human factors management, ranging from case studies to the development of theoretical models.

The chapters are organized into three different topics, which will allow readers to clearly identify the main focus of each chapter.

The first section, comprised of the first seven chapters, is dedicated to occupational safety.

Chapter 1, from Carvalho and Melo describes the matrix-based technique used to perform occupational risk assessment. They claim that this approach has advantages in occupational risk assessments, namely, because it allies the advantages of both the quantitative and qualitative approaches and overcomes some of their limitations. In this chapter, Carvalho and Melo present a study to evaluate the reliability of the matrix-based approach.

Chapter 2, from Debnath et al., discusses regulatory, organizational, and operational issues in road construction safety in Australia. In their study, from the state of Queensland, Australia, they examine how well the tripartite (regulatory, organization, and operational) framework functions. The study identifies several factors influencing the translation of safety policies into practice, including the cost of safety measures in the context of competitive tendering, the lack of firm evidence of the effectiveness of safety measures, and pressures to minimize disruption to the traveling public.

The contribution of Gutiérrez and Sánchez, in Chapter 3, describes the development of an occupational health and safety management system for manufacturing companies in Mexico using factorial analysis. Their research, based on a survey conducted among 32 Mexican manufacturing companies, attempts to give clarity to Mexican manufacturing companies in the creation of a unique management system that covers occupational safety aspects and allows them to accomplish government as well as global clients' requirements.

In Chapter 4, from Rodrigues et al., the authors present a study developed within the Portuguese furniture industrial sector, in which they characterize the safety performance of the sector, namely, by analyzing the corresponding occupational accidents and identifying the key unsafe conditions that can originate these accidents. Using a sample of 14 Portuguese companies of this sector, they also analyzed the applicability of the Safety Climate in Wood Industries as a tool to monitor companies' safety performance and assess the safety climate within those companies. Among other results, they found a strong positive linear correlation between safety climate scores and the companies' safety performance.

Väyrynen et al. present a review about health, safety, environment, and quality (HSEQ) management in Chapter 5. They describe a model used for HSEQ assessment that has been developed and applied within many Finnish company networks. They also focus on small- and medium-sized enterprises (SMEs) and their work systems with outcomes, their HSEQ assessment results, and the concepts of sustainability and safety culture. The authors suggest that such a model can promote productivity and conformity within a work system with more desired outcomes.

In Chapter 6, Vidal et al. develop an analysis of work accidents based on the ergonomic point of view. They present the methodological framework for this analysis, trying also to show its application. They discuss some contemporary visions about work accidents and attempt to cross them with some modern trends of approaches used in ergonomics. They finish by presenting the possible impact of their approach on practice in accident prevention.

Chapter 7, the last chapter of Section I, by Waefler et al., describes a project safety management information system (S-MIS), which aims to develop an information system that supports decisions in safety management. According to the authors, the S-MIS project attempts to provide industry with reliable proactive indicators, as well as a support for decision making in safety management. Based on a pilot project, the S-MIS process has been analyzed for its appropriateness to provide decision makers in safety management with a better quantitative information base. In the authors' opinion, the process still needs to be optimized.

Section II is dedicated to the specific topic of safety and human factors in training and simulation and encompasses four different chapters.

Chapter 8, from Maciel et al., aims to analyze tasks and electrical system operators' potential errors to propose corrective strategies and improvements in the design process and operating systems using hierarchical task analysis and the systematic human error reduction and prediction approach. The results revealed that the method employed is capable of distinguishing the main operator tasks, according to their decision making, to maintain proper system operation.

In Chapter 9, Nazir et al. compare the results of convectional training methods and those based on immersive virtual environments employed in process industries. Two groups of participants are trained according to either a conventional training approach or an immersive virtual environment. The performance of operators is measured in real time by means of suitable and well-defined key performance indicators. The results show that participants trained with immersive virtual environments react significantly more quickly and accurately to a simulated accident scenario than those trained with a conventional approach.

Rocha et al., in Chapter 10, discuss the importance of knowledge management for counterbalancing the process of loss of skills at work, as the social actor responsible for creating the procedures is far from the reality experienced in the field, causing safety problems at work. They argue that the disconnection between what is written and what is real is the absence of spaces of discussion at work that allow the sharing of knowledge or the possibility to externalize strategies and actions that can be used when managing the difficulties in the field.

In Chapter 11, Wagner et al. describe an experimental study with 23 untrained volunteers where they analyzed how the occurrence of an evacuation assistant influences the behavior and the emotional state of evacuees while acting in different conflict situations. Their results give important indications to improve evacuation situations. They have also developed an agent-based simulation model to allow an evacuation, through simulating the cognitive processes of agents in the simulation environment. The authors concluded that the model was capable of reproducing empirically observed human behavior, and it enables simulation scenarios with a high degree of realism.

In Chapter 12, Inaba et al. finish Section II describing an interactive educational game to learn about human error. The aim is the development of a serious game in which individuals can effectively learn the mechanisms of a slip. Using the game, people become immersed in situations that allow them to react to risks and learn about risks without exposing themselves to real danger.

Finally, Section III is dedicated to safety and human factors models and related topics, as well as some other mixed topics, as described briefly in the following paragraphs.

In Chapter 13, Jansen et al. discuss how our daily and work lives are filled with interruptions and transitions from one task to another, resulting in a fragmented workflow. He proposes the transitional journey maps, creating workflow visualizations as a way to produce reflections about interruptions in work activities. He approached two organizations with the request to study human information processing activities at work, the Dutch National and the European Space Operations Centre.

Klockner and Toft talk about the missing links in system safety management in Chapter 14. Their research starts with the premise that organizations have no memory and accidents recur, and that organizations and safety regulators often identify what appear to be reoccurring patterns and themes of the contributing factors identified by safety occurrence investigations. The ongoing frustration is how lessons can be learned from what has already occurred and how that information can be used to identify areas and aspects of organizational safety management systems that are negatively contributing to safety occurrences.

In Chapter 15, Murata uses the Bayesian estimation method to predict the risk of driving drowsiness. The aim of this study was to predict in advance drivers' drowsy states with a high risk of encountering a traffic accident and prevent drivers from continuing to drive under drowsy states. His results indicate that the proposed method could predict in advance the point in time with a high risk of a virtual crash before the point in time of a virtual accident when the participant would surely have encountered a serious accident with a high probability.

Schlacht, in Chapter 16, tries to inspire specialists to use the space missions design, system, and simulation as a model for realizing possible innovation of safety procedures in regular critical and dangerous situations. Assuming the safety-critical systems and space environments share many of the problems regarding the support of human life, the author proposes that space missions can be used as a model to learn how to increase safety and improve user–system interaction.

Chapter 17, from Tappura and Nenonen, proposes a scheme for categorizing effective safety leadership facets, considering that this concept is a key factor for promoting safety performance in organizations. The authors based their work on a literature review, as well as on interviews carried out in a Finnish organization. They concluded that both the transactional and transformational facets of safety leadership should be exercised and developed.

The last chapter of the book, authored by Zixian Yang and Therma Wai Chun Cheung, presents a work on the topic of upper limb repetitive strain injury (RSI) in women involved in housework. The authors have analyzed this problem and confirmed that female homemakers who need to carry out unpaid housework make up a major proportion of patients with upper limb RSI referred to an occupational therapy outpatient clinic in Singapore. According to the authors, their findings provide a logical explanation for the high prevalence of upper limb RSI in women.

On behalf of the entire team that was involved in the development of this book, we are very proud to provide a very broad scope of contributions, which has included some case studies, examples, solutions, models, and challenges presented and proposed here by a broad group of authors from a wide array of disciplines and countries. We greatly enjoyed working with the contributors to this book on the topic of Human Factors in Safety Management. We also want thank the contributors for sharing their findings and insights, as well as the reviewers of the initial versions of these chapters for their essential contribution. We hope that the works presented here can be an inspiration for translating research into useful actions and, ultimately, make a relevant and tangible contribution to the effective improvement regarding the safety of our daily and work settings.

Editors

Pedro Arezes is a full professor of human factors engineering at the School of Engineering of the University of Minho in Guimarães, Portugal. He is also a visiting scholar at the Massachusetts Institute of Technology and Harvard University, University, in Cambridge (MA), USA. At the University of Minho, he coordinates the human engineering research group, and his research interests are in the domains of safety, human factors engineering, and ergonomics. Pedro is also the director of the PhD program, "Leaders for Technical Industries" within the MIT Portugal Program. He has supervised more than 60 MSc theses for several universities and 10 completed PhD theses. He was also the host supervisor of some postdoctorate projects with colleagues from countries such as Brazil, Poland, and Turkey. Dr. Arezes has published in the domains of human factors and ergonomics, safety, and occupational hygiene, as the author or coauthor of more than 50 papers in international peer-reviewed journals, as well as the author or editor of more than 40 books published internationally. He is also the author or coauthor of more than 300 papers published in international conference proceedings with peer review. Dr. Arezes has collaborated, as a member of the editorial board or a reviewer, with more than 15 well-recognized international journals. He is a member of the scientific and organization committees of several international events related to the topics of occupational safety and ergonomics, including being the chair of the 2015 edition of the WorkingOnSafety (WOS) conference and co-chair of the International Conference on Safety Management and Human Factors, an affiliated event of the International Conference on Applied Human Factors and Ergonomics.

Paulo Victor Rodrigues de Carvalho is a researcher at the Nuclear Engineering Institute and a full professor of ergonomics and resilience engineering at the Federal University of Rio de Janeiro in Brazil. He coordinates the Complex Systems Technology research group of the Brazilian National Research Council CNPq, and his research interests are in the domains of safety, human factors engineering and ergonomics, and resilience engineering. He has worked and published in several domains of industrial safety, disaster management, and safety, human factors, and ergonomics, such as safety and accident analysis, resilience modeling, information technology for disaster prevention and response, and naturalistic decision making. Dr. de Carvalho was the supervisor of MSc and DSc theses in the postgraduate programs of informatics, industrial, and environmental engineering at the Federal University of Rio de Janeiro. He has published in the domains of human factors, ergonomics, and safety and is author or coauthor of more than 50 papers published in international peer-reviewed journals, as well as the author of more than 100 papers published in international conference proceedings with peer reviews. Dr. de Carvalho has collaborated, as a member of the editorial board or a reviewer, with more than 10 well-recognized international journals. He is a member of the scientific and organization committees of several international events related to the topics of occupational safety and ergonomics, including co-chair of the International Conference on Safety Management and Human Factors, an affiliated event of the International Conference on Applied Human Factors and Ergonomics.

Contributors

Rodrigo Arcuri
Complex Systems Ergonomics Research
  Unit
Alberto Luiz Coimbra Engineering
  Research and Post-Graduate Institute
Federal University of Rio de Janeiro
Rio de Janeiro, Brazil

Tamara Banks
Centre for Accident Research and Road
  Safety–Queensland
Queensland University of Technology
Queensland, Brisbane Australia

Ana Karla Baptista
Interaction Ergonomia
Belo Horizonte, Brazil

Herbert Biggs
Centre for Accident Research and Road
  Safety–Queensland
Queensland University of Technology
Brisbane, Australia

Simon Binz
School of Applied Psychology
University of Applied Sciences and Arts
  Northwestern Switzerland
Olten, Switzerland

Ross Blackman
Centre for Accident Research and Road
  Safety–Queensland
Queensland University of Technology
Brisbane, Australia

Renato José Bonfatti
Center for Studies on Workers' Health
  and Human Ecology
National School of Public Health
Oswaldo Cruz Foundation
Rio de Janeiro, Brazil

Vamberto Lima Cabral
Companhia Energética do Estado do
  Ceará
Fortaleza, Brazil

Klendson Marques Canuto
Companhia Energética do Estado do
  Ceará
Fortaleza, Brazil

Filipa Carvalho
Research Centre CIAUD
University of Lisbon
Lisboa, Portugal

Therma Cheung Wai Chun
Occupational Therapy Department
Singapore General Hospital
Singapore

Ashim Kumar Debnath
Centre for Accident Research and Road
  Safety–Queensland Queensland
  University of Technology
Brisbane, Australia

Huib de Ridder
Faculty of Industrial Design
  Engineering
Delft University of Technology
Delft, the Netherlands

Nathan Dovan
Centre for Accident Research and Road
  Safety–Queensland Queensland
  University of Technology
Brisbane, Australia

Vitor Figueiredo
Federal University of Itajubá
Itabira, Brazil

Katrin Fischer
School of Applied Psychology
University of Applied Sciences and Arts
Northwestern Switzerland
Olten, Switzerland

Alberto Gallace
Department of Psychology
University of Milano-Bicocca
Milan, Italy

Rosemary Cavalcante Gonçalves
University of Fortaleza
Fortaleza, Brazil

Luis Cuautle Gutiérrez
Industrial and Automotive Engineering
   Faculty
Puebla State Popular Autonomous
   University
Puebla City, Mexico

Shigeru Haga
Department of Psychology College of
   Contemporary Psychology
Rikkyo University
Saitama, Japan

Narelle Haworth
Centre for Accident Research and Road
   Safety–Queensland
Queensland University of
   Technology
Brisbane, Australia

Midori Inaba
Safety Research Laboratory Research
   and Development Center of JR East
   Group
East Japan Railway Company Saitama,
   Japan

Reinier J. Jansen
Faculty of Industrial Design
   Engineering
Delft University of Technology
Delft, the Netherlands

Alessandro Jatobá
Center for Studies on Workers' Health
   and Human Ecology
National School of Public Health
Oswaldo Cruz Foundation
Rio de Janeiro, Brazil

Henri Jounila
Work Science
Industrial Engineering and Management
University of Oulu
Oulu, Finland

Konrad Wolfgang Kallus
Department of Psychology
University of Graz
Graz, Austria

Karen Klockner
Central Queensland University
Brisbane, Australia

Ken Kusukami
Safety Research Laboratory Research and
   Development Center of JR East Group
East Japan Railway Company Saitama,
   Japan

Stefan Ladstätter
Joanneum Research Institute, Digital
Institute for Information and
   Communication Technologies
Graz, Austria

Jukka Latva-Ranta
Work Science
Industrial Engineering and
   Management
University of Oulu
Oulu, Finland

Celina P. Leão
R&D Centro Algoritmi
School of Engineering of the University
   of Minho
Guimaraães, Portugal

Regina Heloisa Maciel
University of Fortaleza
Fortaleza, Brazil

Luciana Maria Maia
University of Fortaleza
Fortaleza, Brazil

Davide Manca
PSE-Lab, Process Systems Engineering
    Laboratory
Dipartimento id Chimica,
Materiali e Ingegneria Chimica "Giulio
    Natta"
Polytechnic University of Milan
Milan, Italy

Rui B. Melo
Research Centre CIAUD
University of Lisbon
Lisboa, Portugal

Atsuo Murata
Department of Intelligent Mechanical
    Systems
Graduate School of Natural Science and
    Technology
Okayama University
Okayama, Japan

Salman Nazir
PSE-Lab
Process Systems Engineering
    Laboratory
Dipartimento di Chimica, Materiali e
    Ingegneria Chimica "Giulio Natta"
Polytechnic University of Milan
Milan, Italy

and

Training and Assessment Research
    Group
Department of Maritime Technology
    and Innovation
University College of Southeast Norway
Horten, Norway

Noora Nenonen
Center for Safety Management and
    Engineering
Department of Industrial
    Management
Tampere University of Technology
Tampere, Finland

Norah J. Neuhuber
Department of Psychology
University of Graz
Graz, Austria

Kjell Ivar Øvergård
Training and Assessment Research
    Group
Department of Maritime Technology
    and Innovation
University College of Southeast
    Norway
Horten, Norway

Lucas Paletta
Joanneum Research Health Institute
    for Biomedical and Health
    Sciences
Graz, Austria

Sami Pikkarainen
Work Science
Industrial Engineering and
    Management
University of Oulu
Oulu, Finland

Martin Pszeida
Joanneum Research Institute,
    Digital
Institute for Information and
    Communication Technologies
Graz, Austria

Raoni Rocha
Federal University of Itajubá
Itabira, Brazil

Matilde A. Rodrigues
Department of Environmental
    Health
Research Centre on Health and
    Environment
School of Allied Health Technology
    of the Institute Polytechnic of
    Porto
Vila Nova de Gaia, Portugal

and

R&D Centro Algoritmi
School of Engineering of the University
    of Minho
Guimaraães, Portugal

Miguel Angel Avila Sánchez
Universidad Popular Autónoma del
    Estado de Puebla
Puebla City, Mexico

Irene Lia Schlacht
Extreme-Design Research Group
Design Department
Polytechnic University of Milan
Milan, Italy

Helmut Schrom-Feiertag
Austrian Institute of Technology
Vienna, Austria

Michael Schwarz
Joanneum Research Institute, Digital
Institute for Information and
    Communication Technologies
Graz, Austria

Ikuo Shirai
Safety Research Laboratory Research and
    Development Center of JR East Group
East Japan Railway Company Saitama,
    Japan

Martin Stubenschrott
Austrian Institute of Technology
Vienna, Austria

Sari Tappura
Center for Safety Management and
    Engineering
Department of Industrial Management
Tampere University of Technology
Tampere, Finland

Yvonne Toft
Central Queensland University
Rockhampton, Australia

René van Egmond
Faculty of Industrial Design Engineering
Delft University of Technology
Delft, the Netherlands

Seppo Väyrynen
Work Science
Industrial Engineering and Management
University of Oulu
Oulu, Finland

Mario Cesar R. Vidal
Complex Systems Ergonomics Research
    Unit
Alberto Luiz Coimbra Engineering
    Research and Post-Graduate Institute
Federal University of Rio de Janeiro
Rio de Janeiro, Brazil

Toni Waefler
University of Applied Sciences and Arts
    Northwestern Switzerland
School of Applied Psychology
Olten, Switzerland

Verena Wagner
Department of Psychology
University of Graz
Graz, Austria

Kaj von Weissenberg
Inspecta Sertifiointi Oy
Helsinki, Finland

Zixian Yang
Occupational Therapy Department
Singapore General Hospital
Singapore

Section I

Occupation Safety

1 Reliability in Occupational Risk Assessment

Stability and Reproducibility Evaluation When Using a Matrix-Based Approach

Filipa Carvalho and Rui B. Melo

CONTENTS

1.1 INTRODUCTION

Every year, millions of people in the European Union are injured at work or have their health seriously harmed in the workplace. With this in mind, we can understand why risk assessment is so important and is considered the key to healthy workplaces. Risk assessment is a dynamic process that allows enterprises and organizations to put in place a proactive policy of managing workplace risks. Risk assessment is the cornerstone of the European approach to prevent occupational accidents and ill health (EU-OSHA, 2009).

The European Agency for Safety and Health at Work (EU-OSHA, 2008) states that risk assessment is the basis for effective management of occupational safety and health (OSH) and the key to reduce both accidents at work and occupational diseases. When properly performed, it can improve safety and health at work and, in general, the performance of companies.

The risk matrix method is probably the most common approach to evaluation in risk analysis. It is a semiquantitative method where probabilities and consequences are categorized, instead of using numerical values (Harms-Ringdahl, 2013).

According to Carvalho and Melo (2015), risk matrices present several advantages, including being generalist, user-friendly, and easy to apply. Nevertheless, they emphasize that we cannot disregard the existing gap in terms of reliability of these applications. In other words, to be useful, these methods must prove to be reliable. To highlight the relevance of reliability, we can recall Kaplan and Goldsen's (Krippendorff, 2004, p. 211): "The importance of reliability rests on the assurance it provides that data are obtained independent of the measuring event, instrument or person. Reliable data, by definition, are data that remain constant throughout variations in the measuring process."

There are very few studies reflecting a concern about risk assessment's outputs when different methods are used, particularly methods relying on risk matrices. In Portugal, the few known studies reinforce the need for further scientific research in this area to ensure the reliability of risk assessments (Branco et al., 2007; Carvalho, 2007).

This chapter reports the results of a study on the reliability of matrix-based methods. This study involved a comparative analysis of four matrices, which were used to estimate and assess six risks identified in two tasks accomplished to produce car airbags.

1.2 RISK ASSESSMENT

Workplaces, work practices, and processes comprise different types of hazards to which many workers are exposed on a daily basis. This exposure frequently results either in workplace accidents or occupational diseases that necessarily must be prevented. There is absolutely no doubt that risk assessment is a fundamental step in the occupational risk management process contributing to the reduction of both types of consequences.

Although the process of risk assessment has no strict established rules to be followed, it should be carried out in a logical and structured manner. The following steps should be observed: identification of the hazards, identification of the possible consequences, estimation of the likelihood of possible consequences, estimation of the possible consequences' severity, estimation of the risk magnitude (i.e., how big is it?), evaluation of the significance of the risk (e.g., is it acceptable?), and recording of the findings. The results will inform on the level and relevance of risks, as well as if the existing control measures are adequate or additional preventive and protective actions are needed.

Risk assessment techniques range from a simple qualitative approach to a detailed quantitative assessment, and each of them presents advantages and inconveniences. While the former approach lacks of objectivity and does not allow cost–benefit analysis, the latter is rather complex and time-consuming and requires well-trained analysts to be applied.

1.2.1 MATRIX-BASED APPROACH

The use of matrices is probably the most common approach to evaluation in risk analysis. It is a semiquantitative method where probabilities and consequences are categorized (Harms-Ringdahl, 2013). The basis for the risk estimate is usually qualitative, although numbers can be used for labeling either the consequences or the frequencies, or both, expressing the hierarchy in both scales. Therefore, these categories are defined either numerically or by a description.

The simplest matrices interpret risk as the combination of consequence (severity) and likelihood (frequency). Therefore, both variables must be coded according to a scale. In this approach, risk level is obtained by either combining the used variables in a preestablished manner or multiplying the attributed values. Once risk level is computed, it is compared to the risk index scale to prioritize actions in terms of preventive or protective countermeasures. This kind of approach is considered very important in occupational risk assessments, because it allies the advantages of both the quantitative and qualitative approaches and overcomes some of their limitations. Plus, it is very effective at promoting audience participation during risk management programs.

Carvalho and Melo (2007) state that this kind of approach has proven to be, in most cases, the only available technique and the most suited to carry out this task. These last evidences assume particular relevance when we think of small and medium enterprises. Most of them do not have the resources to assess risk quantitatively: there is not a permanent OSH practitioner available to perform a risk assessment on a regular basis, and in some cases, it is the employer himself who makes this first approach to risk management, even without experience or adequate knowledge to do so.

Despite the benefits referred, the validity and reliability of risk matrices and other evaluation techniques have not been studied enough (Harms-Ringdahl, 2013).

1.2.2 RELIABILITY

In general, reliability refers to the extent to which a test, experiment, or measuring procedure gives the same results on repeated trials or applications (Olsen, 2013). Matrix-based risk assessment methods rely on coding categories, which are considered to be reliable if separate coding attempts end up with the content coded in a similar way.

Reliability and agreement are still generally broad terms and require further definition to ensure their correct application to measurements within the OSH risk assessment domain. *Reliability* refers to a proportional consistency of variance among coders and is correlational in nature, while *agreement* refers to the interchangeability among coders, and addresses the extent to which coders make essentially the same coding (Olsen, 2013). Therefore, coders can be reliable in the coding process when the range of codes assigned by one coder is consistent with the range of codes assigned by another coder, even if the codes assigned to each individual event do not meet with consensus. There is agreement among coders when codes assigned to each individual event are the same between coders (i.e., consensus is attained on the codes assigned to each individual event).

Then, high reliability can be obtained even when there is low agreement and the opposite is true.

According to Krippendorff (2004), there are three types of reliability: stability, reproducibility, and accuracy. These are distinguished not by how agreement is measured, but by the way the reliability data are obtained (Table 1.1). Without information about the circumstances under which the data for reliability assessments have been generated, agreement measures remain uninterpretable.

Stability is the degree to which a process is unchanging over time. It is measured as the extent to which a measuring or coding procedure yields the same results on repeated trials. The data for such assessments are created under test–retest conditions; that is, one observer rereads, recategorizes, or reanalyzes the same text, usually after some time has elapsed, or the same measuring device is repeatedly applied to one set of objects. Under test–retest conditions, unreliability is manifest in variations in the performance of an observer or measuring device. There is stability when one person is consistent with himself or herself; for example, coding categories present intracoder reliability if a coder can categorize the same content, on a later occasion, similarly to how he or she coded it previously.

Reproducibility is the degree to which a process can be replicated by different analysts working under varying conditions, at different locations, or using different

TABLE 1.1

Types of Reliability

Reliability	Designs	Causes of Disagreement	Strength
Stability	Test–retest	Intracoder inconsistencies	Weakest
Reproducibility	Test–test	Intracoder inconsistencies + intercoder disagreements	Medium
Accuracy	Test–standard	Intracoder inconsistencies + intercoder disagreements + deviations from a standard	Strongest

Source: Krippendorff, K., *Content Analysis: An Introduction to Its Methodology* (2nd ed.), Thousand Oaks, CA: Sage, 2004.

but functionally equivalent measuring instruments. Demonstrating reproducibility requires reliability data that are obtained under test–test conditions; for example, two or more individuals, working independent of each other, apply the same recording instructions to the same units of analysis. There is reproducibility when two or more persons are consistent with each other; for example, coding categories have intercoder reliability if one coder can categorize a set of content similarly to how a second or more coders categorize it.

Accuracy is the degree to which a process conforms to its specifications and yields what it is designed to yield. There is accuracy when one, two, or more persons are consistent with a standard value, which is taken to be correct. This is the strongest reliability test available, but the most difficult to accomplish. To establish accuracy, analysts must obtain data under test–standard conditions (they must compare the performance of one or more data-making procedures with the performance of a procedure that is taken to be correct).

1.3 OBJECTIVES OF THE STUDY

This study involved a comparative analysis of different matrices, which were used to estimate and assess six risks identified in two tasks accomplished to produce car airbags. We have assessed intermethod, intercoder (reproducibility), and intracoder (stability) reliability of four matrices applied within the OSH scope. As the risk-level estimation depends on the intermediate variables used by each matrix, both types of variables were analyzed.

In synthesis, this study was performed to pursue three main objectives, which are described in Figure 1.1:

- Intermethod reliability
- Intercoder reliability or reproducibility
- Intracoder reliability or stability

1.4 METHODOLOGY

1.4.1 STAGES OF THE STUDY AND PROCEDURES

This study presents a structure similar to any risk assessment as part of the risk management process (BSI, 2004; ISO/IEC, 1999; ISO, 2009; Suddle, 2009; van Duijne et al., 2008). Therefore, it comprises four fundamental stages: characterization of work situations, hazard identification, risk estimation, and risk evaluation (Figure 1.2).

The first stage, *characterization of work situations*, includes the characterization of both the operators and the company, followed by a task analysis, for example, task identification and characterization in terms of prescribed objectives, as well as in terms of executing conditions.

The second stage, *hazard identification*, integrates a list of identified hazards, potential risks, and eventual consequences. In this stage, we also identify the potentially exposed people.

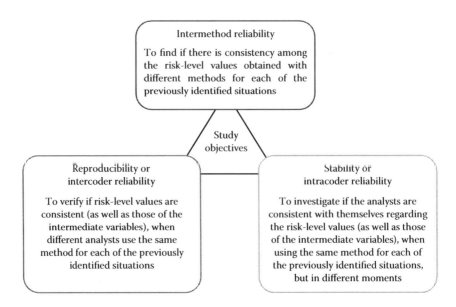

FIGURE 1.1   Identification and description of the three objectives of the study.

The third stage, *risk estimation*, includes risk characterization in terms of the variables required by the four tested matrices, namely, *likelihood* (L) and *severity* (G), and *risk magnitude* (R) estimation.

In the fourth stage, *risk evaluation*, we identify the risk level and, consequently, set up a hierarchy of intervention needs. Risk acceptability was established by the comparison between *risk magnitude* (R), obtained in the former stage, and a *risk index*, proposed by each matrix. A synthesis of these four stages of the study is represented in Figure 1.3.

1.4.2   Sample: Analyzed Tasks and Identified Risks

A previous inspection to a car airbag production unit allowed us to identify six risky work situations in two tasks, which were fully described and illustrated with pictures and videos. The two tasks were *fabric cutting press* and *turning and folding airbags*.

Table 1.2 shows the relationship between the nature of the assessed risks, the respective task or situation, and the adopted codes to identify the situation being analyzed. For the characterization of work situations, additional data were collected: noise exposure levels, lighting conditions, and the risk of developing musculoskeletal disorders, whenever relevant.

1.4.3   Data Collection

For data collection, we used different methods, tools, and equipment, according to the specificity of each stage of the study. Although this study focuses on the results obtained in stages 3 and 4, a global overview is provided.

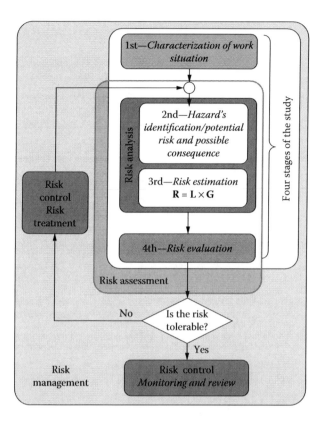

FIGURE 1.2 Flowchart illustrating the four stages of the study, as a part of the risk management process. (Adapted from BSI, BS8800—Occupational health and safety management systems—Guide, London: BSI, 2004; ISO/IEC, Guide 51—Safety aspects—Guidelines for their inclusion in standards [2nd ed.], Geneva: ISO/IEC, 1999; ISO, ISO 31000—Risk management—principles and guidelines [1st ed.], Geneva: ISO, 2009; Suddle, S., *Safety Science*, 47(5), 668–679, 2009; Van Duijne, F. H., et al., *Safety Science*, 46(2), 245–254, 2008.)

FIGURE 1.3 Synthesis of the four stages of the study.

In the first two stages, data collection relied on free and systematized observations and made use of video recording, documental research, analysis grids, and a questionnaire specifically developed for this purpose. A single analyst conducted these stages and interviewed the workers based on the former questionnaire. We tried to collect and analyze useful data in order to offer the analysts a complete

TABLE 1.2

Relationship between the Nature of the Assessed Risks, the Respective Task or Situation, and the Adopted Codes to Identify the Situation Being Analyzed

Task	Nature of the Assessed Risk	Code
	Slipping, tripping, and falling	TA1
A	Noise exposure	TA2
	Objects falling on the feet	TA3
	Mechanical contact—crushing	TB2
B	Physical overload	TB4
	Hit by a forklift truck	TB6

characterization of the analyzed work situations. In this way, all analysts could perform stages 3 and 4—risk estimation and risk evaluation—relying on the same exact information. For the second stage, a worksheet was created taking into account the identified hazards, the potential risks, and the possible consequences. An online questionnaire was developed for the third stage.

To provide a global overview of the study, we present a summary of the used methods and tools and equipment, their purposes, and some of their particular characteristics.

1.4.3.1 Documental Research

Hazards associated with work activities can be present as a result of any or a combination of the following: substances, machinery and processes, work organization, tasks, procedures, people, and circumstances in which the activities take place, including the physical aspects of the plant or premises (Gadd et al., 2003). The characterization of work situations began with the analysis of the information included in relevant sources, such as legislation and standards with relevant impact on this particular activity sector; internal workplace accident reports; European data on workplace accidents within this activity sector; environmental risk assessment reports; work procedures; occupational disease reports; internal ill health and incident data; and tools, equipment, and materials characteristics.

These analyses resulted in relevant information to be considered in both the questionnaire and the analysis grid specifically developed for characterization of the workstations. At this level, the analyst focused on the sociodemographic and organizational aspects, as well as on the possible causes of workplace accidents.

1.4.3.2 Tools

In order to get better-organized information about the work and the workplace, we developed an *analysis grid*, which was used throughout the systematized observations. This tool allowed gathering all relevant information in a unique document, organized by topics that represent the principal items influencing the working conditions. The following list summarizes the main topics considered: (1) organizational conditions; (2) equipment, machinery, and tools; (3) manual materials handling; (4) postures and related issues; (5) personal protective equipment (PPE); (6) building characteristics; (7) electricity hazards; and (8) complementary information used with the quick exposure check (QEC) tool.

Considering that the participation of the workers is seen as a useful way to make a risk analysis, because they are the ones that know how the work is actually performed, we decided to develop a *first questionnaire* to collect data on how safely or risky the workers work. It was organized in the following three parts:

- Part A: Focused on the operator's characterization, which involved indicators such as age, gender, experience, safety knowledge, life habits, and health complaints (identification of body regions, pain and discomfort levels, etc.).
- Part B: Concerned with the operator's perception of his or her working conditions. It involved questions about PPE characteristics, equipment use, and characteristics of hand tools.
- Part C: Involved a set of questions to collect the information required for the QEC tool (a method to assess musculoskeletal disorders risk).

Some of the questions we included in the questionnaire were:

- Have you had OSH training? If yes, in which domains?
- Do you usually feel pain or physical discomfort? If yes, the workers should indicate the affected regions on the body map figure available in the questionnaire.
- Which postures do you adopt more often to get the job done?
- How do you evaluate the available PPE?
- How do you evaluate the available tools?
- How do you evaluate the working environment?

Whenever the workers were required to evaluate an item about their working conditions, a 5-point scale was available.

The QEC observational tool was integrated into our tools (topic 8 of the analysis grid and Part C of the questionnaire) because it was our intention to compare the results with those obtained with the four matrices applied. However, we do not report this part of the study here.

To accomplish the second stage, namely, identify hazards, the potential risks, and the possible consequences, a *worksheet* was developed (Figure 1.4). It presents nine fields: workstation components, description of the situation or task, risk scenarios, pictures to illustrate the situations, identified hazards, associated risks, possible consequences, safety measures, and individual susceptibility. To complete this stage, this sheet was filled out with the information collected in the previous stage. This procedure was very useful because it allowed us to gather all information in a unique document.

As we said previously, a second questionnaire was developed to perform the third stage. It is an online questionnaire, named QuePOLPER, which was developed with LymeSurvey® software. With this online questionnaire, the analysts could perform

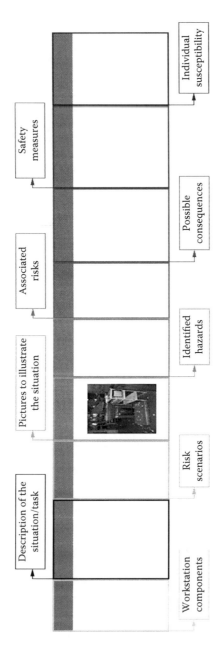

FIGURE 1.4 Illustration of the worksheet developed for the second stage of the study.

risk estimation from wherever and whenever they considered it appropriate. The QuePOLPER was organized in three parts:

- The first part presented the rules and conditions to take part in the study.
- The second part described the tasks under analysis and the situations to be assessed resorting to the four matrices. All variables estimated by each matrix were well described so that the analysts would just select the option they considered adequate.
- The third part addressed the analysts' characteristics, namely, age, gender, academic background, OSH professional experience, knowledge about the matrix-based approach, perception about the ease or difficulty when using this kind of approach, and the analysts' perception of matrices' reliability.

With this questionnaire, we ensured that all analysts would possess the same information about the situation under analysis to estimate the variables involved in each method (likelihood, severity, and so on).

The QuePOLPER included 42 questions, of which 6 depended on previous answers and the estimated time to fill it out was around 30 minutes.

Finally, the Statistical Package for the Social Sciences (SPSS© version 20) was used for the four stages. This tool was used to automatically generate, for all assessments, the main studied variable—risk level.

Considering that we intend to accomplish an intracoder reliability assessment, the two last stages were repeated with a lag time of 5 months.

1.4.3.3 Equipment

Video recording was the chosen technique to collect images related to work activity, and previous verbal consent of the workers involved was obtained. A digital camera SONY Handycam, HDR-SR10 model, was used.

Noise, light, and thermal variables were measured to provide a better characterization of the workplace. Noise was measured with a Bruel & Kjaer sound meter, 2260 model. The device was carefully placed near the operator's ear and was subjected to verification in the workplace before each series of measurements. Both continuous A-weighted sound pressure level (dB [A]) and maximum peak level (dB [C]) were measured. The illuminance level was assessed with a digital Krochmann lux meter, 106E model. The device was strategically put on the surface of the workstations. Dry and wet air temperatures were assessed with a THIES sling psychrometer, 450 model. Air humidity was computed from these two variables.

1.4.4 Participants

To complete the last two stages of a risk assessment (risk estimation and risk evaluation), 81 analysts were invited. All analysts were familiarized with the matrices used (either theoretically or in practice).

The selection criteria to include them in the study comprised their academic background, OSH practice experience, and expertise in using this type of method.

TABLE 1.3

Analysts' Groups, Respective Dimension (N) in Each Round of the Study, and Associated Code

	Code	Sample Group	N First Round	N Second Round
Academic background	SA_A1	Ergonomists	19	16
	SA_A2	OSH practitioners	5	5
	SA_A3	Engineers	14	12
OSH practice experience	SA_B1	With practical experience	28	23
	SA_B2	With no practical experience	16	16
Expertise	SA_C1	Experts	15	2
	SA_C2	Nonexperts	29	27

With these specifications, we organized the analysts in seven groups as presented in Table 1.3.

Only 44 (26 women and 18 men) out of the 81 invited analysts agreed to participate in the first risk assessment round (response rate 54%). Five months later, in the second risk assessment round, only 39 analysts (23 women and 16 men) responded in good time to integrate the intracoder reliability assessment.

1.4.5 Used Matrices

A set of four different matrix-based methods were selected from those described in international references and those used in particular organizations. These four matrices can be divided into two categories (Carvalho and Melo, 2007):

- Simple matrix methods (MMS 3 × 3 and BS8800), which resort to the use of two single variables (likelihood/frequency and severity/consequence) to compute risk magnitude.
- Complex matrix methods (MMCP and WTF), which rely on three or more variables. In addition to the previous variables, we can find exposure factor, procedures and safety conditions, or number of persons exposed or affected.

The selection of these particular methods was based on the following criteria: all methods should have a risk index scale with five levels, two methods belonging to each of the above-described categories should be included, risk magnitude (R) should result from a preestablished combination of the intermediate variables (e.g., likelihood and severity), and use of the same label to identify variables taking part in risk estimation was not compulsory (e.g., likelihood or frequency, severity or consequence).

Below is a brief summary of the characteristics of each method.

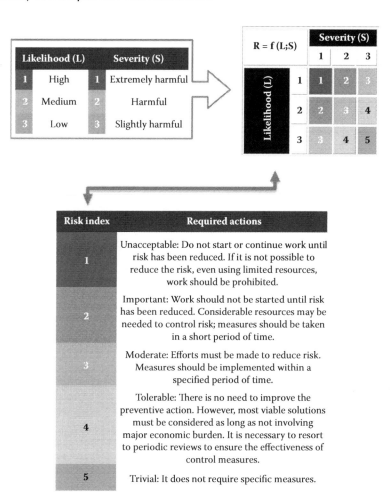

FIGURE 1.5 Risk assessment matrix used in MMS 3 × 3 method. (From Carvalho, F., and Melo, R. B., *WORK: A Journal of Prevention, Assessment, and Rehabilitation*, 51(3), 591–600, 2015.)

1.4.5.1 3 × 3 Simple Matrix Method or MMS 3 × 3

It is a simple method using a symmetrical (3 × 3) risk estimation matrix resorting to two variables, gravity/severity (G/S) and likelihood (L), both expressed in scales with three levels. It integrates a risk index scale with five levels, to prioritize intervention. Figure 1.5 shows the risk assessment matrix used in this method.

1.4.5.2 BS8800 Simple Matrix Method or BS8800

It is a method that was introduced in BS8800 (2004). An asymmetrical matrix is used in this method. Its matrix resorts to two variables: gravity/severity (G/S) expressed on a three-level scale and likelihood (L), presenting a scale of four levels. The risk index integrates five levels of intervention priority. Figure 1.6 shows the risk assessment matrix used in this method.

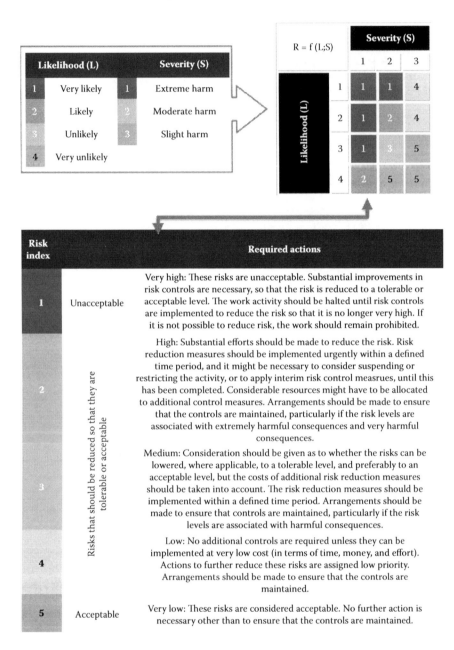

FIGURE 1.6 Risk assessment matrix used in BS8800 method. (From Carvalho, F., and Melo, R. B., *WORK: A Journal of Prevention, Assessment, and Rehabilitation*, 51(3), 591–600, 2015.)

As we can see in both methods, risk magnitude results from a preestablished combination of the used variables (e.g., likelihood and severity) and not from their product, which is commonly expressed by the relation R = S × L, where R = risk magnitude, L = likelihood, and S = severity, as described by Marhavilas et al. (2011a). This kind of method is similar to the methods described by Ni et al. (2010), BSI (2004), and Reniers et al. (2005a). Once risk magnitude (R) is determined, it is compared to the risk index scale, which helps us to define the priority of the required interventions.

1.4.5.3   P: Complex Matrix Method or MMCP

This method computes risk considering the frequency (F) of an accident or exposure to a hazard, the severity (S) of the potential consequences, the adopted procedures and safety conditions (Ps), and the number of people exposed or affected (N). The corresponding risk score (R) is obtained using Equation 1.1:

$$R = F * S * Ps * N$$

(1.1)

All variables (F, S, Ps, and N) are expressed on a five-level scale. On its turn, risk magnitude varies between 1 (very bad) and 625 (very good). This method integrates a risk index scale with five intervention priority levels and is illustrated in Figure 1.7.

1.4.5.4   William T. Fine Method or WTF

This method takes the potential consequences (C) of an accident, the exposure factor (E), and the probability factor (P) into account and computes the risk score (R) using Equation 1.2:

$$R = C * E * P$$

(1.2)

Each variable (C, E, and P) is assessed on a six-level scale. The risk magnitude scale varies between 0.05 (optimal situation) and 10,000 (worst situation). This method integrates a risk index scale with five levels of intervention priorities and is illustrated in Figure 1.8.

Considering the classification given by Carvalho and Melo (2007), the two last methods—WTF and MMCP—use a complex matrix similar to the proportional risk assessment technique (PRAT) technique described by Marhavilas et al. (2011a) and Marhavilas and Koulouriotis (2008), and to the Kinney and Fine technique described by Reniers et al. (2005a). The idea behind these methods is the same as in the risk matrix; however, some differences are present as a formula for calculating risks due to a particular hazard is presented.

1.4.6   STATISTICAL ANALYSIS

For data processing, we resorted to SPSS version 20. The nonparametric Friedman test and Krippendorff's alpha coefficient (α_K) were the statistical techniques used.

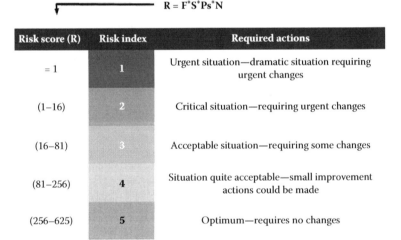

Frequency (F)		Severity (S)		Procedures and safety conditions (Ps)		Number of people exposed or affected (N)	
1	Frequent	1	Catastrophic	1	Do not exist or are not known	1	≥51
2	Likely	2	Critical	2	Serious deficiencies	2	31–50
3	Occasional	3	Marginal	3	Some weaknesses in procedures and lack of implementation of others	3	11–30
4	Remote	4	Negligible	4	Sufficient, but can be improved	4	4–10
5	Unlikely	5	Insignificant	5	Very good—enough and well implemented	5	1–3

$$R = F^*S^*Ps^*N$$

Risk score (R)	Risk index	Required actions
= 1	1	Urgent situation—dramatic situation requiring urgent changes
(1–16)	2	Critical situation—requiring urgent changes
(16–81)	3	Acceptable situation—requiring some changes
(81–256)	4	Situation quite acceptable—small improvement actions could be made
(256–625)	5	Optimum—requires no changes

FIGURE 1.7   Illustration of MMCP method. (From Carvalho, F., and Melo, R. B., *WORK: A Journal of Prevention, Assessment, and Rehabilitation*, 51(3), 591–600, 2015.)

Whenever the Friedman test outputs led to H_0 rejection, a posterior analysis was performed to identify the pairs that rendered those results.

A significance level of 0.05 was adopted as a criterion to reject the null hypothesis. For the pairs of methods not revealing differences between risk-level values, α_K was used to find if there was consistency.

Krippendorff's alpha coefficient (α_K) was chosen to assess intermethod reliability, reproducibility (intercoder reliability), and stability (intracoder reliability) because it allows uniform reliability standards to be applied to a great diversity of data (Krippendorff, 2004, 2007):

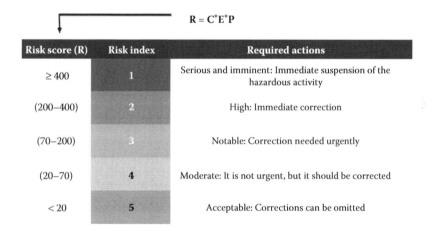

Potential consequences (C)		Exposure factor (E)		Probability factor (P)	
100	Catastrophic	10	Continuous—several times a day	10	Very likely
50	Multiple fatalities	6	Frequent—about 1 time a day	6	Possible
25	Fatality	5	Occasional—from 1 time/week to 1 time/month	3	Rare
15	Extremely serious injury-permanent disability, amputation	4	Unusual—from 1 time/month to 1 time/year	1	Repeat unlikely
5	Disabling injury	1	Rare—it is known that it occurs, but frequency is very low	0.5	Never happened
1	Minor cuts, bruises, minor damage	0.5	Remotely possible—it is not known whether it occurs, but it is possible to occur	0.1	Practically impossible

$$R = C^*E^*P$$

Risk score (R)	Risk index	Required actions
≥ 400	1	Serious and imminent: Immediate suspension of the hazardous activity
(200–400)	2	High: Immediate correction
(70–200)	3	Notable: Correction needed urgently
(20–70)	4	Moderate: It is not urgent, but it should be corrected
< 20	5	Acceptable: Corrections can be omitted

FIGURE 1.8 Illustration of WTF method. (From Carvalho, F., and Melo, R. B., *WORK: A Journal of Prevention, Assessment, and Rehabilitation*, 51(3), 591–600, 2015.)

- It is applicable to any number of values per variable. Its correction for chance makes α_K independent of this number.
- It is applicable to any number of coders ($k \geq 3$), not just the traditional two.
- It is applicable to small and large sample sizes alike. It corrects itself for varying amounts of reliability data.
- It is applicable to ordinal variables.
- It is applicable to data with missing values in which some coders do not attend to all coding units.

TABLE 1.4
Guidelines for Interpreting α_K and %Agr Values

	α_K	%Agr
Good consistency/agreement level	>0.8	>80%
Acceptable consistency/ agreement level	$0.6 \leq \alpha_K < 0.8$	$60\% \leq \%Agr < 80\%$
Low consistency/agreement level	<0.6	<60%

Source: Adapted from Krippendorff, K., *Content Analysis: An Introduction to Its Methodology* (2nd ed.), Thousand Oaks, CA: Sage, 2004; Krippendorff, K., Testing the reliability of content analysis data: What is involved and why, 2007, http://www.asc.upenn.edu/usr/krippendorff/dogs.html.

α_K is a statistical measure of the extent of consistency among coders, and is regularly used by researchers in the area of content analysis (Gwet, 2011).

Coders were grouped according to Table 1.3 in order to investigate the relative importance of academic background, OSH practice experience, and expertise level regarding matrix-based methods' use, on both intercoder and intracoder reliability. Therefore, we had eight groups of analysts, including the complete set.

The software application KALPHA© (macro for SPSS) was used to compute α_K (Hayes, 2005). The KALPHA macro was set up to bootstrap 1000 samples of hypothetical values of α_K, leading to the estimation of α_K true, for a 95% confidence interval.

Although the agreement percentage (%Agr) does not account for agreements that would be expected by chance, and therefore overestimates the level of agreement, it was also computed in some cases, to provide a better interpretation of the results.

Cutoffs for both α_K and %Agr are presented in Table 1.4.

Bearing in mind that this study attempts to assess individual risk, we also tried to identify the most protective methods, as far as the worker's protection is of concern. Mean rank values were used to rank methods, and it was assumed that more protective methods present lower values.

1.5 RESULTS AND DISCUSSION

Risk levels obtained with each of the four methods for the six work situations, in both rounds of the study, are presented in Figure 1.9. The Friedman test revealed significant differences among those values ($p < 0.05$). However, one may observe that methods MMS 3 × 3 and MMCP present a weaker ability to differentiate risks, which may jeopardize the main goals of a risk assessment, namely, ranking intervention priorities toward risk control. In the particular case of method MMS 3 × 3, it may have happened due to a phenomenon that Ni et al. (2010) call risk tie, which

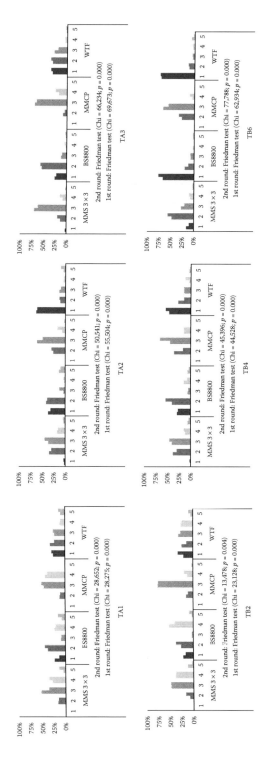

FIGURE 1.9 Risk-level frequency (%) and Friedman test outputs obtained in each situation (TA1, TA2, TA3, TB2, TB4, and TB6) with each method in each round of the study.

is related to the use of short scales to rank both severity and likelihood (both present three levels in this particular method). Consequently, when using this method to assess different risks, they may end with the same risk magnitude, not allowing a correct allocation of resources. As for the MMCP method, we believe that the subjectivity of the categorization labels used in scales may explain this fact.

From the point of view of the workers' protection, WTF and BS8800 seem to be the most powerful methods, as they assume the first position of the mean rank all the time (Figure 1.10), which means that both methods computed the higher-risk magnitude values.

Despite the differences previously reported, the posterior analysis revealed no significant differences ($p > 0.05$) between the results of some pairs of methods. Consequently, Krippendorff's alpha coefficient was used to assess reliability. Regarding the intermethod reliability (Figure 1.11), Krippendorff's alpha coefficient results revealed low consistency ($\alpha_K < 0.6$) in most cases (only three cases revealed $\alpha_K > 0.6$). It appears that the assessed risk type or nature influences the results, and reinforces that the choice of a risk assessment method should be a rigorous process (Carvalho, 2007).

It is worth noting that consistency increased from round 1 to round 2, which may be due to a deeper knowledge of the situations under analysis.

In addition to intermethod reliability, this study also analyzed inter- and intra-coder consistency.

Considering that for intercoder assessment $0 \le \alpha_K < 0.3$ (Figure 1.12), it is possible to verify that there is a low intercoder consistency level ($\alpha_K < 0.6$). It seems that Krippendorff's alpha coefficient values are a consequence of the systematic inconsistencies found among coders (type II error). Therefore, it is appropriate to state that risk assessments made by different analysts with the same method may lead to different intervention priorities.

The analysts' academic background (SA_A1, SA_A2, SA_A3), their OSH practice experience (SA_B1, SA_B2), and their level of expertise in using these methods (SA_C1, SA_C2) did not influence the results. This fact contradicts the disadvantage pointed out for this type of approach, particularly its results' dependency on these last two factors (Marhavilas et al., 2011a; Marhavilas and Koulouriotis, 2008; Reniers et al., 2005b). Likewise, it is possible to highlight that the method referenced as the most popular (MMS 3×3) among the experts' group (SA_C1) did not show relevant differences in the results of α_K.

Intercoder reliability evaluation for the intermediate variables used in each method revealed the same trend over the two evaluation rounds (Figure 1.13). It is worth highlighting (1) the similarity of the obtained results for all groups of analysts and (2) the fact that three variables (gravity [G], severity [S], and consequence factor [C]) delivered α_K values within the range [0.25, 0.5]. For these variables, the %Agr revealed high-level consensus (%Agr > 0.8) when using methods MMS 3×3 and BS8800 to assess TA2, TA3, and TB4. These results suggest that the clarity of descriptors may minimize the uncertainties of coding, which results in higher levels of consensus.

Despite the disadvantages pointed out to %Agr, we have used it to confirm the results obtained with Krippendorff's alpha because this coefficient may suffer from

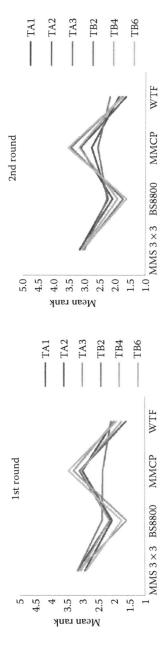

FIGURE 1.10 Mean rank obtained with each method in each situation (TA1, TA2, TA3, TB2, TB4, and TB6) in both rounds.

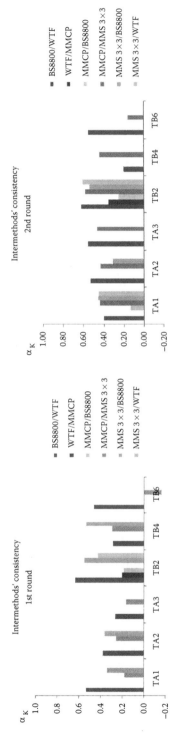

FIGURE 1.11 Results of Krippendorff's alpha coefficient to evaluate intermethod consistency, in both rounds.

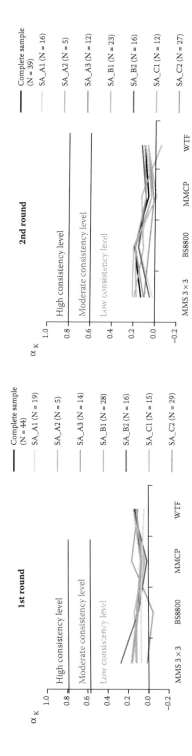

FIGURE 1.12 α_K values obtained for the intercoder reliability assessment of risk-level variable: left, first round of the study; right, second round of the study.

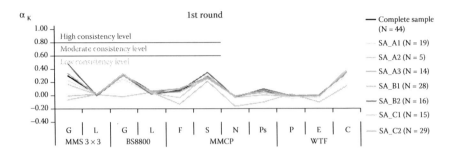

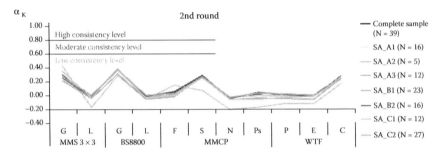

FIGURE 1.13 α_K values obtained for the intercoder reliability assessment of the intermediate variables: top, first round of the study; bottom, second round of the study.

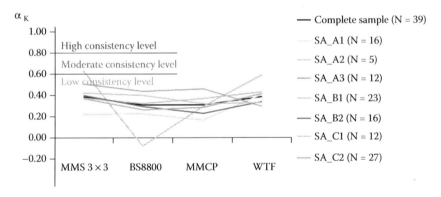

FIGURE 1.14 α_K values obtained for the intracoder reliability assessment of the risk-level variable.

type I errors when small-dimension samples are used. This may have happened when the analysts' sample was split in several sets and a pair of analysts was compared.

For the other variables (likelihood [L], procedures and safety conditions [Ps], number of people exposed or affected [N], exposure factor [E], and probability factor [P]), the α_K values were slightly lower, which may be related to the subjectivity associated with the labels used in these variables' scales.

The obtained values of α_K for the intracoder reliability evaluation (Figure 1.14) concerning the risk-level variable show slightly higher consistency levels

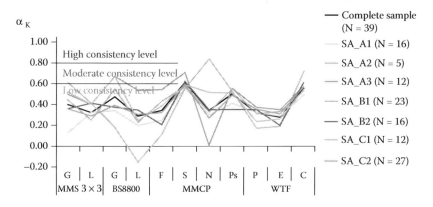

FIGURE 1.15 α_K values obtained for the intracoder reliability assessment of the intermediate variables.

$(0.2 \leq \alpha_K < 0.6)$ than the intercoder reliability evaluation, highlighting methods MMS 3 × 3 and WTF.

According to Figure 1.14, the analysts' professional experience and their level of expertise did not seem to produce relevant differences in the results of α_K. As for their academic background, some inconsistency $(-0.07 \leq \alpha_K \leq 0.62)$ was found within the OSH group (SA_A2), whereas greater consistency $(0.3 \leq \alpha_K \leq 0.45)$ and a higher agreement level were found within the engineers' group (SA_A3).

Intracoder reliability evaluation (Figure 1.15) for the intermediate variables used in each method came to reinforce the trend already found, highlighting gravity (G), severity (S), and the consequence factor (C) as the ones with higher levels of consistency. The coding process of these particular variables seems to be facilitated due to the objectivity associated with the labels applied to the scale categories.

1.6 CONCLUSIONS

This chapter reports the results of a study in which reliability was estimated for matrix-based methods used to assess occupational risks. Six working situations were assessed with four different matrices by a set of qualified OSH practitioners. Intracoder and intercoder reliability assessments were also accomplished for both the final risk magnitude and the intermediate variables of each method.

Statistically significant differences $(p < 0.05)$ were found among the risk levels obtained for the six working situations with all four methods, and low intermethod reliability levels $(\alpha_K < 0.6)$ were confirmed. Therefore, it seems reasonable to conclude that despite each method's intrinsic validity, its choice should not be arbitrary. Nevertheless, there was some consensus in the power revealed by the methods regarding the protection they confer to the workers. The WTF and BS8800 methods proved to be the most powerful.

It was shown that assessing occupational risks with matrix-based methods could be performed with higher reliability if done by the same, rather than different, coders. However, the results revealed that the use of a matrix-based approach to accomplish the requirements of the legislation should be done with caution, as most inter- and intracoder assessments showed low levels of consistency. This means that the use of risk matrices within the OSH scope revealed low reliability, either in terms of stability or in terms of reproducibility.

Apparently, there are no relevant differences among risk assessment results obtained by individuals with different levels of OSH practice experience and different levels of expertise in using risk matrices. Such evidence seems to be opposite to that found in several scientific references on the use of this type of approach.

In most cases, inter- and intracoder reliability proved to be low ($\alpha_K < 0.6$) for both the risk level and the intermediate variables of each assessment method. Consequently, different priority interventions may be suggested, depending on the analyst involved or the moment of the risk assessment accomplishment.

It was still possible to register similar α_K values among the analyses concerning the intermediate variables, which reveals that some of them seem to be easier to code than others. In fact, gravity, severity, and consequence factors were the three variables for which the agreement was higher (%Agr > 80%), and therefore variables such as likelihood, frequency, and probability factor may require improvements regarding their coding process.

The results of this study can be used to improve risk assessment methods based on matrices by eliminating misleading or confusing issues related to the variables in use and providing guidance on the coding process. Modification of labels used to categorize these variables or detailed descriptions of each category may be helpful.

Considering the advantages attributed to matrix-based methods, we consider that further research regarding their reliability is of major importance to establish adequate criteria to support the risk assessment methods' selection process. The soundness of such assessment instruments remains a concern given the variety of matrices available and the lack of studies regarding their validity (accuracy), as reliability comprises more than stability and reproducibility.

REFERENCES

Branco, J. C., Baptista, J. S., and Diogo, M. T. (2007). Comparação da avaliação dos riscos por dois métodos correntemente utilizados na Industria Extractiva. In P. Arezes, J. S. Baptista, M. Barroso, A. Cunha, R. Melo, A. S. Miguel, and G. P. Perestrelo (Eds.), *Colóquio Internacional Segurança e Higiene Ocupacionais—SHO2007* (pp. 177–181). Guimarães, Portugal: Sociedade Portuguesa de Segurança e Higiene Ocupacionais (SPOSHO).
BSI [British Standard Institution]. (2004). BS8800—Occupational health and safety management systems—Guide. London: BSI.
Carvalho, F. (2007). Estudo comparativo entre diferentes métodos de avaliaçao de risco, em situaçao real de trabalho. Master thesis in Ergonomia na Segurança no Trabalho, Cruz-Quebrada.
Carvalho, F., and Melo, R. B. (2007). Comparação entre diferentes métodos de avaliação de risco, em situação real de trabalho. In C. Guedes Soares, A. P. Teixeira, and P. Antão (Eds.), *Riscos Públicos e Industriais* (Vol. 2, pp. 853–868). Lisbon: Edições Salamandra.

Carvalho, F., and Melo, R. B. (2015). Stability and reproducibility of semi-quantitative risk assessment methods. *WORK: A Journal of Prevention, Assessment, and Rehabilitation*, 51(3), 591–600.

EU-OSHA [European Agency for Safety and Health at Work]. (2008). Facts 81 (EN)—Risk assessment—The key to healthy workplaces. Bilbao, Spain: EU-OSHA.

EU-OSHA [European Agency for Safety and Health at Work]. (2009). Síntese da campanha: Locais de trabalho seguros e saudáveis: Bom para si. Bom para as empresas. Bilbao, Spain: Agência Europeia para a Segurança e Saúde no Trabalho.

Gadd, S., Keeley, D., and Balmforth, H. (2003). *Good Practice and Pitfalls in Risk Assessment*. Sheffield, UK: Health and Safety Executive.

Gwet, K. L. (2011). On The Krippendorff's alpha coefficient. http://agreestat.com/research_papers/onkrippendorffalpha.pdf.

Harms-Ringdahl, L. (2013). *Guide to Safety Analysis for Accident Prevention* (1st ed.). Stockholm: IRS Riskhantering AB.

Hayes, A. F. (2005). An SPSS procedure for computing Krippendorff's alpha [Computer software]. http://www.afhayes.com/spss-sas-and-mplus-macros-and-code.html (accessed January 12, 2012).

ISO [International Organization for Standardization]. (2009). ISO 31000—Risk management—principles and guidelines (1st ed.). Geneva: ISO.

ISO/IEC [International Organization for Standardization and International Electrotechnical Commission]. (1999). Guide 51—Safety aspects—Guidelines for their inclusion in standards (2nd ed.). Geneva: ISO/IEC.

Krippendorff, K. (2004). *Content Analysis: An Introduction to Its Methodology* (2nd ed.). Thousand Oaks, CA: Sage.

Krippendorff, K. (2007). Testing the reliability of content analysis data: What is involved and why. http://www.asc.upenn.edu/usr/krippendorff/dogs.html.

Marhavilas, P. K., and Koulouriotis, D. E. (2008). A risk-estimation methodological framework using quantitative assessment techniques and real accidents' data: Application in an aluminum extrusion industry. *Journal of Loss Prevention in the Process Industries*, 21(6), 596–603.

Marhavilas, P. K., Koulouriotis, D. E., and Gemeni, V. (2011a). Risk analysis and assessment methodologies in the work sites: On a review, classification and comparative study of the scientific literature of the period 2000–2009. *Journal of Loss Prevention in the Process Industries*, 24(5), 477–523.

Marhavilas, P. K., Koulouriotis, D. E., and Mitrakas, C. (2011b). On the development of a new hybrid risk assessment process using occupational accidents' data: Application on the Greek Public Electric Power Provider. *Journal of Loss Prevention in the Process Industries*, 24(5), 671–687.

Ni, H., Chen, A., and Chen, N. (2010). Some extensions on risk matrix approach. *Safety Science*, 48(10), 1269–1278.

Olsen, N. S. (2013). Reliability studies of incident coding systems in high hazard industries: A narrative review of study methodology. *Applied Ergonomics*, 44(2), 175–184.

Reniers, G. L. L., Dullaert, W., Ale, B. J. M., and Soudan, K. (2005a). Developing an external domino accident prevention framework: Hazwim. *Journal of Loss Prevention in the Process Industries*, 18(3), 127–138.

Reniers, G. L. L., Dullaert, W., Ale, B. J. M., and Soudan, K. (2005b). The use of current risk analysis tools evaluated towards preventing external domino accidents. *Journal of Loss Prevention in the Process Industries*, 18(3), 119–126.

Suddle, S. (2009). The weighted risk analysis. *Safety Science*, 47(5), 668–679.

Van Duijne, F. H., van Aken, D., and Schouten, E. G. (2008). Considerations in developing complete and quantified methods for risk assessment. *Safety Science*, 46(2), 245–254.

2 Regulatory, Organizational, and Operational Issues in Road Construction Safety

Ashim Kumar Debnath, Tamara Banks,
Ross Blackman, Nathan Dovan,
Narelle Haworth, and Herbert Biggs

CONTENTS

2.1 INTRODUCTION

The safety of roadworkers is a high priority for occupational health and road safety authorities, and especially for the workers themselves and the organizations that employ and represent them. There are substantial risks involved in undertaking construction and maintenance in close proximity to moving traffic, and compromises between traffic flow, acceptable risk exposure levels, and equipment and resource levels associated with such tasks. While road construction and maintenance works (commonly known as roadworks) are essential for maintaining and improving the mobility and safety of all road users, the process of building safer roads and roadsides needs to be managed to minimize risks to both the motorists and roadworkers.

Reports from highly motorized countries, including the Netherlands, the United States, and Great Britain, show that around 1%–2% of road fatalities occur at roadworks (NWZSIC, 2012a, 2012b; SWOV, 2010). Numerous studies have found that crash rates increase significantly during roadworks compared with prework periods (Doege and Levy, 1977; Garber and Zhao, 2002; Khattak et al., 2002; SWOV, 2010; Whitmire et al., 2011). Roadwork crashes are also reported to be more severe than other crashes (Pigman and Agent, 1990), possibly associated with the relatively high proportional involvement of trucks (Bai and Li, 2006; Krux and Determan, 2000; SWOV, 2010). Compared to some other countries, relatively little is known about roadwork crashes across Australia, primarily because it is difficult to identify roadwork crashes in official records (Haworth et al., 2002; Debnath et al., 2013). Thus, it is difficult to obtain accurate comparative information on crash rates, crash severity, and other variables of interest. Based on New South Wales data (RTA, 2008), it is estimated that nationally each year at least 50 deaths and 750 injuries occur to workers and the public in crashes at roadworks, with a cost of more than $400 million (Debnath et al., 2012).

Because of multiple floods in recent years in the state of Queensland in Australia, significant maintenance and rehabilitation works are being undertaken on the state road network. This sharp increase in roadwork activities has been accompanied by a number of roadworker fatalities and injuries in recent times. In addition, it has become a cause of driver frustration—resulting from frequent stopping at roadwork sites and the associated increase in travel time—which might influence driver behavior and compliance with roadwork signage and traffic rules. For example, a recent study of driver speeds in a Queensland worksite (Debnath et al., 2014a) showed that almost all (97.8%) vehicles drove above the posted speed limit in the activity area of the roadwork site. This high rate of noncompliance with roadwork signage poses significant threat to roadworkers, as well as to the traveling public themselves. To improve safety at Queensland roadwork sites for roadworkers and the traveling public, a large research project is being undertaken using a multidisciplinary approach because the factors ultimately determining safety outcomes at roadworks are complex and have not been systematically explored before.

Conceptually, roadwork operation and management of safety can be seen in a three-level framework (see Figure 2.1 for a description of each of the levels and the associated tasks undertaken in the research project). First, at the *regulatory level*, roadworks operate at the interface between the work environment, governed by workplace health and safety regulations, and the road environment, which is subject to road traffic regulations and practices. This may lead to a less than optimal compromise between measures to improve worker safety and measures to maintain traffic flow.

Second, at the *organizational level*, national, state, and local governments plan and purchase road construction and maintenance, which are then delivered in-house or tendered out to large construction companies (or alliances). Many smaller companies then supply services and labor. The lack of a nationally accepted framework for safety management tasks and competencies, in combination with a largely subcontracted workforce that shifts regularly between organizations, projects, and sites, means that workers receive different messages relating to safety performance when they change projects or primary contractors. Reports highlight the confusion caused by the plethora of documents that provide information regarding requirements for traffic control at roadworks (Austroads, 2009). Third, at the *operational*

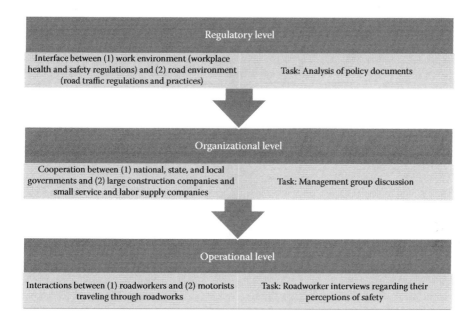

FIGURE 2.1   Three-level framework of roadwork operation and safety management.

level, roadworks are difficult to isolate from the general public, and so it is difficult to implement effective occupational health and safety controls. The credibility and appropriateness of signage and roadworks' speed limits are consistently questioned by drivers (MCR, 2009; Highways Agency, 2006) and compliance is low (Debnath et al., 2014b).

This multidisciplinary project involves partners at each of the regulatory, organizational, and operational levels working together to enhance the safety of roadworkers by investigating the real and perceived dangers at roadworks, strengthening organizational policies and practices for roadworker safety, testing innovative on-road treatments and educational initiatives to improve driver behavior at roadworks, and developing models of safety management in complex systems that span different regulatory frameworks. The research project is designed to be conducted in two settings: within the organizations that are involved in purchasing and delivering roadworks, and the actual roadwork sites. This ongoing project has completed multiple studies to date in order to understand the various safety issues related to roadworks in light of the three-level framework explained above.

This chapter presents the critical safety issues in roadworks identified by taking a structured approach to synthesizing outputs derived from three studies: analyzing (1) regulatory policy documents, (2) opinions of roadwork subject matter experts, and (3) roadworkers' perceptions regarding the common roadwork incidents and their causes, hazards, and effective measures for improving safety. The findings obtained from these three studies, along with their methodologies, are discussed in the next section. Findings from each study were then synthesized to discuss the common safety issues identified.

2.2 IDENTIFYING SAFETY MANAGEMENT ISSUES IN ROADWORKS

2.2.1 ANALYSIS OF POLICY DOCUMENTS

The purpose of the review of policy documents was to analyze and critique the documentation specifically pertaining to safety culture and occupational health and safety to more accurately identify the common policies and processes currently being implemented in Australia. Generally, construction projects are cooperative ventures between a number of organizations, each bringing their own management process, procedures, and safety culture to the project. The extent to which agreement is reached between project partners on safety-critical positions, and the allocation of responsibility for safety management tasks are critical to the development and maintenance of the project's safety culture. Industry partners contributing to this phase of the research project include a major purchaser of roadworks, a project manager, a construction company, a workers' union, and many road construction worker representatives.

To accurately identify and review the relevant documentation, inquiries were made by the researchers to the partner organizations, requesting any formal or informal policies or procedures currently being implemented with regard to a best practice risk management list developed by the researchers. The list was developed based on best practice risk management approaches identified through a combination of academic and government publications and the Construction Safety Competency Framework (Anderson et al., 1998; Biggs and Biggs, 2012; Dingsdag et al., 2008; Haworth et al., 2000, 2008; Health and Safety Executive, 2000). Table 2.1 details the list of selection criteria for the review of policy and procedure documentation. Given the comprehensive nature of this research, a broad search list was deliberately employed in order to identify all potential documents related to the subject matter. No exclusion criteria were applied in the policy documentation gathering stage.

Based on the selection criteria, a total of 22 documents across three organizations were obtained. One organization reported that it did not have any policies or procedures regarding the safety of roadworkers. Instead, it stated that it accepted that it was the responsibility of the organization and contracting company to ensure roadworker safety by abiding by the minimum occupational health and safety requirements. An initial screening of the obtained documents confirmed that all supplied documents related to safety culture and occupational health and safety management in roadworks. Therefore, no document exclusions were required and all documents proceeded to the full review stage.

Rigorous analysis of documents against the best practice risk management list detailed in Table 2.1 identified that development of policy and procedural documentation varied considerably across the 13 approaches to occupational health and safety management in roadworks. More specifically, most of the researched construction partners demonstrated mature development of policy and procedural documentaion with regard to setting work schedules, providing safe equipment, competency assessment, training, risk identification and control, and communication of safety commitments and responsibilities.

TABLE 2.1

List of Selection Criteria for the Review of Policy and Procedure Documentation

Work schedules are set with consideration of fatigue, high-risk periods, adverse weather conditions, etc.

Equipment is fit for purpose and maintained appropriately.

Employees and subcontractors are competent and capable.

Employees and subcontractors are appropriately inducted or trained and sufficiently fit and healthy to perform roadworker tasks.

Risks are identified and controlled.

Stakeholders are communicated and consulted with in regard to risk management.

Safety program effectiveness is evaluated.

Safety responsibilities and the organization's commitment to safety are communicated to employees and subcontractors.

Employees and subcontractors are recruited and selected based on safety records and awareness of safety issues.

Employees and subcontractors are recognized for good and poor safety behaviors through an official scheme of incentives and disincentives.

Incident involvement is recorded and monitored and identified high risks are managed.

Relevant components of the occupational health and safety and workers' compensation management system are implemented.

Ongoing safety culture is developed through either formal training, conferences, or internal processes.

Risk management policies should be updated regularly.

Maturity of documentation of occupational health and safety management in roadworks varied between the construction partners with regard to the monitoring, evaluating, and recognizing of employee and departmental safety performance and incident involvement. To enhance roadworker safety, the construction industry should consider developing policies that clearly articulate how safety performance and program effectiveness are to be evaluated, recognizing both the good and poor safety behaviors of employees and subcontractors, and collaboratively across the industry recording and monitoring risks and developing a consistent approach to the management of high risks.

The analysis identified a gap between the best practice risk management list developed by the researchers and the supplied risk management documentation with regard to consideration of safety in employee selection decisions. More specifically, none of the provided policy documents indicated that employees and subcontractors are recruited and selected based on an awareness of safety issues and a consideration of safety records. Organizations that proactively recruit and select safe employees may achieve better safety performance records than organizations that do not consider safety when recruiting and selecting employees. When considering this gap, it is important to recognize that this analysis was limited to the documents supplied to the researchers. Therefore, it is possible that some construction partners are informally adopting risk management approaches but have not demonstrated evidence of this through their supplied policy documents.

Additionally, the policy documents analysis identified a substantial overlap between the reviewed policies of the construction partners. While it was encouraging to identify consistent policies and procedures between the organizations, such as the unified use of the risk management hierarchy of controls, it is also important to recognize that confusion may exist in the industry due to the identified gaps in policies and discrepancies, such as the process for reporting incidents. Where discrepancies and gaps in organizational policies exist, confusion may occur for roadworkers, with regard to which organization's policies and procedures should be adhered to in any given situation.

Preliminary discussions suggested that when a project alliance occurs, roadworkers are expected to follow the policies and procedures of the leading project company. Examination of the supplied policy documents from each organization identified that wherever there was a difference between the standard represented in an organization's policy and that same policy of the relevant jurisdiction or client, the more stringent standard applied. Furthermore, given that the policy analysis identified gaps between the best practice risk management list developed by the researchers and the risk management approaches included in the supplied policy documents, the construction industry may benefit from including all aspects of the best practice risk management list detailed in Table 2.1.

Finally, it is important to note that the review of the organizational policies and procedures identified risk management approaches that were documented by the construction partners who participated in this research. It was unclear from this analysis the extent to which policy affected workplace practices. Subject matter interviews were conducted to better understand how the documented approaches translated to workplace risk management practices in roadworks.

2.2.2 SUBJECT MATTER EXPERT INTERVIEWS

Subject matter expert interviews focused on the identification of safety-critical positions and safety management tasks required to be undertaken on projects involving roadworkers. This process involved roadworker safety discussions with employees across various levels and job titles from each of the partner organizations. The development of these discussions was structured around the identified information obtained from the preliminary review of organizational policies, and a previously developed generic safety competency framework for the construction industry by Dingsdag et al. (2006).

In total, 24 employees across the three partner organizations participated in the discussions. Subject matter experts' job positions ranged from safety managers, project leaders, project managers, safety coordinators, supervisors, and engineers to health and safety representatives. Obtaining this diversity in the participants' job roles was considered essential to achieving an accurate representation of the organizations' safety-critical positions. Responses in the discussions were considered to be reflective of the current processes and procedures undertaken to ensure the safety of roadworkers in the researched construction partners. It is important to note that while the job positions were similar to those of employees interviewed in other phases of this research, the purpose of these discussions was to investigate the organizational level, taking a top-down approach to roadworker safety. In groups of up to 10 participants, subject matter experts were asked to identify within their own organizations the safety-critical positions and the processes and strategies currently implemented by their organizations to better assist in the shared responsibility of the safety management tasks across these positions. They were also asked to identify the safety management tasks required to be undertaken on worksites within their organization. The experts were then prompted to identify any processes and strategies implemented between organizations for the improvement of safety to their roadworkers. Finally, the subject matter experts were asked to identify what could be done at an organizational level to better improve the safety of roadworkers.

The subject matter experts identified many safety-critical positions, demonstrating a shared understanding and belief that safety in roadwork environments is everyone's responsibility. The following safety-critical positions within organizations were identified: general managers, directors, managers and supervisors, team leaders, workplace health and safety officers, work health and safety (WHS) consultants, health and safety representatives, rehabilitation and return-to-work coordinators, principal contractors, designers, plant suppliers, and operators. Additionally, several committees and teams were identified as being required to perform a critical role in roadworker safety management. These included the Executive Management Team, Safety Health and Environment Board, Group Safety Health and Environment Teams, Shared Services Teams, and WHS Governance Committee.

Based on the discussions with the subject matter experts, key safety management themes emerged. The main initial concern identified was that the cost of improving the safety of roadworkers needed to be considered against the reality of the risk associated with executing the task. A major concern raised by participants also involved training and development. It was a commonly held notion that not only was this needed for staff and workers who were present on site, but also there would be great benefit in further public education campaigns. Recurring themes for training for the organizations' employees included performing gap analyses of what they were currently missing and tailoring the appropriate training to meet their needs. There was also an increased demand for more education and training for employees to better understand how to properly utilize safety essentials for worksites, including physical barriers and proper road signage. In addition, there was consensus that there needed to be greater accountability and empowerment from the general worker with regard to safety.

The need for greater accountability and responsibility from the workers for safety extended to the initial project planning stages, with a common belief that there needed to be better incorporation of safety management tasks during this process, with an increased demand for more safety management stipulations to be inserted into initial contracts with clear and transparent terminology. These tasks needed to then be clearly communicated to all personnel on the worksite, with start-up and toolbox meetings critical to the communication of safety expectations on site. The subject matter experts believed that enhanced communication would better assist with the governance of the project, ensuring that work was carried out to an agreed-upon standard. Again, these standards needed to be clearly communicated and imposed on all workers on site during initial inductions.

Additionally, it was made clear that it was important for workers to have a thorough understanding of industry standard manuals to better assist with drawing up the site and helping with resourcing. It was stated that safety on site could be further improved by continued supervision and inspections of sites, with regular risk assessments to be carried out. There was also a belief that assessing key performance indicators for safety would be important.

2.2.3 Interviews with Roadworkers

As mentioned previously, accurate and reliable data on incidents and injuries in work zones are difficult to obtain in official records in Australia, and the limited available data are often fragmented and incomplete. In such cases, supplementary data are particularly valuable, including data sourced from surveys and interviews. A specific phase of the current study sought to gain an understanding of workplace hazards and mitigation measures from the perspective of people directly involved in road construction, maintenance, and traffic control activities (collectively referred to as roadworkers). Subsequently, 66 roadwork personnel were engaged in semistructured interviews, including 25 traffic controllers; 15 laborers and machinery operators; 21 managers, engineers, or supervisors; and 5 directors, planners, or designers. Interviews were conducted individually and face-to-face over an average period of 20 minutes, with the exception of three interviews, which were conducted by telephone.

As their descriptions suggest, the interviewed personnel were occupied in distinctly different roles and spent different proportions of their working time in the field where they could be exposed to work zone hazards. According to their general level of exposure to traffic, participants were grouped as fully exposed ($n = 25$), semiexposed ($n = 15$), or nonexposed ($n = 26$). About 50% ($n = 32$) of respondents worked throughout the work zone, while 11 respondents (all traffic controllers) worked only at either end of the work zone. Another 11 respondents (nonexposed to traffic) worked mostly from offices. Twelve respondents (eight nonexposed and four semiexposed) had an approximately 50–50 split of office and on-site work. About 40% of respondents reported that they work predominantly on foot, and another 47% reported to be mostly on foot and sometimes in a vehicle. Only nine participants (including seven nonexposed) reported staying predominantly inside vehicles when working on site. The high proportion of participants who work on foot in work zones indicates that there should have been a thorough collective understanding of the common hazards and mitigation measures. Most of the participants worked during daytime ($n = 49$), while seven respondents (including five traffic controllers) worked only at night. The remaining 10 respondents had both day and night work experience.

2.2.3.1 Common Roadwork Incidents and Their Causes

Participants were asked to recall and discuss any safety-critical incidents that they had experienced or witnessed personally, or heard about directly through colleagues. A range of incidents were subsequently recounted in varying amounts of detail. The most frequently recounted type of incident involved a public vehicle intruding into the work area in a work zone. This was mentioned by 38% ($n = 25$) of the 66 participants. The next most reported incident type was a traffic controller being hit by a vehicle, recalled by one-third ($n = 22$) of participants. Rear-end crashes were the third most frequently reported incident type, mentioned by 29% ($n = 19$) of participants. Reversing incidents (a work vehicle or machinery reversing into another work vehicle, machinery, object, or worker) were also mentioned relatively frequently, with 23% (15) of participants having witnessed or heard directly about this type of event.

A public vehicle was typically involved in the three most commonly mentioned types of incident (work zone intrusion, hit traffic controller, and rear-end crash). The fourth most frequently noted incident, reversing incidents, usually involved a work vehicle or mobile machinery. Whether involving the traveling public, workers only, or a combination of motorists and workers, human error was the main causative factor cited in the majority of these incidents.

The most common causes of incidents according to participants were vehicles ignoring signage and traffic controllers ($n = 26$), distracted driving ($n = 14$), driver error ($n = 6$), and drunk driving ($n = 5$). With regard to drivers ignoring signage, this behavior results in speeding through work zones and, as such, is directly related to arguably the most common and problematic hazard in work zones accommodating public traffic. Ignoring traffic controller instructions (e.g., stop or slow) can result in vehicles driving into work areas or closed lanes, rear-end crashes with vehicles stopping or stopped near the traffic controller, or head-on crashes with oncoming vehicles when violating a stop instruction. Distracted driving, often due to drivers observing work zone activities or using mobile phones and in-vehicle devices, is likely to cause rear-end crashes with preceding vehicles. Drunk driving was reported to be associated with speeding and consequently not complying with stop traffic controls. Interested readers are referred to Debnath et al. (2013), where the common work zone incidents and their causes are described in detail.

2.2.3.2 Common Roadwork Hazards

In addition to the common incidents and their associated causes, participants were asked to describe the situations in which they feel unsafe in order to gain a better understanding of the work zone hazards. The commonly reported hazardous situations (see Table 2.2) include excessive vehicle speeds in work zones, working in wet weather, driver frustration and aggression toward roadworkers, working close to live traffic lanes, working during night, dawn and dusk hours, and drivers on mobile phones leading to distracted driving. While these situations include factors related to the working environment (e.g., weather, visibility, and traffic), drivers' actions (e.g., speeding, aggression, and distraction) are again reported to create a significant share of the unsafe situations in work zones. Although some hazards were universally reported by all groups of roadworkers, in general, roadworkers' perceptions regarding the hazards were found to vary according to their exposure to traffic.

Participants were asked to specify the nature of the hazards in the situations they reported to feel unsafe. Excessive speeding, particularly in the absence of enforcement, poses significant hazards to roadworkers, as speeding is directly related to severity of incidents. Roadworkers believed that drivers deliberately violate the posted speed limits and drive at speeds that they perceive to be appropriate for the

TABLE 2.2
Commonly Reported Hazards in Work Zones

Hazards at Roadworks	Total Frequency[a]	Frequencies by Respondent Groups[a]		
		Fully Exposed	Semiexposed	Nonexposed
Excessive vehicle speeds	40	18	9	13
Working in wet weather	20	12	2	6
Driver frustration and aggression toward workers	18	11	1	6
Working close to traffic stream	14	0	3	11
Distracted driving	11	4	4	3
Working during night hours	9	5	2	2
Working close to machinery	7	1	1	5
Highway works (more unsafe than local street works)	7	5	0	2
Setting up signage	6	4	1	1
Working on hilly and curved roads with restricted escape path	6	3	1	2
Working during dawn and dusk hours	5	2	3	0

[a] Frequency values represent the number of respondents who reported a particular hazard.

road environment. Driver frustration resulting from travel time delay at roadworks was also believed to contribute to the prevalence of speeding.

The hazards of working in wet weather include reduced visibility and slippery surfaces, which reduce skid resistance and require greater stopping distances, so the chances of not noticing signage or traffic controllers and underestimating required stopping distances are higher. These eventually could lead to failing to stop under stop or slow directions and being involved in rear-end crashes with vehicles stopped ahead. Although working under rainy conditions is not common, sometimes workers need to continue working in order to meet deadlines or to reopen the road to traffic as soon as possible.

Driver frustration and aggression were reported as hazardous mainly by the traffic controllers. It was reported that forms of aggression can range from verbal abuse to throwing objects, spitting, and threatening roadworkers. The hazards in working close to traffic lanes, mostly reported by the nonexposed roadworkers, include throwing of loose materials from pavement by passing traffic, inability to see oncoming traffic properly (often on the hilly and windy roads), and not having an adequate escape path. Reduced visibility and higher numbers of fatigued drivers were the common hazards reported for working during night, dawn, and dusk hours. Distracted driving due to mobile phone use, which was also reported as a common cause of work zone incidents, was reported mostly by the exposed and semiexposed groups as a significant hazard, as this often results in motorists disobeying or not noticing traffic lights and signage. The signage setup process, particularly on highways, was perceived as a hazard by some roadworkers (mostly from the exposed group) because of their exposure to high-speed traffic. Further details about the hazards are available in Debnath et al. (2015).

In summary, findings from the commonly reported work zone incidents, their causes, and unsafe situations at work show that driver actions are responsible for creating most of the hazards in work zones. Speeding, noncompliance with traffic signage and traffic controller instructions, and distracted driving were the common hazardous behaviors in work zones. Other sources of hazards include improper working environment (e.g., working in wet weather and inadequate escape path) and not maintaining safety practices (e.g., tampering with reversing beepers). It is believed that construction companies and workplace safety regulators have the potential to control workers' compliance with safety practices and to treat hazards related to improper work environments. However, controlling driver behavior in work zones is a more difficult challenge, responsibility for which falls to a broader range of organizations and agencies, including police, road authorities, and road users themselves.

2.2.3.3   Measures to Improve Safety

A wide range of safety measures, including physical, informational, regulatory and enforcement, and educational measures, are used to control driver behavior in work zones (Debnath et al., 2012). No single measure has been proven to change driver behavior to the desired extent, although some measures appear useful in particular situations, while multiple measures in combination are more effective generally. To examine this issue, roadworkers were asked about the effectiveness of current safety measures with which they had experience or specific knowledge. The measures most commonly seen as effective by participants are summarized in Table 2.3.

Roadworkers perceived the available measures to be generally effective ($n = 24$), but in something of a contradiction also emphasized that some (such as regulatory and advisory signs) are only effective to the extent that motorists comply. As noted above, motorists' propensity to ignore or not notice signs, and their frustration regarding excessive signage, was frequently reported. This implies that although a particular safety measure may be theoretically appropriate for its intended purpose, poor compliance can render it ineffective. Thus, while many respondents considered the current safety measures to be generally effective, there was a lack of universal agreement.

Active police enforcement was considered the most effective safety measure, based on the frequency with which it was mentioned. Most motorists exceed speed limits in the absence of enforcement according to about 60% of respondents. Further, approximately half of all respondents claimed that visible police presence substantially reduces speeding even in the absence of enforcement. While the perceived presence of police, regardless of actual enforcement, was reportedly effective in encouraging speed limit compliance, respondents believed that the effect diminished quickly once the police presence was no longer apparent. The use of speed feedback systems in conjunction with police presence was also reported by some respondents ($n = 5$) to reduce speeds.

Apart from police presence, enforcement, and speed feedback systems, several other speed reduction measures were noted. These include portable temporary speed

TABLE 2.3

Commonly Reported Effective Countermeasures

Safety Measures	Total Frequency[a]	Frequencies by Respondent Groups[a]		
		Fully Exposed	**Semiexposed**	**Nonexposed**
Active police enforcement	34	11	4	19
Driver education and work zone–oriented licensing	11	8	1	2
Public awareness of roadwork safety	10	6	2	2
Improving communication among and between workers and traffic controllers	8	3	4	1
Physical separation of workers from traffic	6	0	2	4
Spotters to alert workers of incoming hazards	6	1	0	5
Speed feedback system	5	0	0	5
Speed humps ahead of traffic controller	5	3	0	2

[a] Frequency values represent the number of respondents who reported a particular safety measure.

humps positioned near traffic control points and use of pilot cars to guide traffic through work zones at safe speeds. Some also suggested introducing the lowest work zone speed limit (typically 40 km/h in the study area) in advance of, rather than at, the traffic control point, so that motorists may pass the traffic control at lower speeds.

Respondents perceived a lack of public awareness and education about work zone hazards and driving conditions. Traffic controllers in particular thought that driver education, including addition of work zone issues to licensing program materials, was a potentially effective safety improvement measure. Respondents also proposed refresher courses to advise and update drivers about work zone signage and related regulations. Some posited that driver training lacks any practical work zone component, leading to novice drivers lacking supervised experience with work zone hazards and driving conditions. Media campaigns, including television, radio, and newspaper advertisements, were also a well-supported measure for improving public awareness.

Some respondents from the exposed and semiexposed groups felt that communication among and between workers and traffic controllers could be improved. It was claimed that workers without a (two-way) radio can often be unaware as to what is happening behind them, which could potentially result in reversing-related crashes. Additional to improved communication, some respondents claimed that reversing should be limited as far as possible, and where unavoidable, both radio and reversing cameras should be used. It was argued that mandatory single-channel radio use by all work zone personnel would improve communication and also hazard awareness among them.

Greater or more frequent physical separation of roadworkers from live traffic emerged as another potential safety improvement measure. Some respondents claimed that portable barriers should be available in long-term work zones, and that truck-mounted attenuators should offer both rear and side impact protection. It was noted that the installation and removal processes typically restrict barrier use to long-term work zones only. Further, while barriers undoubtedly improve the safety of workers working behind them, traffic controllers most often remain exposed to live traffic.

As noted previously, continuous alerts from work vehicles and plant (e.g., reversing beepers) sometimes blend with background noise and may not be heard sufficiently by roadworkers. Some respondents argued that a "spotter" could be employed to prevent crashes in such situations; however, it was also reported that some workers disobey the alerts and prefer not to rely on others.

Some respondents argued that there was scope for improving the conspicuity of currently available signage, specifically by adding flashing lights or just simply attaching flags to the signage.

Several other potential measures for mitigating work zone hazards were mentioned. These include antigawk screens (i.e., roadside safety screens) to address driver distraction, more conspicuous and comfortable safety vests, portable traffic lights to reduce exposure of traffic controllers, increased penalties for work zone traffic violations, and periodically alternating electronic messages on variable message signs (VMSs) to limit driver desensitization to repetitive messages.

Although most respondents perceived that safety improvement could be achieved with appropriate additional measures, a significant number felt that there was little scope for further improvement. They perceived that compliance with current rules and signage would be effective enough for ensuring their safety, which was somewhat at odds with their stated experience of poor compliance among motorists. Only 19% of the nonexposed group suggested that there was no scope for substantial safety improvements, compared with 42% and 60% of the exposed and semiexposed groups, respectively. The nonexposed respondents therefore appeared to have a better understanding than the other respondents of the potential for further work zone safety improvement. Additionally, younger respondents indicated that they were more positive about the potential for further improvement than those aged 54 years or older.

2.3 DISCUSSION

This chapter identifies safety issues in road construction and mitigation approaches from the regulatory, organizational, and operational levels. At the regulatory level, the examination of policy documents showed that there were indeed two explicit foci: worker safety and the safety of members of the public traveling through worksites. In general, the policies of the road construction companies focused on worker safety, while those of the road authorities had a greater emphasis on road safety for the public.

At the organizational level, the factors influencing the impact of the policy documents on practices were examined in the management discussion groups. These group discussions revealed that the cost of safety measures was an important consideration in their adoption. Road construction companies are competing on price in tendering for work, wherein safety (particularly traffic control) is a significant component of the cost, and therefore subject to downward pressure. The lack of good data on the relative effectiveness of different approaches to achieving safety objectives was identified as a barrier to selecting and justifying what were judged to be the best safety measures. The evaluations of the effectiveness of on-road safety measures that we will be conducting in a later part of the program of research will provide valuable input into addressing the concerns expressed at the organizational level of needing more objective information to assess the cost-effectiveness of particular safety measures.

The interviews with workers revealed how the goal of the road authorities to minimize disruption to the traveling public influenced operational decisions with potential safety consequences. To maintain traffic flow, many works on high volume roads are undertaken at night, with workers consequently less visible to drivers and more exposed to drunk drivers and fatigued drivers.

Driver frustration and aggression, ranging from verbal abuse to throwing objects and physical assault, were reported by roadworkers as a common cause of feeling unsafe. This was not mentioned as a safety issue in any of the policy documents or in the interviews with management. Thus, it seems that some safety issues experienced at the operational level are not feeding back into programs and policies to mitigate their harm at the organizational or regulatory levels.

Several important conclusions can be drawn from the perspective of people at the operational level who are directly involved in construction, maintenance, and traffic control activities. First, public vehicles and related driver behavior issues pose the greatest perceived challenge with regard to minimizing roadwork risks and hazards. The perceptions of workers are supported to a large extent by the work zone safety literature. While the considerable risks and hazards directly associated with construction activities, such as mobile plant, tools and machinery, hazardous materials, and other environmental factors, are clearly recognized, they are seen to be more within the control of workers and also management. Speeding, driver distraction, lack of awareness, driver fatigue, impairment, and driver aggression are thus all key concerns of those who are exposed to public vehicle traffic in and around work zones.

Many of the incidents that workers described in the interviews occurred in the approach to roadworks, where traffic controllers are the group most exposed to risk of injury. The traffic controllers, who are commonly employed by subcontractors, were also reported to be subjected to driver aggression. While the regulatory framework states that the prime contractor is responsible for their safety, the occurrence of ongoing incidents involving injuries to traffic controllers suggests that more attention is needed to improve their safety. This was supported by the comments from management of the need for continued inspection and supervision of sites. The management comment also was consistent with worker interviews reporting that some workers ignored spotters or tampered with reversing alarms.

In both the discussions with management and the interviews with workers, respondents proposed that public education campaigns should be undertaken to improve driver behavior at roadworks. Similar suggestions have been made in earlier studies (Haworth et al., 2002; Pratt et al., 2001), but there remains a need for formal and reliable program evaluations (Arnold, 2003; Haworth et al., 2002; MVA Consultancy, 2006; Ross and Pietz, 2011). While other categories of safety measures have been evaluated objectively in terms of the extent of speed reduced, evaluations of educational measures have typically relied on public perceptions of their effectiveness obtained from surveys. For example, after a 5-year advertising and awareness campaign in Queensland, almost all drivers surveyed (97%) agreed that the campaign encourages drivers to slow down and 93% agreed that the campaign helped them to realize the potential consequences of speeding at roadworks and of disregarding traffic control signals and directions (TMR, 2009). Yet no measures of the effects on speeds at roadworks were collected.

The transportation and construction sector stands to benefit substantially from increasing collaboration on the development of a more standardized approach to increasing the safety around roadworks. The information identified in this chapter will be used to inform trials of a series of safety interventions, collaboratively agreed upon by our industry partners. The findings may also be applied internationally to inform the development of policies and procedures to enhance road construction safety globally.

2.4 CONCLUSIONS

Attempts to control the full suite of roadwork risks and hazards are largely developed, refined, and formalized at the organizational and regulatory levels, with the implicit (and often explicit) objective of achieving "zero-harm" outcomes. At the operational level, however, there appears to be widespread acceptance that working on roadways is inherently dangerous, largely due to the unpredictability of traffic, to the extent that not all risks and hazards can be controlled absolutely. In this sense, it could be argued that there is something of a divide between organizational and operational perspectives, and that much of the resultant gap is bridged by the regulatory frameworks, procedures, and requirements provided by the appropriate organizations. A range of factors influence the translation of safety policies into practice, including the cost of safety measures in the context of competitive tendering, lack of firm evidence of the effectiveness of safety measures, and pressures to minimize disruption to the traveling public.

REFERENCES

Anderson, W., Plowman, B., Leven, B., and Fraine, G. 1998. Workplace fleet safety system. Brisbane, Australia: Queensland Transport.

Arnold Jr., E.D. 2003. Use of police in work zones on highways in Virginia: Final report. Charlottesville, VA: Virginia Transportation Research Council.

Austroads. 2009. National approach to traffic control at work sites. Report No. AP-R337/09. Sydney: Austroads.

Bai, Y., and Li, Y. 2006. Determining major causes of highway work zone accidents in Kansas. Lawrence: University of Kansas/Kansas Department of Transport.

Biggs, H.C., and Biggs, S.E. 2012. Interlocked projects in safety competency and safety effectiveness indicators in the construction sector. *Safety Science* 52:37–42.

Debnath, A.K., Blackman, R., and Haworth, N. 2013. Understanding worker perceptions of common incidents at roadworks in Queensland. In *Proceedings of the 2013 Australasian Road Safety Research, Policing and Education Conference*, Brisbane, Australia.

Debnath, A.K., Blackman, R., and Haworth, N. 2014b. A Tobit model for analyzing speed limit compliance in work zones. *Safety Science* 70:367–377.

Debnath, A.K., Blackman, R., and Haworth, N. 2015. Common hazards and their mitigating measures in work zones: A qualitative study of worker perceptions. *Safety Science* 72:293–301.

Debnath, A.K., Blackman, R.A., and Haworth, N.L. 2012. A review of the effectiveness of speed control measures in roadwork zones. In *Proceedings of the Occupational Safety in Transport Conference*, Gold Coast, Australia.

Debnath, A.K., Blackman, R.A., and Haworth, N. 2014a. Effectiveness of pilot car operations in reducing speeds in a long-term rural highway work zone. In *Proceedings of the Transportation Research Board Annual Meeting 2014*, Washington, DC.

Dingsdag, D.P., Biggs, H.C., and Sheahan, V.L. 2008. Understanding and defining OH & S competency for construction site positions: Worker perceptions. *Safety Science* 46:619–633.

Dingsdag, D.P., Biggs, H.C., Sheahan, V.L., and Cipolla, D.J. 2006. *A Construction Safety Competency Framework: Improving OH&S Performance*. CRC for Construction Innovation. Brisbane, Australia: Icon.Net Pty Ltd.

Doege, T.C., and Levy, P.S. 1977. Injuries, crashes and construction on a superhighway. *American Journal of Public Health* 67:147–150.

Garber, N.J., and Zhao, M. 2002. Crash characteristics at work zones. Charlottesville: Virginia Transportation Research Council.

Haworth, N., Greig, K., and Wishart, D. 2008. Improving fleet safety—current approaches and best practice guidelines. Sydney, Australia: Austroads.

Haworth, N., Symmons, M., and Mulvihill, C. 2002. Safety of small workgroups on roadways. Clayton, Australia: Monash University Accident Research Centre.

Haworth, N., Tingvall, C., and Kowadlo, N. 2000. Review of best practice road safety initiatives in the corporate and/or business environment. Clayton, Australia: Monash University Accident Research Centre.

Health and Safety Executive. 2000. Driving at work: Managing work-related road safety. Sudbury, UK: Health and Safety Executive.

Highways Agency. 2006. *Roadworkers Safety Focus Groups*. Publication HA79/06. Guildford, UK: Highways Agency.

Khattak, A.J., Khattak, A.J., and Council, F.M. 2002. Effects of work zone presence on injury and non-injury crashes. *Accident Analysis and Prevention* 34:19–29.

Krux, W., and Determan, D. 2000. Pre-crash warning system for temporary road maintenance sites. Berlin: Siemens AG, Traffic Guidence Systems.

MCR [Marketing and Communications Research]. 2009. Roadwork Safety Campaign 2009. Quantitative research report. Report prepared for DTMR by MCR Pty Ltd.

MVA Consultancy. 2006. Roadworkers' safety research—phase two: A report to the Highways Agency by MVA. London: Highways Agency Great Britain.

NWZSIC [National Work-Zone Safety Information Clearinghouse]. 2012a. Fatalities in motor vehicle traffic crashes by state and work zone 2010. http://www.workzonesafety. org/crash_data/workzone_fatalities/2010 (acessed March 12, 2014).

NWZSIC [National Work-Zone Safety Information Clearinghouse]. 2012b. Motor vehicle traffic fatalities by year, construction/maintenance zone and the highest "driver or motorcycle operator" BAC in the crash. http://www.workzonesafety.org/crash_data/ workzone_fatalities/alcohol_fatalities (accessed March 12, 2014).

Pigman, J.G., and Agent, K.R. 1990. Highway accidents in construction and maintenance work zones. *Transportation Research Record* 1270:12–21.

Pratt, S.G., Fosbroke, D.E., and Marsh, S.M. 2001. Building safer highway work zones: Measures to prevent worker injuries from vehicles and equipment. Cincinnati, OH: National Institute for Occupational Safety and Health.

Ross, J.H., and Pietz, A.J. 2011. Maximizing investments in work zone safety in Oregon: Final report. Salem: Oregon Department of Transportation.

RTA [Roads and Traffic Authority]. 2008. Road traffic crashes in New South Wales. Statistical statement for the year ended 31 December 2007. Sydney: New South Wales RTA.

SWOV. 2010. Roadworks and road safety: SWOV fact sheet. Leidschendam, the Netherlands: SWOV Institute for Road Safety Research.

TMR [Queensland Department of Transport and Main Roads]. 2009. 2009 Queensland road safety awards nomination. Roadwork safety: Making safety around roadworks everyone's responsibility. Brisbane: TMR.

Whitmire II, J., Morgan, J.F., Oron-Gilad, T., and Hancock, P.A. 2011. The effect of in-vehicle warning systems on speed compliance in work zones. *Transportation Research Part F: Traffic Psychology and Behaviour* 14:331–340.

3 Development of an Occupational Health and Safety Management System for Manufacturing Companies in Mexico Using Factorial Analysis

Luis Cuautle Gutiérrez and
Miguel Angel Avila Sánchez

CONTENTS

3.1   INTRODUCTION

Because of mass production, several departments and key functions of firms were created. First, a clear division between management and the operative area was distinguished, each one with well-defined duties. However, in order to be successful, both parts were made conscious of synergy elements. The companies started to create systems to control their activities and indicators that show day-to-day operation performance.

On an organizational level, top management took a leading role in the business destiny, and its activities were focused on resource optimization and survival.

In 1 year, there were 1,412 decreases in Mexico for labor hazards and 411,000 job accidents. These two matters represent 81% of the threats registered by the Social Security Mexican Institute (IMSS). The main harm to the body occurs in the hands and wrists, as well as the ankles and feet, followed by the abdomen, backbone, and pelvis. This situation causes injuries, trauma, burns, and mutilation. The accidents occur at a rate of 62% in men and 38% in women, but due to the increasing demand of women in assembly processes, the rate is growing backward.

According to official numbers, accidents in workplaces appear during the productive ages of 25–29 years old, with seniority in a job of 1–4 years. The employments with the greatest danger from the point of view of prevention are loaders in self-service stores, machinery and tools operators, managers, and hospital and hotel personnel.

On the other hand, and due to market globalization, the implementation and maintenance of a formal management system has become a regular practice in process assurance and even a contractual requirement.

Mexico is characterized by its manufacturing sector. The main areas are the automotive, petroleum, aerospace, chemical, and metallurgical industries. These types of factories are used for work in hazard situations all the time, increasing accident probability.

Currently, many international standards deal with quality, environment, technology, social responsibility, safety, and occupational health, among others. Some of them have common points but different goals.

Therefore, and in order of their priorities, the companies have adopted one or more management systems to deal with this situation. These systems work independently and in a simultaneous way, creating chaos among the personnel because of the inexistence of a central axis that provides the organization with certainty and conviction. The need to have an integral management system according to the specific operations of the factory has become a crucial issue.

Mexican companies are not far from this situation, and they have tried to reproduce world-class practices mostly in the quality field. Nevertheless, the continuous demands of society and international markets have forced Mexican companies to work with management systems such as occupational health and safety.

This research emerges with the intention to give clarity to Mexican manufacturing companies in the creation of a unique management system that covers those aspects and allows them to accomplish government as well as global clients' requirements.

3.2 LITERATURE REVIEW

Table 3.1 shows a report from IMSS with an important 0.27 rise in accidents, disabilities, and deaths from 2004 to 2013. In 2013, there were 69,089 accidents of production staff workers, 33,749 of lifting workers, 16,389 of machinery operators, 12,057 of truck drivers, and 9,689 of packaging and manufacturing workers.

In terms of gender, men exceed women in injuries due to job accidents. The most common injuries and their respective amounts during the period of 2011–2013 in Mexico are presented in Table 3.2.

Meanwhile, in 2013, the most frequent job illnesses were hearing loss (1489 cases), lung disease (1184), pulled back muscle and lower back strain (434), carpal tunnel

TABLE 3.1

Rates of Accidents, Illnesses, Handicaps, and Deaths in Workplaces in Mexico: 2004–2013

Year	Owners	Average Workers	Job Accidents per 100 Workers	Job Illnesses per 10,000 Workers	Job Handicaps per 100 Cases	Job Deaths per 10,000 Workers
2004	804,389	12,348,259	2.29	6.01	7.16	0.87
2005	802,107	12,735,856	2.32	5.73	6.51	0.87
2006	810,181	13,578,346	2.28	3.47	5.77	0.79
2007	823,999	14,424,178	2.50	1.87	4.51	0.73
2008	833,072	14,260,309	2.88	2.58	4.22	0.79
2009	825,755	13,814,544	2.86	2.97	4.69	0.80
2010	829,500	14,342,126	2.81	2.42	5.50	0.78
2011	821,572	14,971,173	2.82	2.74	5.72	0.82
2012	824,823	15,671,553	2.77	3.10	5.57	0.74
2013	833,105	16,224,336	2.56	3.92	6.07	0.61

Source: Secretaría del Trabajo y Previsión Social, Accidentes de trabajo según ocupación y sexo, 2011–2013, Tlalpan: Mexico, Secretaría del Trabajo y Previsión Social, 2013, http://www.stps.gob.mx/bp/secciones/dgsst/estadisticas/Nacional%202004-2013.pdf (accessed December 15, 2014).

TABLE 3.2

Job Accidents in Terms of Injury and Gender

Injury Type	2011 Men	2011 Women	2012 Men	2012 Women	2013 Men	2013 Women
Superficial trauma	71,643	35,282	82,524	40,972	77,377	39,893
Dislocations and sprains	63,004	35,096	64,642	37,224	61,222	36,002
Wounds	63,083	16,174	65,173	16,554	60,597	15,687
Fractures	30,647	5,896	31,914	6,315	31,587	6,285
Trauma	12,208	3,580	11,813	3,499	11,747	3,409
Burns	7,094	2,731	7,155	2,853	6,755	2,826
Foreign bodies	4,353	487	3,985	441	3,360	390
Mutilation	2,993	478	3,073	475	3,243	472
Poisoning	756	330	744	309	643	333
Minor frequencies/other	44,239	17,496	37,846	14,584	38,178	15,654

Source: Secretaría del Trabajo y Previsión Social, Accidentes de trabajo según ocupación y sexo, 2011–2013, Tlalpan: Mexico, Secretaría del Trabajo y Previsión Social, 2013, http://www.stps.gob.mx/bp/secciones/dgsst/estadisticas/Nacional%202004-2013.pdf (accessed December 15, 2014).

syndrome (men, 58; women, 278), shoulder damage (281), Quervain's tenosynovitis (men, 40; women, 216), and bursitis (227).

The self-management program of health and safety in the workplace was founded in 1995 with the creation of preventive programs, which were implemented by mandate in factories with 100 or more workers that export products in the northern region of Mexico, Jalisco, and Mexico City, in order to promote a health and safety culture and achieve the commitments derived from the North American Agreement on Labor Cooperation.

In 1997, the Safety, Health, and Labor Environment Regulation was published in the Official Diary. Article 12 states that the secretary will form assessment and certification programs through voluntary engagements from those companies that need them.

The Mexican Social Prevention and Labor Secretary established a technical program in occupational health and safety management. Five steps are used to represent the basic security scheme: planning, resource creation, voluntary commitment, audits, and recognition. The accomplishment of those requirements led the Mexican firms to obtain the "Empresa Segura" Award. The elements are in Figure 3.1.

For this last point, there are three types of awards in accordance with the maturity level of the occupational health and safety management system (OHSMS): (1) for the accomplishment of the national standard, (2) for continuous improvement actions, and (3) for results in management. However, only 15 Mexican manufacturing factories have been recognized for their safety due to their great documentation work and lack of commercial benefits. The national oil company Petroleos Mexicanos (PEMEX) is the pioneer in implementing this kind of management.

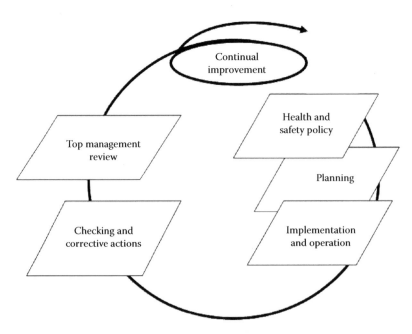

FIGURE 3.1   Safety and health successful management elements.

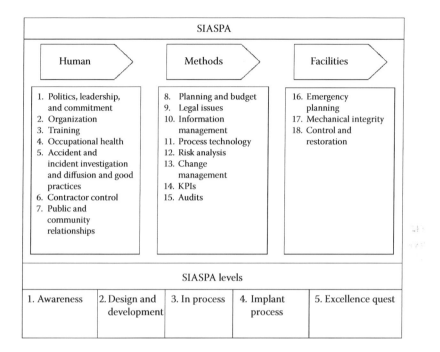

FIGURE 3.2 SIASPA components and levels. KPIs, key performance indicators.

The health programs deal with the identification of important alarms and the promotion of actions to eliminate them. These issues consider job environments, as well as pandemics.

In Mexico, occupational health and safety have been treated individually, creating different management systems and confusion among Mexican manufacturing companies. One of the first attempts to combine both approaches was developed by PEMEX.

In 1997, SIASPA (Sistema Integral de la Administracion de la Seguridad y Proteccion Ambiental) was implemented by PEMEX. This system is shown in Figure 3.2 and is integrated by five levels of progress and 18 elements related to the human factor, work methods, and operations. Until 2005, there were 159 PEMEX facilities with this management system.

Unfortunately, the SIASPA method is not known among manufacturing industries. In addition, the investment involved for its implementation, as well as the size of the companies, represents a huge problem for its consideration.

In terms of safety, the twentieth century has witnessed the worldwide birth and growth of the safety movement, from machinery risks prevention and compensatory laws to state-of-the-art perspectives.

Throughout history, there has been several safety focuses (Figure 3.3).

In the construction industry, through the use of the balanced scorecard and quality function deployment (QFD) tools, a safety management system was developed (Figure 3.4).

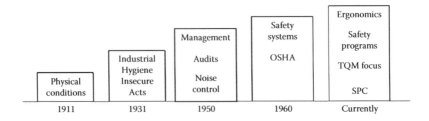

FIGURE 3.3 Safety focuses. OSHA, Occupational Safety and Health Administration; TQM, total quality management; SPC, statistical process control.

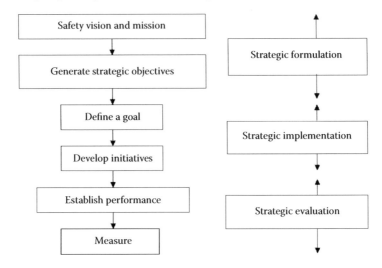

FIGURE 3.4 Safety system structure proposal. (From Gunduz, M., and Simsek, B., *Canadian Journal of Civil Engineering*, 34(5), 623, 2007. Copyright © Canadian Science Publishing or its licensors.)

To implement a safety management system, Petersen (2000) notes six criteria:

1. Daily actions assurance by the supervisor and teams to demonstrate that safety is an organization value
2. Involvement of top management as key participants
3. Requirement of actions and not only commitments
4. Demand for involvement in routine activities
5. Allowance of flexibility
6. Positive awareness by the employees

The benefits for the certified companies in safety programs, as noted by Carrillo and Garcia, are

1. Reduction in social security premium payments
2. Safety and hygiene culture
3. Teamwork in the improvement of safety procedures

Due to a safe environment, personnel's sense of belonging increases. This situation promotes productivity and quality. Worldwide, the OHSAS 18000 series deals with OHSMSs. OHSAS is a specification that allows organizations to control risk, improve performance, and achieve internal and external requirements.

OHSAS 18001 was created with the intention to be an International Organization for Standardization (ISO) standard, such as BS 5750 for ISO 9000 and BS 7750 for ISO 14000.

The requirements established by OHSAS 18001 are similar to those for quality management systems. However, they include an occupational health and safety policy, planning to identify risks, hazard assessment and control, emergency preparation and response, and corrective and preventive actions.

3.3 RESEARCH

In this study, an OHSMS is developed by using structural equation models (SEMs) to help Mexican manufacturing companies conduct their efforts. The data in regard to the actions followed by the firms have been collected through an email survey, and their results have been considered for the creation of the OHSMS.

Based on extensive research on the different standards and best practices related to occupational health and safety in Mexico and around the world, a construct was created for the identification of the elements of an OHSMS (Table 3.3).

Using this construct, a questionnaire is designed. This instrument consists of 13 questions using a 5-point Likert scale for each of the items, from 1, strongly disagree, to 5, strongly agree. In order to measure the reliability of the survey, the total population and sample of Mexican manufacturing companies were calculated in Table 3.4.

TABLE 3.3
OHSMS Construct

	Levels	
1	**2**	**3**
Occupational health and safety management system	Risks	Assessment (VAR00029)
		Control (VAR00030)
	Management	Occupational health and safety policy (VAR00031)
		Planning (VAR00032)
		Inspection (VAR00033)
	Process	Emergency preparation and response (VAR00034)
	Improvement	Corrective actions (VAR00035)
		Preventive actions (VAR00036)
		Audits (VAR00037)
		Education and training (VAR00038)
	Documentation	Occupational health and safety manual (VAR00039)
		Procedures (VAR00040)
		Registers (VAR00041)

TABLE 3.4

Stratified Random Sample according to Management System

Stratum	Management System	Total Population (N)	Sample (n)
1	Clean industry	376	32
2	Safe company	15	8

FIGURE 3.5　Mexican states considered in the total sample.

The trial consisted of an email inquiry directed to the people in the manufacturing firms in charge of any certified management system. In the first stage, nine Mexican companies were contacted: 70% from the state of Puebla, 10% from the state of Mexico, 10% from the state of Morelos, and 10% from the state of Tlaxcala.

With their data, Cronbach's alpha was developed. From the results, Cronbach's alpha equals 0.9440. Therefore, the instrument is confirmed as reliable.

In the second phase of the research, 23 different Mexican manufacturing companies were included. In this new sample, northern states were considered, such as Nuevo León, Tamaulipas, San Luis Potosi, and Guanajuato, as well as southern states such as Morelos, Guerrero, and Quintana Roo (Figure 3.5).

The authors emphasize the inconvenience of obtaining answers from Mexican manufacturing companies because of their lack of trust and interest in this kind of research.

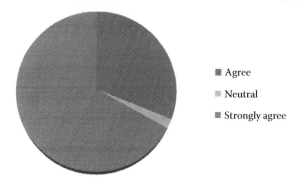

FIGURE 3.6 Total qualification of OHSMS items.

The survey was conducted in the same way, and the contributions of the items proposed are shown in Figure 3.6. It is obvious that the elements suggested were well accepted among the participating factories.

3.4 FACTORIAL ANALYSIS

Factorial analysis is a procedure that is employed in the reduction and synopsis of the data. This method makes no distinction between the dependency or independency of the variables. In addition, factorial analysis is a multivariate interrelation technique that is useful in the construction of scales under a theoretical concept.

At present, there are two types of factorial analysis: exploratory (EFA) and confirmatory (CFA). Both techniques create p linear combinations with p interrelated variables. The goal of both methods is only to keep the k first combinations to replace the p variables in the principal component analysis (PCA).

This research employs EFA and CFA to determine the attributes that integrate the OHSMS.

3.4.1 EXPLORATORY FACTORIAL ANALYSIS

The main components of the OHSMS were defined with the findings of the survey and the aid of LISREL software. Table 3.5 illustrates the three factors obtained from the 13 items studied.

An alternative perspective consists of the use of statistical criteria. In this case, using the varimax criterion as the orthogonal rotation method, the components of

TABLE 3.5
OHSMS Components

Factor	Variance
1	6.578
2	2.440
3	1.037

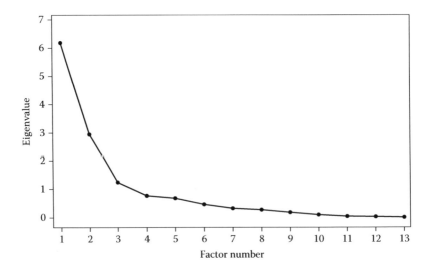

FIGURE 3.7 OHSMS sedimentation chart.

each factor are defined in Figure 3.7. Generally, the highest point of the slope (above an eigenvalue equal to 1) shows the quantity of significant factors.

Both procedures conclude the same: three factors must be considered for occupational health and safety.

Then, the PCA and varimax were used as extraction and rotation methods, respectively. The interpretation was realized identifying the variables a high load in the same factor. Table 3.6 shows the factors obtained with their respective value for each element.

To verify the discoveries, EFA was performed with different kinds of extraction methods: underweighted least squares, generalized least squares, maximum likelihood, principal axis factoring, alpha factoring, image factoring, and two rotation methods—varimax and oblimin. All tests showed the same conclusions.

According to their nature and based on ISO standards, the factors are named health and safety focus (e.seguri), health and safety management (g.seguri), and continual improvement (mejora). Therefore, the three main factors and their components are established in Table 3.7.

3.4.2 Confirmatory Factorial Analysis

The regression techniques, as well as the path analysis, are types of SEMs that analyze the causal and noncausal relationships between the variables considered in the constructs. It is necessary to develop a measurement for the construct that validates the relationship between itself and its indicators. Considering the information of the EFA, the three main factors were validated using CFA, and the results are shown in Figure 3.8.

The root mean square error of approximation (RMSEA) calculated for each system is less than 0.08 (recommended value) and the Tucker–Lewis index is higher

TABLE 3.6
Orthogonal Rotated Component Matrix

	Component		
	1	2	3
Assessment	0.919	0.123	0.079
Risk control	0.806	0.279	–0.218
OHS policy	0.170	0.762	0.313
Planning	0.144	0.334	0.830
Inspection	0.270	0.702	0.237
Emergency preparation and response	0.853	0.191	0.295
Corrective actions	0.865	0.157	0.221
Preventive actions	0.790	0.039	0.396
Audits	0.232	0.434	0.743
Education and training	0.151	0.433	0.763
OHS manual	0.001	0.820	0.172
Procedures	0.139	0.815	0.294
Registers	0.249	0.797	0.232

Note: Extraction method, PCA; rotation method, varimax.

TABLE 3.7
OHSMS Factors

Health and Safety Focus (e.seguri)	Health and Safety Management (g.seguri)	Continual Improvement (mejora)
Assessment	OHS policy	Planning
Risk control	Inspection	Audits
Emergency preparation and response	OHS manual	Education and training
Corrective actions	Procedures	
Preventive actions	Registers	

than 0.90 (recommended value). Thus, the three main factors can be considered for a SEM.

3.5 STRUCTURAL EQUATION MODEL

The variables for the OHSMS are defined as

Occupational health and safety focus (e.seguri). Commitment to prevent industrial risks and labor diseases should meet national and international standards and strive to exceed employees' expectations. It includes risk

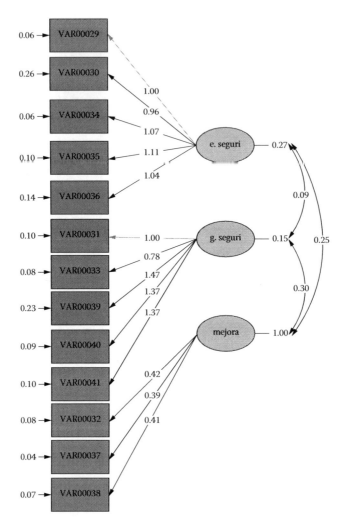

FIGURE 3.8 OHSMS variables' confirmatory analysis.

assessment and control, emergency preparation and response, and corrective and preventive actions.

Occupational health and safety management (g.seguri). This is a system approach to identify, understand, and manage occupational health and safety issues. This factor considers the occupational health and safety policy and the documental aspect of the management system.

Continual improvement (mejora): This is the willingness to increase the current health and safety prevention measures. Factors include planning, audits, education, and training.

The model considers the focus and management of the OHSMS as the first stage that causes the second stage, continual improvement.

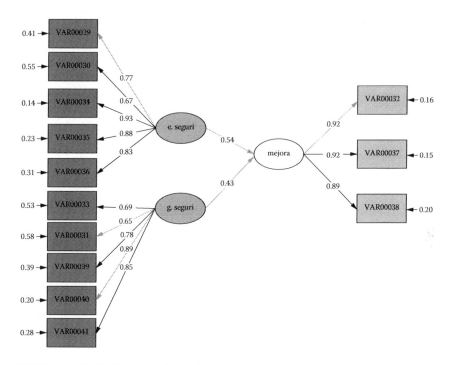

FIGURE 3.9   OHSMS structural equation model.

Using LISREL, a SEM was developed (Figure 3.9). The results are shown in Table 3.8.

3.6   CONCLUSIONS

This chapter shows a survey conducted among 32 Mexican manufacturing companies with at least one management system certified in terms of occupational health or safety. Forty-three percent of the companies are located in the central part of Mexico, while 32% and 25% come from the northern and southern states, respectively. In terms of the business sectors, the automotive, metallurgical, and chemical results are the most representative.

Regarding the EFA and based on the results generated by the software Minitab, three main components of the OHSMS were calculated, which were confirmed employing different extraction methods (unweighted least squares, unweighted maximum likelihood, and principal axis factoring, among others) and two principal methods of rotation (varimax and oblimin).

Related to the confirmatory analysis, the relationship among the proposed elements of the OHSMS was confirmed. An OHSMS is proposed in Figure 3.10 for the Mexican manufacturing companies in order to facilitate the implementation of both issues in working areas. This proposal denotes an approach parallel to the one proposed by Mexican standard IMNC NMX-SAST-001-IMNC-2008.

TABLE 3.8
OHSM Goodness-of-Fit Results

Goodness-of-Fit Statistics	Goodness-of-Fit Acceptable Levels	Result	Approval
Absolute Fitness Measures			
Estimated noncentrality parameter (NCP)	In terms of χ^2	0.669	N/A
Goodness-of-fit index (GFI)	[0–1]	0.709	Marginal
Root mean square residual (RMSR)	In terms of the entrance matrix	0.0269	Marginal
Root mean square error of approximation (RMSEA)	Acceptable levels under 0.08	0.0162	Acceptable
Expected cross-validation index (ECVI)	N/A	2.943	Marginal
Incremental Fitness Measures			
Adjusted goodness-of-fit index (AGFI)	0.90	0.573	Medium
Tucker–Lewis index (TLI/NNFI)	0.90	0.999	Acceptable
Normed fit index (NFI)	0.90	0.921	Acceptable
Parsimony Fitness Measures			
Parsimony goodness of fit (PGFI)	Higher values indicate biggest parsimony of the model	0.483	Low
Normed χ^2	Lower limit: 1.0 Upper limit: 5	1.0	Acceptable
Parsimony normed fit index (PNFI)	Higher values indicate smaller fitting	0.732	Marginal
Akaike information criterion (AIC)	Smaller values indicate parsimony	818.577	Low

Note: N/A, not applicable.

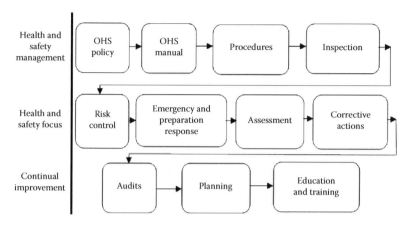

FIGURE 3.10 OHSMS model.

It is expected that this OHSMS will allow Mexican firms to control their operations in terms of occupational health and safety, and to prevent and reduce accidents in their facilities.

This proposal needs to be implemented in Mexican manufacturing companies and get the feedback of top management. Also, it can be used in an integrated management system that considers OHSM.

The authors suggest the creation of more SEMs using other types of relationships and employing other estimation methods and different correlation matrices, as well as other management systems, such as quality, environment, and social responsibility, among others, to create more integrated management systems.

REFERENCES

Petersen, D., 2000, Safety management 2000: Our strengths and weaknesses, *Professional Safety*, 45(1), 16–19.

Secretaría del Trabajo y Previsión Social, 2013, Accidentes de trabajo según ocupación y sexo, 2011–2013, Tlalpan: Mexico, Secretaría del Trabajo y Previsión Social, http://www.stps.gob.mx/bp/secciones/dgsst/estadisticas/Nacional%202004-2013.pdf (accessed December 15, 2014).

BIBLIOGRAPHY

Arkins, P., 2003, Solving the mystery: A summary of OHSAS 18000, ISO 18000 & ISO. IEC JTC-1/SC31, *Professional Safety*, 12(3), 13.

Bolton, F. N., and Kleinsteuber, J. F., 2001, A perspective on the effectiveness of risk assessment by first-line workers and supervisors in a safety management system, *Human and Ecological Risk Assessment*, 7(7), 1777–1786.

Camacho, M., 2007, Implantación de un sistema de administración de seguridad, salud y protección ambiental en petróleos Mexicanos, in *Proceedings of the Reunión Nacional de Seguridad Industrial y Protección Ambiental*, Mérida, Yucatán, December, 5–30.

Carrillo, J., and García, H., 2002, Evolución de las maquiladoras y el rol del gobierno y del mercado en la seguridad en el trabajo, *Papeles de Población*, 8(33), 173–186.

Casas, M., 2002, Los modelos de ecuaciones estructurales y su aplicación en el índice Europeo de satisfacción del cliente, http://www.uv.es/asepuma/X/C29C.pdf (accessed March 19, 2010).

Cuautle, L., and Cholula, C., 2014, Integrated management system proposed for manufacturing companies in Mexico, *European Journal of Business and Social Sciences*, 3(3), 39–53.

Gundz, M., and Simsek, B., 2007, A strategic safety management framework through balanced scorecard and quality function deployment, *Canadian Journal of Civil Engineering*, 34(5), 622–630.

Hair, J., Anderson, R., Tatham, R., and Black, W., 2000, *Análisis multivariante*, Vol. 491, Madrid, Prentice Hall.

Instituto Mexicano de Normalización y Certificación, 2000, Norma Mexicana IMNC NMX-SAST-001-IMNC-2008, Sistemas de administración de seguridad y salud en el trabajo, Especificación, Cuauhtémoc, Mexico, Instituto Mexicano de Normalización y Certificación.

Koehn, E. E., and Datta, N. K., 2003, Quality, environmental, and health and safety management systems for construction engineering, *Journal of Construction Engineering and Management*, 129(5), 562–569.

Landero, R., and González, M., 2006, *Estadística con SPSS y metodología de la investigación*, Mexico: Trillas.

Mairani, J., 2007, No matter the plant size, quality management systems measure up, *Plant Engineering*, 61(4), 25–26.

Malhotra, N. K., 1997, *Investigación de mercados*, 2da. Edición, editorial Pearson Educación, Mexico.

Pheng, L. S., and Pong, C. Y., 2003, Integrating ISO 9001 and OHSAS 18001 for construction, *Journal of Construction Engineering and Management*, 129(3), 338–347.

Programa de Autogestion en Seguridad y Salud en el Trabajo, 2009, Lineamientos generales de operación, Tlalpan: Mexico, Secretaría del Trabajo y Previsión Social, http://autogestion.stps.gob.mx:8162/pdf/Lineamientos%20Generales%202008.pdf (accessed January 13, 2015).

Schaechtel, D., 1997, How to build a safety management system, *Professional Safety*, 42(8), 22–23.

Shah, G. C., 2003, Strategic business elements of ISO9001:2000, *Hydrocarbon Processing*, 82(9), 109–110.

Wilkinson, G., and Dale, B. G., 1999, Integration of quality, environmental and health and safety management systems: An examination of the key issues, *Proceedings of the Institution of Mechanical Engineers, Part B, Journal of Engineering Manufacture*, 213(3), 275–283.

4 Characterization of the Portuguese Furniture Industry's Safety Performance and Monitoring Tools

Matilde A. Rodrigues, Pedro Arezes, and Celina P. Leão

CONTENTS

4.1 INTRODUCTION

Portugal has a large cultural tradition of using forestry resources for construction and furniture. An important milestone in Portuguese history was the use of wood for shipbuilding in the era of the discoveries. Nowadays, wood from Portuguese forests, as well as imported wood, continues to be an important source of raw materials for industry (ITTC, 2008). In fact, the industry linked to wood products, particularly those that are in the wood and mattress manufacturing sector (Code 310 of the Classification of Economic Activities), underpins a significant part of Portugal's economy. The furniture subsector (Code 3109) represents 70.4% of the entire sector; it is the largest one, particularly in the north and center of Portugal, where most of the industries are located.

Despite the importance of the furniture industry, this subsector has been strongly affected by the crisis that Portugal has faced in recent years. Its economic and organizational constraints may have led it to become more vulnerable to the impact of the crisis. Several of these industries are dependent on the domestic market (Cardoso, 2014), which has been in decline. Furthermore, most of the workforce is unqualified, and the professionalization of their management, marketing, and trade policies is low (EGP, n.d.), which puts the companies in this sector in a fragile situation, particularly when they are trying to improve exports. This scenario is reflected in the number of accidents in the sector, which remains high (Eurostat, 2012; GEE, 2015) and is expected to increase in the remaining companies due to a lack of investment by the companies in occupational safety and health (OSH) during this period of crisis (Loureiro et al., 2014).

To reverse this trend, these industries have made a great effort, increasing exports. In 2014, this sector was expected to achieve a record 2.2 billion euros in exports, representing an increase of 10% compared to 2013 (Cardoso, 2014). In fact, the furniture industry is trying to increase its competitiveness not only through modernization, but also by improving its safety performance. Understanding the principal sources of accidents, monitoring their safety performance, and identifying critical areas of intervention will enable the companies to implement a more efficient OSH management strategy, and consequently improve their safety performance and their image to the outside.

Safety climate measures can be important tools to monitor the companies' safety performance (together with other measures and strategies), by warning of problems related to safety, preferably before injuries occur, and allowing for safety interventions and management programs to be designed (Arezes and Miguel, 2005; Vinodkumar and Bhasi, 2009; Rodrigues et al., 2015a). They have been linked to important safety-related outcomes, such as risk perception, safety management systems, accident rates, and safe behavior (Varonen and Mattila, 2000; Rundmo, 2000; Johnson, 2007; Huang et al., 2007; Arocena et al., 2008; Nielsen et al., 2008; Tharaldsen et al., 2008; Vinodkumar and Bhasi, 2009; Lu and Yang, 2011; Fugas et al., 2012). However, for companies to achieve good results, suitable instruments, which include the specificities of the sector of activity, need to be used. In the case of the furniture industry, the Safety Climate in Wood Industries (SCWI) measure, developed by Rodrigues et al. (2015b), can be a suitable instrument for measuring safety climate. This instrument was based on a multilevel construct, including three levels of analysis: organizational, group, and individual levels. Furthermore, the instrument includes items specific to the furniture industry, which enhances the instrument's sensitivity (Zohar, 2008).

In light of the points above, knowledge about the sources of accidents in the furniture industry appears to be important to help both competent authorities and the companies define an intervention strategy to reduce occupational accidents, particularly during this period of crisis. Moreover, it is important to understand how furniture companies can monitor their safety performance and identify the most critical areas. In this regard, new strategies beyond the traditional ones, as safety climate measures, can give companies important new insights to define priorities of intervention.

This study is an extension of the work presented by Rodrigues et al. (2015a). It aims, as a first stage, to analyze the occupational accidents in Portugal in the furniture subsector and to identify the key unsafe conditions that can originate these accidents, providing OSH practitioners with badly needed information to define an intervention strategy. Furthermore, an analysis of the SCWI as a tool to monitor the furniture subsector safety performance will be also performed.

4.2 METHODOLOGY

4.2.1 Sample

For the present study, 14 Portuguese furniture companies were analyzed. They varied in size from micro- to medium-sized companies and were all located in northern Portugal. A total of 403 blue-collar workers, divided into 33 workgroups, were considered for this study. The formation of the groups took into consideration the department or sector of activity, supervisors, and physical boundaries. The participants' age was on average 39.49 years (standard deviation [SD] = 10.09, interval range 18–63 years) and the majority were males (86.6%). They had been employed by their companies for an average of 10.47 years (SD = 7.27, interval range 0–37 years) and had been engaged in manual labor for an average of 17.49 years (SD = 12.06, interval range 1–50 years).

4.2.2 Identification of Unsafe Conditions and Behaviors

A safety audit was performed for each of the analyzed companies to identify unsafe conditions and behavior. To support that, a checklist was developed by the researchers based on Portuguese legislation and other specific guidelines for the furniture subsector (e.g., Decreto Lei n° 103/2008; Decreto Lei n° 24/2012; Decreto Lei n° 347/1993; Portaria n° 987/93; Miguel et al., 2005). The checklist comprised a set of 112 items related to the safety conditions of workplaces, equipment and machinery, safe behavior, and procedures. A 5-point Likert scale adapted from Reese (2012) was used to characterize the level of deficiency of each feature or behavior under analysis (1 = very deficient and 5 = excellent). An option for "not applicable" situations was also considered in the checklist. This option was considered in situations where a risk factor was not found to be applicable to the specific situation under analysis. At the end of each safety audit, the results were discussed with the companies' management and supervisors.

4.2.3 ACCIDENT REPORTS

Data about occupational accidents in the year 2012 gathered by the Portuguese Office of Strategy and Studies (GEE) for the "furniture manufacturing for other purposes" subsector (Code 3109 of the Classification of Economic Activities) were used for the purposes of this study. They derived from the accidents reported by the companies and were presented in the format of the European Statistics on Accidents at Work Methodology (ESAW-III) (Eurostat, 2013). A total of 2546 accidents were included in this analysis.

Since the only data available from GEE were not specified by categories, the data analysis was limited.

4.2.4 SAFETY CLIMATE ANALYSIS: SCWI TOOL

The analysis of the safety climate was performed by using the SCWI tool, previously developed and validated by the authors in Rodrigues et al. (2015b). The first part of SCWI includes workers' demographic questions: age, gender, department or sector, professional activity, number of years working at the company, number of years in the applicable professional activity, and previous involvement in work accidents. The second part includes 34 items for measuring safety climate, analyzing three different levels: organizational (13 items), group (12 items), and individual (9 items) levels. The dimensions analyzed at the organizational level included management investment in safety issues, continuous improvement of safety systems, and safety communication. At the group level, workers were asked about supervisor concerns regarding workers' safety practices, involvement in safety issues, and effort in regard to compliance with rules and the use of safety protection. The individual level comprised a single dimension related to workers' commitment to safety. A description of the items included at each level of analysis can be found in Table 4.1. The level of agreement with each item was assessed using a 5-point Likert scale that ranged from 1 = strongly disagree to 5 = strongly agree.

4.2.5 DATA ANALYSIS

Descriptive analysis of the accident reports and unsafe conditions and behaviors was performed.

For the analysis of the relationship between safety climate and safety performance, separate analyses of both variables were performed first. To determine the workgroups' safety performance, the total number of items checked on the checklist and their corresponding score (according to the 5-point scale) were shown in percentages of the maximum value. For safety climate, scores were determined by adding up the level of respondents' agreement with each item on the scale. After that, bivariate correlation analysis was performed using both variables. The non-parametric

TABLE 4.1
Description of the Items Included in the SCWI Tool, by Level of Analysis

Organizational Level	Item Description
The management of this company …	Reacts quickly when a dangerous situation is detected, or an accident or incident occurs
	Insists on thorough and regular safety audits and inspections
	Is not interested in continually improving safety levels in each department
	Does not invest in modernizing work machines
	Invests in the implementation of measures to minimize the manual handling of loads
	Provides all the equipment needed to do the job safely
	Is strict about working safely when we are working under pressure
	Requires each supervisor or team leader to help improve safety in his or her sector or department
	Invests much time and money in safety training for workers
	Uses all available information to improve existing safety rules
	Promotes the development of appropriate work procedures for the tasks performed by workers
	Does not consider workers' suggestions about improving safety
	Provides workers with sufficient information on safety issues
Group Level	Item Description
My supervisor or team leader …	Makes sure we receive all the equipment needed to do the job safely
	Does not check frequently to see whether we are all obeying the safety rules
	Discusses how to improve safety with us
	Rather than using explanations, compels us to act safely
	Worries that I fulfill the regulations and work procedures
	Worries that I use all of the machines' protections
	Lets safety rules and procedures be ignored when we are working under pressure
	Frequently tells us about the hazards of our work
	Makes sure we follow all the safety rules, not just the most important ones
	Is strict about safety at the end of the shift, when we want to go home
	Spends time helping us learn to see problems before they arise
	Insists that we wear our personal protective equipment even if it is uncomfortable
Individual Level	Item Description
I …	Believe that safety is the main priority when I do my work
	Report dangerous situations immediately to one of my superiors whenever I see them
	Try to always follow the rules and work procedures when I run my work
	Do not use the personal protective equipment necessary for performing tasks
	Do not always use the machines' protections

(Continued)

TABLE 4.1 (Continued)
Description of the Items Included in the SCWI Tool, by Level of Analysis

Individual Level	Item Description
	Refuse to ignore safety rules, even when the work is delayed and production must be increased
	Disregard safety rules at the end of the shift, when I want to go home
	Clarify all my questions about the risks to which I am exposed
	Do not bring it to my colleagues' attention when I see them violating some rule or safety procedure

Source: Adapted from Rodrigues, M.A., et al., *Theor. Issues Ergon. Sci.* 16(4): 412–28, 2015.

Kruskal-Wallis H Test (KW), alternative to one-way ANOVA, was used to analyze significant differences (or not) between the distributions of the workgroups' safety climate level, considering a significance level of .05.

All statistical analysis was done using the statistical software package IBM SPSS® version 20.

4.3　RESULTS

4.3.1　ANALYSIS OF OCCUPATIONAL ACCIDENTS IN THE "FURNITURE MANUFACTURING FOR OTHER PURPOSES" SUBSECTOR

The ESAW-III (Eurostat, 2013) is used in Portugal to analyze the causes and circumstances of occupational accidents. However, only six of the variables proposed by ESAW-III are mandatory: working environment, specific physical activity, deviation, contact mode of injury, material agent associated with the deviation, and material agent associated with the contact mode of injury. The results found for these variables, by nonspecific category, are presented in Tables 4.1 through 4.6.

TABLE 4.2
Distribution of Accidents according to the Place Where the Worker Was at the Moment of the Accident

Description of the Site	Frequency	Percentage
Industrial site	2303	90.42
Construction site, construction, opencast quarry, opencast mine	19	0.76
Tertiary activity area, office, amusement area, miscellaneous	37	1.47
Health establishment	9	0.35
Public area	29	1.12
In the home	37	1.45
In the air, elevated, excluding construction sites	40	1.58
Ignored	73	2.85

4.3.1.1   Working Environment

Table 4.2 presents the distribution of accidents according to workplace, work premises, or general environment where the accident happened. According to the results, the great majority of accidents occurred when the victim was present or working at an industrial site (90.42%).

4.3.1.2   Specific Physical Activity

The specific physical activity refers to what the victim was doing just before the accident (Eurostat, 2013). The results in Table 4.3 show that most work accidents occurred when carrying something by hand (31.95%), working with handheld tools (23.48%), and operating machines (19.47%).

4.3.1.3   Deviation

Table 4.4 describes the distribution of accidents according to deviation from normal working process. It is noticed that loss of control is particularly important among the factors that cause accidents, since they form the most frequent type of deviation

TABLE 4.3
Distribution of Accidents according to the Specific Physical Activity Performed

Description of the Specific Physical Activity	Frequency	Percentage
Operating machine	496	19.47
Working with handheld tools	598	23.48
Driving or on board a means of transport or handling equipment	18	0.71
Handling of objects	295	11.58
Carrying by hand	813	31.92
Movement	241	9.46
Ignored	86	3.38

TABLE 4.4
Distribution of Accidents according to the Deviation

Description of the Deviation	Frequency	Percentage
Deviation by overflow, overturn, leak, flow, vaporization, or emission	214	8.38
Breakage, bursting, splitting, slipping, fall, or collapse of material agent	79	3.10
Loss of control (total or partial) of machine, means of transport or handling equipment, handheld tool, object, or animal	1126	44.19
Slipping, stumbling and falling, or fall of persons	184	7.21
Body movement without any physical stress (generally leading to an external injury)	84	3.28
Body movement under or with physical stress (generally leading to an internal injury)	749	29.41
Shock, fright, violence, aggression, threat, or presence	1	0.04
Ignored	112	4.39

TABLE 4.5

Contact Mode of Injury

Description of the Contact Mode of Injury	Frequency	Percentage
Contact with electrical voltage, temperature, or hazardous substances	32	1.27
Horizontal or vertical impact with or against a stationary object (the victim is in motion)	280	10.99
Struck by object in motion, collision with	607	23.82
Contact with sharp, pointed, rough, or coarse material agent	641	25.16
Trapped, crushed, etc.	132	5.20
Physical or mental stress	736	28.90
Bite, kick, etc. (animal or human)	1	0.04
Ignored	118	4.62

(44.19%). Body movement under or with physical stress, such as, for example, lifting, carrying, pushing, pulling, or putting down, was also found to be an important event that triggers accidents.

4.3.1.4   Contact Mode of Injury

Table 4.5 describes the circumstances of the accident, that is, how the victim was hurt. Results indicate that most injury occurrences were from workers' exposure to physical or mental stress (28.90%); from contact with sharp, pointed, rough, or coarse material agents (25.16%); or because they were struck or collided with an object in motion (23.82%).

4.3.1.5   Material Agent Associated with the Deviation

The results in Table 4.6 describe the tools, objects, and instruments involved in the deviation from normal activity. The category of materials, objects, products, machine or vehicle components, debris, and dust is most frequently related to the accidents (30.09%). Fixed machines and equipment, as well as office equipment, personal equipment, sports equipment, weapons, and domestic appliance categories, are also related to a significant proportion of the accidents.

4.3.1.6   Material Agent Associated with the Contact Mode of Injury

In relation to the material agent with which the victim came into contact, the results in Table 4.7 also emphasize the category of materials, objects, products, machine or vehicle components, debris, and dust (43.43%). Results are slightly different from those seen regarding the material agent connected to the deviation. In the case of contact, fixed machines and equipment have less impact.

4.3.2   Analysis of Unsafe Conditions and Unsafe Behavior

The main unsafe conditions and unsafe behavior were identified. Table 4.8 presents the percentage of companies where specific unsafe conditions and behavior were observed. In general, results show higher percentages of unsafe conditions and behavior for ergonomic issues and the use of machines.

TABLE 4.6

Distribution of Accidents according to Material Agent Associated with the Deviation

Description	Frequency	Percentage
Buildings, structures, and surfaces—at ground level (indoor or outdoor, fixed or mobile, temporary or not)	103	4.06
Buildings, structures, and surfaces—above ground level (indoor or outdoor)	45	1.75
Hand tools, not powered	152	5.98
Handheld or hand-guided tools, mechanical	99	3.90
Hand tools—without specification of power source	33	1.31
Machines and equipment—fixed	401	15.76
Conveying, transport, and storage systems	46	1.83
Land vehicles	14	0.56
Materials, objects, products, machine or vehicle components, debris, and dust	766	30.09
Chemical, explosive, radioactive, and biological substances	28	1.12
Office equipment, personal equipment, sports equipment, weapons, and domestic appliances	396	15.56
Living organisms and human beings	1	0.04
No material agent or no information	460	18.06

TABLE 4.7

Distribution of Accidents according to Material Agent Associated with the Contact Mode of Injury

Description	Frequency	Percentage
Buildings, structures, and surfaces—at ground level (indoor or outdoor, fixed or mobile, temporary or not)	201	7.91
Buildings, structures, and surfaces—above ground level (indoor or outdoor)	22	0.88
Motors and systems for energy transmission and storage	5	0.20
Hand tools, not powered	126	4.95
Handheld or hand-guided tools, mechanical	64	2.50
Hand tools—without specification of power source	29	1.15
Machines and equipment—fixed	184	7.23
Conveying, transport, and storage systems	51	1.99
Land vehicles	29	1.15
Materials, objects, products, machine or vehicle components, debris, and dust	1107	43.46
Chemical, explosive, radioactive, and biological substances	28	1.12
Office equipment, personal equipment, sports equipment, weapons, and domestic appliances	362	14.21
Living organisms and human beings	1	0.04
No material agent or no information	361	14.19

TABLE 4.8
Principal Unsafe Conditions and Unsafe Behaviors Identified at the Companies Analyzed

Unsafe Conditions and Unsafe Behaviors	Percent of Companies
Floor, Stairs, and Storage	
Uneven or poorly maintained flooring	36.4
Passageway obstructed or not marked	43.3
Vertical ladder without adequate protection	27.3
Slippery stairs	18.2
Cables and wires scattered on the floor	54.5
Loose material such as waste wood scattered on the floor	45.5
Materials stored near machines and work areas	54.5
Materials, tools, and auxiliary equipment unsafely stored	36.4
Chemicals	
Chemicals incorrectly stored	54.5
Spraying tasks performed without appropriate ventilation or extraction systems	18.2
Ergonomics	
Tasks require execution of repetitive work for long periods of time (e.g., polishing)	100
Tasks require excessive force or uncomfortable or awkward postures	100
Workplaces with inadequate dimensions	72.7
Heavy or bulky loads handled by only one worker	54.5
Materials are moved over long distances	36.4
Routes where loads are moved with steep slopes or stairs	36.4
Loads difficult to grasp or carry close to the middle of the worker's body	90.9
No mechanical devices for moving loads	45.5
No mechanical devices to help feed the wood into the machine	81.8
Machines—General Features	
Wood pieces blocking machines removed by hand while saw is running	27.3
Nonuse of push sticks to cut small pieces	45.5
Nonuse of pressure pads for fixing wood to machines	54.5
Machines remain running when they are not being used	18.2
Tools are handled in the direction of the worker's body	45.5
Little space around machines, hindering access and movement of people and materials	27.3
Machines have sharp edges, sharp angles, or rough surfaces	9.1
Machines do not have an efficient system for collecting wood dust	27.3
Areas for machines' moving parts are not marked	45.5
Machines running without guards or with guards in incorrect positions	100
Saws, blades, and drills have cracks or other defects	18.2
Specific Machines	
Spindle molder without protections for cutters, cutter block, and spindle behind the fence (e.g., pressure pads to form a tunnel guard, a guard covering the top of spindle and back of cutters)	90.9

TABLE 4.8 (Continued)
Principal Unsafe Conditions and Unsafe Behaviors Identified at the
Companies Analyzed

Unsafe Conditions and Unsafe Behaviors	Percent of Companies
Spindle molder without the use of a power feed unit for straight cuts	36.4
Spindle molder without the use of a guiding device and adjustable guard for curved work	45.5
Bandsaw without the correct use of adjustable saw protection height	90.9
Bandsaw without protection from the moving parts of the machine	54.5
Self-adjusting guard for crosscut	63.6
Panel dimension saws without a transparent protective cover for the circular saw	81.8
Panel dimension without mechanisms to adjust saws in height	27.3
Edge banding with guard open or nonexistent	27.3
Mechanical presses without two-handed controls or similar protections	63.6
Gluer without protection to prevent access to the area of the cylinder	18.2
Planer thicknesser without the cutter head enclosed or otherwise guarded	72.7
Planer thicknesser without antikickback fingers on thicknesser feed	45.5
Double-side thickness, side planer molder, and thicknesser without protection to prevent access to dangerous parts of the machine	45.5
Round-end tenoner without adequate protections for cutters and saws	63.6
Mortiser, drillings, and multirip saw without adequate protection from moving parts (current, drill, mill, driving mechanism)	90.9

Among ergonomic issues, repetitive work for long periods of time and tasks requiring excessive force or uncomfortable or awkward postures were found to be common to all companies. Furthermore, workers handling heavy loads, loads that are difficult to grasp or carry, workplaces with inadequate dimensions, and workers feeding the wood into the machine by hand were unsafe situations identified at most of the companies analyzed.

The results in regard to machines emphasize the presence of machines without guards or with the guard in an incorrect position at all the companies, including machines such as the spindle molder, panel dimension saw, bandsaw, mortiser, drill, and multirip saw. Some procedures adopted by workers—beyond the removal of the safety guard—were also found at several companies, such as not using pressure pads to fix wood to machines or push sticks to cut small pieces, as well as manual tools being moved in the direction of the worker's body.

4.3.3 ANALYSIS OF SCWI AS AN INDICATOR OF WORKGROUP SAFETY PERFORMANCE

Table 4.9 summarizes the descriptive results of the safety climate scores by group and level of analysis. Groups from the cutting department presented the lowest safety climate level (7, 11, 15, 18, 22, 27, and 30) and groups from the storage and assembly departments had the highest safety climate level (8, 10, 12, 16, 20, 23, 25, 26, and 29). An analysis to

TABLE 4.9

Safety Climate Average Scores, by Group and Level of Analysis

Company	Group	Organizational Scale ($x \pm$ SD)	Group Scale ($x \pm$ SD)	Individual Scale ($x \pm$ SD)	Total Average Score ($x \pm$ SD)
A	1	35.0 ± 2.5	28.4 ± 3.0	28.4 ± 1.0	91.8 ± 4.96
B	2	37.1 ± 6.9	35.4 ± 4.9	34.4 ± 4.1	106.9 ± 8.0
	3	32.9 ± 3.3	35.7 ± 4.9	31.6 ± 2.2	100.1 ± 7.8
	4	46.2 ± 4.7	41.8 ± 3.8	33.8 ± 3.3	121.8 ± 8.5
	5	41.3 ± 1.4	40.3 ± 2.1	31.0 ± 1.6	112.7 ± 2.8
	6	49.0 ± 2.5	46.7 ± 5.5	33.0 ± 7.5	128.8 ± 4.8
C	7	37.8 ± 6.6	45.6 ± 3.2	36.2 ± 4.4	119.6 ± 8.1
	8	46.4 ± 7.1	41.5 ± 4.3	33.2 ± 3.0	121.1 ± 8.2
	9	44.4 ± 3.8	41.6 ± 3.0	33.4 ± 4.3	119.4 ± 6.5
	10	43.7 ± 3.8	40.1 ± 3.0	40.5 ± 4.1	124.2 ± 4.8
D	11	45.6 ± 4.8	41.4 ± 4.1	32.8 ± 4.4	119.8 ± 8.0
	12	47.4 ± 5.2	43.3 ± 2.7	31.45 ± 3.2	122.1 ± 9.6
E	13	50.8 ± 1.6	41.0 ± 2.2	34.8 ± 1.0	126.6 ± 2.9
F	14	44.0 ± 3.1	38.1 ± 2.5	31.3 ± 3.0	113.4 ± 8.8
G	15	35.5 ± 1.9	31.2 ± 3.6	26.2 ± 2.9	93.0 ± 6.5
	16	33.5 ± 1.5	40.8 ± 5.0	30.5 ± 2.2	104.7 ± 7.7
H	17	55.4 ± 4.0	44.7 ± 2.7	36.1 ± 2.9	136.3 ± 7.7
	18	53.9 ± 2.1	45.8 ± 2.0	33.9 ± 3.4	133.6 ± 5.0
	19	54.0 ± 4.8	45.3 ± 3.6	36.3 ± 3.9	135.7 ± 8.0
	20	56.1 ± 4.6	48.7 ± 4.5	37.4 ± 5.1	142.3 ± 9.2
I	21	24.7 ± 1.6	27.6 ± 2.5	18.3 ± 1.1	70.7 ± 1.9
J	22	45.4 ± 3.6	37.6 ± 3.8	31.6 ± 2.9	114.6 ± 6.2
	23	48.6 ± 2.7	40.8 ± 3.3	35.0 ± 1.5	124.4 ± 3.0
	24	45.9 ± 1.4	43.0 ± 1.4	33.4 ± 0.9	122.3 ± 0.9
	25	48.8 ± 1.8	47.2 ± 2.2	36.4 ± 2.6	132.4 ± 4.5
K	26	51.1 ± 4.5	42.9 ± 4.6	35.1 ± 2.0	129.1 ± 7.1
	27	42.8 ± 3.8	38.5 ± 1.5	30.2 ± 1.1	111.5 ± 5.0
L	33	51.2 ± 1.8	39.2 ± 1.8	30.25 ± 1.0	120.7 ± 1.0
M	28	49.8 ± 3.8	41.2 ± 1.7	32.5 ± 3.2	123.7 ± 6.1
	29	56.0 ± 3.1	43.8 ± 3.8	38.5 ± 1.8	137.7 ± 6.4
	30	51.0 ± 3.6	41.2 ± 4.1	31.4 ± 2.8	123.6 ± 7.7
	31	53.5 ± 3.5	46.9 ± 4.7	36.7 ± 3.1	137.1 ± 8.4
N	32	40.4 ± 1.8	39.4 ± 1.7	27.7 ± 1.74	107.5 ± 5.3

Source: Adapted from Rodrigues, M.A., et al., *Theor. Issues Ergon. Sci.* 16(4): 412–28, 2015.

verify if these differences are statistically significant was performed. The results show differences in safety climate between workgroups, even in groups belonging to the same company, which were significant at the organizational level ($KW(32) = 237.32, p < 0.001$), group level ($KW(32) = 169.83, p < 0.001$), and individual level ($KW(32) = 154.01, p < 0.001$) and for the total safety climate score ($KW(32) = 209.16, p < 0.001$).

The relationship between safety climate level and the workgroup' safety performance was analyzed. Aggregated mean scores of the safety climate were used for this analysis. A strong positive linear correlation was found at all levels under analysis (organizational level, $r = 0.639$, $p < 0.01$; group level, $r = 0.686$, $p < 0.01$; individual level, $r = 0.564$, $p < 0.01$), meaning that with higher safety climate scores, safer behavior and workplaces with better safety conditions are expected.

4.4 DISCUSSION

The results of this study indicated that the work done by furniture workers exposes them to numerous ergonomic risk factors. The most important ones include repetitive work; working in forced, uncomfortable, and awkward body postures; handling heavy loads and loads that are difficult to grasp or carry; workplaces with inadequate dimensions; and workers feeding the wood into the machine by hand. The painting and polishing sectors are particularly critical, since several workers perform highly repetitive movements, which are considered an important source of musculoskeletal disorders (MSDs) (Coury et al., 2002). Workers make the same movements, with the same arm, for long periods of time, and in most cases in workplaces with unsuitable dimensions. Furthermore, in all sectors, workers handle heavy or bulky loads, or carry loads along difficult paths (e.g., those with stairs), mostly alone. In fact, only a few companies have rollers to transport wood or mechanical devices to help feed the wood into the machine. All these risk factors are important sources of MSDs.

The ergonomic risk factors identified are also related to the occurrence of other occupational accidents beyond those linked to physical stress. García-Herrero et al. (2012) found a strong direct correlation between occupational accidents and ergonomic risk factors, particularly the handling of harmful substances, the need to move heavy loads using significant force, working on unstable or uneven surfaces, working in areas of difficult access, or stretching to reach items that are out of reach from the usual working body position. This kind of effect is also reflected in accidents in the furniture industry. The majority of occupational accidents occurred when workers were carrying something by hand and working with handheld tools (e.g., polisher). For example, because they are carrying heavy and bulky loads alone, there is a higher risk of dropping the load.

Particularly revealing are the results indicating that the furniture industry in Portugal deals with high risks of worker contact with saws and blades or being squeezed. The presence of machines without guards or with the guard in an incorrect position was found at all of the companies analyzed, and the spindle molder, panel dimension saw, bandsaw, mortiser, drills, and multirip saw were the most problematic machines. These results are in accordance with previous studies made in Portugal. Miguel et al. (2005) have previously identified saws, drills, and milling cutters without any protection or with their protection compromised by workers as the most common hazards in the furniture subsector. Furthermore, results of accidents in the furniture subsector showed that operations with machines are the most important physical activities related to the occurrence of accidents, and contact with sharp, pointed, rough, or coarse material agents is the second most frequent contact mode of injury.

Another important risk related to the absence of safety guards is the projection of machine parts (e.g., saws or saw parts) and other objects. Without suitable protection and maintenance, saws, milling cutters, drills, and particles can be projected and cause severe injuries to workers (Miguel et al., 2005). This is particularly critical for the panel dimension saw, where the transparent protective cover for the circular saw was only found to be used at two companies. However, even in these companies, protection was not always used, and in some cases, it was adjusted to the highest limit and did not carry out its intended protection function. Furthermore, only these two companies had a plan for saw maintenance. However, despite results showing that several companies do not use saw protection, most of them were found to adjust the saw in accordance to the wood height, thereby reducing the risk of accidents.

In addition, situations involving materials and cables stored in passageways and materials stored near machines and work areas were observed at several of the companies analyzed. This increases the risk of falling at the same level and colliding with or against stationary objects.

In light of the results obtained, the cutting sector, where several cutting machines are used and workers are exposed to ergonomic risk factors, is one of the most problematic sectors of activity. These results are reflected in the safety climate achieved for the workgroups linked to the cutting sector. Additionally, although workers in the cutting sector must comply with more rules and safety procedures than other workers due to the risk to which they are exposed, they are more likely not to comply with any of them. Cutting is an important sector for production because the other sectors depend on its results. If production in the cutting sector is delayed, other sectors are left with no material for their work. Thus, workers from this department may be subjected to greater pressure from supervisors; this may explain the tendency to value production over safety (Reese, 2012), which may lead workers to ignore some of the safety rules and procedures. In fact, important unsafe behavior was found in this sector, particularly regarding the risk of contact with running saws and blades. Not using safe guards or push sticks to cut small pieces, removing by hand pieces blocked in machines, and not using pressure pads to fix wood to machines were critical types of unsafe behavior identified at several companies. Also, workers always use the driving bar for cutting small pieces at only two companies. Taking this into account and considering the operator's constant exposure, the probability of a cut to the hand when operating these machines is high. For the other machines, such as manual crosscuts, the round-end tenoner, and the spindle molder, protection was used in some cases.

The high number of unsafe conditions and amount of behavior identified at the companies analyzed can be linked to their low professionalization (EGP, n.d.), as well as the type of preventive services that the companies analyzed have. According to Portuguese legislation (Decreto Lei n° 109/2000), all companies are required to organize a system for OSH with a preventive organization model of some kind. Therefore, they need to implement a risk prevention plan, including an organizational structure, the definition of staff competencies, procedures, and all the necessary resources for preventive actions (Suárez-Cebador et al., 2014). Most of the companies analyzed have external advisory services, which makes it more difficult to perform a deeper analysis of risks and how to control them. Suárez-Cebador et al.

(2014) found, for electrical accidents in the construction industry, a relationship of dependence for this type of preventive organization adopted by companies and the accidents.

The low safety performance observed in most of the workgroups, which have a high number of unsafe conditions and behavior, is clearly reflected in the level of safety climate. These results are in accordance with previous studies, particularly the studies of Varonen and Mattila (2000) in wood-processing companies and of Cooper and Phillips (2004), which suggested that differences in safety climate level among workgroups may be related to workplace safety conditions. All the hierarchical levels of SCWI were strongly related to safety performance. At the organizational level, this reflects the importance of the policies and procedures defined by the companies' management to increase workplace safety (Guldenmund, 2007). The group and individual levels also appeared to be important. Supervisor concerns regarding workers' safety practices, involvement in safety issues and effort with regard to compliance with rules and the use of safety protection, and the workers' commitment to safety are all factors that improve workgroups' safety performance. According to these results, safety climate—particularly the SCWI tool—is an interesting measure for companies to monitor their safety performance (together with other traditional measures), allowing them to identify critical sectors when the analysis is performed by workgroup. Furthermore, a multilevel instrument can better allow the companies to identify faults in the involvement of hierarchical levels.

It is also important to emphasize that the level of safety climate does not depend on the company's size. According to a study by the Portuguese Management School of Porto (EGP, n.d.), most furniture companies in the country are small, with low professionalization and unqualified and undifferentiated workforces. Bearing this in mind, in this study, companies of different sizes were included, with special attention paid to include some small companies, as they are the most representative of this sector. The results showed that companies with fewer than 12 workers have the same chance of having a high safety climate level as they do a very low safety climate level.

4.5 CONCLUSION

This study highlighted that companies in the furniture industry still have many problems related to safety, particularly with the use of unsafe machines, unsafe behavior, and manual work. However, most of them can be prevented through correct work procedures. To achieve this goal, companies need to change priorities, particularly by providing adequate training, promoting high involvement of all hierarchical levels in safety, and paying careful attention to the design and selection of tools, equipment, material, and usage conditions. However, this is not an easy task. Some evidenced risk factors can be related to the current pressure that companies are suffering, denoting a lower interest in safety than in production. In fact, it was observed that the tasks are being performed with a minimum number of workers (e.g., furniture handling) and the workers are susceptible to unsafe behavior, most of which can be related to an attempt to increase production (e.g., removing safe guards). Therefore, the intervention in this sector needs to be performed not only at the worker level, but also at the management level, in order to avoid unsafe behavior and promote safe procedures with a better and more efficient management of resources. In this way, companies' safety climate will be improved.

The study also showed that SCWI is a good tool to be used by furniture companies as an indicator of workgroups' safety performance. It was shown that the higher safety climate scores are, the safer behavior and the better safety conditions in workplaces will be.

ACKNOWLEDGMENT

The authors are grateful to the Portuguese Office of Strategy and Studies, for supplying the national accident data pertinent to the furniture industry.

REFERENCES

Arezes, P.M. and A.S. Miguel. 2005. Individual perception of noise exposure and hearing protection in industry. *Hum. Factors* 47(4): 683–92.
Arocena, P., I. Núñez, and M. Villanueva. 2008. The impact of prevention measures and organizational factors on occupational injuries. *Safety Sci.* 46(9): 1369–84.
Cardoso, M. 2014. Indústria do mobiliário espera ano recorde. *Expresso*, November 18, p. 7.
Cooper, M.D., and R.A. Phillips. 2004. Exploratory Analysis of the Safety Climate and Safety Behavior Relationship. *J. Safety Res.* 35(5): 497–512.
Coury, H.J.C.G., I.A. Walsh, M. Alem, and J. Oishi. 2002. Influence of gender on work-related musculoskeletal disorders in repetitive tasks. *Int. J. Ind. Ergonom.* 29: 33–39.
Decreto Lei n° 347/1993 de 1 de outubro. *Diário da República, I Série A—N° 231.* Lisbon: Ministério do Trabalho e da Solidariedade.
Decreto Lei n° 109/2000 de 30 de junho. *Diário da República, I Série A—N° 149.* Lisbon: Ministério do Emprego e da Segurança Social.
Decreto Lei n° 103/2008 de 24 de junho. *Diário da República, 1ª Série—N° 120.* Lisbon: Ministério da Economia e da Inovação.
Decreto Lei n° 24/2012 de 6 de fevereiro. *Diário da República, 1ª Série—N° 26.* Lisbon: Ministério da Economia e do Emprego.
EGP [Escola de Gestão do Porto]. (n.d.). Estudo estratégico das indústrias de madeira e mobiliário. Porto, Portugal: Associação das Indústrias de Madeira e Mobiliário de Portugal (AIMMP). http://norteemrede.inescporto.pt/rede-informacao-regional/publicacoes/estudo-estrategico-da-industria-de-mobiliario.
Eurostat. 2012. Annual detailed enterprise statistics for industry (NACE Rev.2 B-E). Luxembourg: Publications Office of the European Union, April. http://appsso.eurostat.ec.europa.eu/nui/setupModifyTableLayout.do.
Eurostat. 2013. *European Statistics on Accidents at Work (ESAW)—Summary Methodology.* Luxembourg: Publications Office of the European Union.
Fugas, C.S., S.A. Silva, and J.L. Meliá. 2012. Another look at safety climate and safety behavior: Deepening the cognitive and social mediator mechanisms. *Accident Anal. Prev.* 45: 468–77.
García-Herrero, S., M.A. Mariscal, J. García-Rodríguez, and D.O. Ritzel. 2012. Working conditions, psychological/physical symptoms and occupational accidents. Bayesian network models. *Safety Sci.* 50: 1760–74.
GEE [Gabinete de Estratégia e Estudos]. 2015. *Acidentes de Trabalho 2002–2012—Séries.* Lisbon: GEE.

Guldenmund, F.W. 2007. The use of questionnaires in safety culture research—An evaluation. *Safety Sci.* 45(6): 723–43.

Huang, Y.-H., J.-C. Chen, S. DeArmond, K. Cigularov, and P.Y. Chen. 2007. Roles of safety climate and shift work on perceived injury risk: A multi-level analysis. *Accident Anal. Prev.* 39(6): 1088–96.

ITTC [International Tropical Timber Council]. 2008. Draft annual report, 2007. Yokohama, Japan: ITTC.

Johnson, S.E. 2007. The predictive validity of safety climate. *J. Safety Res.* 38(5): 511–21.

Loureiro, I.F., C. Vale, M. Rodrigues, and R. Azevedo. 2014. Can the external environment affect the occupational safety conditions and unsafety behaviours? In *Occupational Safety and Hygiene*, ed. P. Arezes, J.S. Baptista, M. Barroso, P. Carneiro, P. Cordeiro, N. Costa, E. Melo, A.S. Miguel, and G. Perestrelo, 423–27. Boca Raton, FL: CRC Press/Taylor & Francis.

Lu, C.-S., and C.-S. Yang. 2011. Safety climate and safety behavior in the passenger ferry context. *Accident Anal. Prev.* 43(1): 329–41.

Miguel, A.S., G. Perestrelo, J.M. Machado, M. Freitas, F. Campelo, F.J. Lopes, J.M. Silva, and C. Braga. 2005. *Manual de segurança higiene e saúde no trabalho para as industrias da fileira da madeira.* Porto: Associação das Indústrias de Madeira e Mobiliário de Portugal (AIMMP).

Nielsen, K.J., K. Rasmussen, D. Glasscock, and S. Spangenberg. 2008. Changes in safety climate and accidents at two identical manufacturing plants. *Safety Sci.* 46(3): 440–49.

Portaria nº 987/93 de 6 de outubro. *Diário da República, I Série B—Nº 234.* Lisbon: Ministério do Emprego e da Segurança Social.

Reese, C.D. 2012. *Accident/Incident Prevention Techniques.* 2nd ed. Boca Raton, FL: CRC Press/Taylor & Francis.

Rodrigues, M.A., P. Arezes, and C.P. Leão. 2015a. Safety climate and its relationship with furniture companies' safety performance and workers' risk acceptance. *Theor. Issues Ergon. Sci.* 16(4): 412–28.

Rodrigues, M.A., P. Arezes, and C.P. Leão. 2015b. Multilevel model of safety climate for furniture industries. *Work* 51: 557–70.

Rundmo, T. 2000. Safety climate, attitudes and risk perception in Norsk Hydro. *Safety Sci.* 34(1–3): 47–59.

Suárez-Cebador, M., J.C. Rubio-Romero, and A. López-Arquillos. 2014. Severity of electrical accidents in the construction industry in Spain. *J. Safety Res.* 48: 63–70.

Tharaldsen, J.E., E. Olsen, and T. Rundmo. 2008. A longitudinal study of safety climate on the Norwegian continental shelf. *Safety Sci.* 46(3): 427–39.

Varonen, U., and M. Mattila. 2000. The safety climate and its relationship to safety practices, safety of work environment and occupational accidents in eight wood-processing companies. *Accident Anal. Prev.* 32(6): 761–69.

Vinodkumar, M.N., and M. Bhasi. 2009. Safety climate factors and its relationship with accidents and personal attributes in the chemical industry. *Safety Sci.* 47(5): 659–67.

Zohar, D. 2008. Safety climate and beyond: A multi-level multi-climate framework. *Safety Sci.* 46(3): 376–87.

5 HSEQ Assessment Procedure for Supplying Network
A Tool for Promoting Sustainability and Safety Culture in SMEs

Seppo Väyrynen, Henri Jounila, Jukka Latva-Ranta, Sami Pikkarainen, and Kaj von Weissenberg

CONTENTS

5.1   INTRODUCTION

The scope of our chapter is to review health, safety, environment, and quality (HSEQ) management models based on both the literature and the empirical contexts. A model used for HSEQ assessment has been developed and applied within many Finnish company networks, especially in the process, other manufacturing, and energy industries. Further, our main emphasis of the review is to examine the role of small- and medium-sized enterprises (SMEs) and even microenterprises as partners in networks. Especially, we focus on their work systems (WSs) with outcomes, their HSEQ assessment results, and the concepts of sustainability and safety culture.

5.1.1 General Backgrounds

Human behavior and technology are generally interrelated, and they are also inter-related at work. Changes in technologies affect social relationships, attitudes, and feelings about work (Hatch and Cunliffe, 2006), and so do changes in organization and management. A WS is a natural concept to address the above and other relevant elements and aspects. Organizations tailor WSs to fit various areas of work and act as enablers of business (e.g., health care and hospitals) (see Carayon et al., 2014; Hignett et al., 2013).

Experts have traditionally viewed a WS as a microergonomics system that focuses only on persons and technologies (i.e., often on an individual person and his or her tools or some other technological artifact). When implementing new devices, tech-nologies, or ways of work, the WS evolves continuously even though planning and education are involved. For example, users may explore new ways of using the tech-nology, or the demands on the WS with outcomes from its environment may con-tinue to change (Eason, 2009).

Kleiner and Hendrick (2008) discussed the same concepts within a broader socio-technical WS framework (i.e., macroergonomically). They described a WS as a com-bination of

- A technological subsystem (things needed to perform the work)
- A personnel subsystem (people needed to do the work)
- An environmental subsystem (elements outside of the WS focused on)
- An internal environmental subsystem (e.g., cultural and physical characteristics)
- An organization subsystem (e.g., organizational structure and processes)

All these separate WSs operate within a larger "system of systems." Systems are also engaged in transactions with other systems. Managing this complexity is a challenge (Eason, 2005; Kleiner and Hendrick, 2008). This chapter understands WSs with outcomes as a combination of the microergonomics and macroergonomics approaches with a high level of complexity. The level of complexity is high because many employers are involved, and they all contribute to the same production, intra-organizationally and interorganizationally. All contributions of collaborative enter-prises are accomplished at the same site and premises.

It is reasonable to make enhancements to all levels of the WS to gain maxi-mum profit. The WS needs to be in good shape and balance. Some parts of the WS might be affected very easily, but some parts are not so definite. Holistic ergonom-ics (micro and macro) aim to optimize WSs, for performance and effectiveness, including in a key role people without detriment to their health, safety, or other factors of well-being at work, or as we can best express well-being in a WS. A WS "produces" both desired and undesired outcomes, such as accidents and material and environmental losses (Figure 5.1). For example, desired outcomes are promoted by applying ergonomics knowledge to guarantee a high level of usability of tools and workstations. Therefore, it is important to discuss and analyze how these ele-ments should be balanced and managed so that production can be satisfactory for

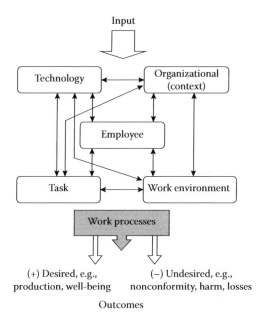

FIGURE 5.1 So-called balanced WS with outcomes model modified to represent a general systemic structure and processes, enablers, "behind" desired and undesired outcomes at work. In short, a WS comprises a combination of people (i.e., human workforce, employee), technology, and tasks within a space and other work environment (tangible and intangible), as well as the interaction of these components within a managed goal-oriented organization with its processes and outcomes (Carayon and Smith, 2000; Väyrynen, 2005; Väyrynen and Nevala, 2010; Carayon et al., 2014).

the person doing it (Carayon and Smith, 2000; Smith and Carayon, 1995). In addition, these elements should be as productive, as safe, and of as good quality as possible (cf. International and European Standard, 2004; Väyrynen, 2005; Väyrynen et al., 2006; Väyrynen and Nevala, 2010). When a balance is not achievable by minimizing the negative aspects of an element, the whole system balance should be improved by enhancing the positive aspects of other elements or their interaction (Smith and Carayon, 2000).

All the components in the WS itself are potential objects of losses. Human beings can be hurt through accidental injuries and occupational diseases. Absences from work and early retirements cause considerable losses to individuals, enterprises, and society. According to the principles of occupational risk prevention, the person has to be protected within the whole entity. On the other hand, the person often plays a role when undesired outcomes occur within the system, causing losses to the person, outside persons, or other components, including the environment (Väyrynen, 2005; Väyrynen et al., 2006, 2008; Väyrynen and Nevala, 2010; Reiman and Väyrynen, 2011). Harms-Ringdahl (2013) addressed a bundle of negative issues in this way: "Quite often, auditing also includes safety, health and environmental aspects, since the management of these is similar and sometimes also integrated. Variation concerning which elements should be included is large." (p. 220) Carayon et al. (2014) examined WSs and noted that they may result in employee and organizational outcomes, as well as quality and safety ones.

On the whole, it is wise to link issues so that one can speak about a holistic safety, health, environment, and quality (SHEQ) system, as do many modern enterprises (Hutchison, 1997). Väyrynen et al. (1997) described a holistic pilot model of a steel mill's collaboration with a supplying company. In the field of safety, Levä (1998), one of the Finnish pioneers of this area, presented three reasons why safety, quality, and environmental issues should be managed as an integrated entity: First, this approach ensures the conformity of techniques, meaning a common set of tools and techniques, which enables them to be used in handling problems of other areas. Second, it promotes the structural conformity of systems, which refers to building one comprehensive management system for all three areas. Third, this approach guarantees the conformity of politics, aiming for shared strategic objectives and goals for all the areas.

Safety management accentuations and practices are most efficient in a comprehensive management system. In this kind of total quality management (TQM) (Hutchison, 1997; Zink, 2000, 2011), system, quality management, safety management, and environmental management are all connected by the general management of the enterprise (Väyrynen, 2003; Väyrynen et al., 2012; Zülch et al., 1998). These management areas should, however, be discussed as separate entities, still seamlessly belonging to the TQM system.

The above can imply the need for integrated management systems (IMSs) (Wilkinson and Dale, 2007). An IMS assures purchasers that products and services satisfy quality requirements. Further, responsible organizations also have to be concerned about the well-being of their employees, their working environment, the impact of operations on the local community, and the long-term effects of their products while in use and after they have been discarded.

One way of thinking and explaining integration for improved risk management is connected with a definition of the term *accident* by the UK Health and Safety Executive that supports an integrated approach. By this definition, an accident is "any unplanned event that results in injury or ill health of people, or damage or loss to property, plant, materials or the environment or a loss of business opportunity" (Hugnes and Ferrett, 2003).

Maybe a little bit narrower scope is behind the goals of (occupational health and) safety culture. The European Agency for Safety and Health at Work has produced (Eeckelaert et al., 2011) a collection of central aspects of safety and health culture. Specifically, it explained that the concept of safety culture has been used more and more in safety research, particularly in high-risk industries such as the nuclear and petrochemical industry, and (public) mass transportation (railway and aviation), recognizing the importance of the human element and soft organizational aspects in accident and risk prevention. The nuclear energy sector, within which the concept has predominantly developed, describes it according to the following two extractions. First, they have noted that "the safety culture of an organization is the product of individual and group values, attitudes, competencies and patterns of behaviour that determine the commitment to, and the style and proficiency of, an organization's health and safety programmes" (p. 3) (HSE, 2005). In addition, HSE (2005) stated, "Organizations with a positive safety culture are characterised by communications founded on mutual trust, by shared perceptions of the importance of safety, and by confidence in the efficacy of preventive measures" (p. 3).

WSs together with management systems should be seen as important and recognized elements in enterprises' strategy development (Cecich and Hembarsky, 1999; Dzissah et al., 2000). Development actions are needed in order to succeed in strategy work in small enterprises (Rajala and Väyrynen, 2011b; Reiman and Väyrynen, 2011). A totality of aims for development activities in enterprises is to increase productivity, shorten time to market, simplify processes, facilitate information and knowledge sharing, and increase employee well-being. In organizational development activities, the characteristics of the organization are not always fully taken into account, and developmental processes are implemented without a deeper understanding of the culture (Järvenpää and Eloranta, 2000). Usually, it takes time to see what kinds of benefits and cost savings are gained through different developmental actions and improvements.

5.1.2 Current Issues

For instance, the global World Steel Association (2013) emphasizes strongly the importance of sustainability within the steel industry, and so do the global International Organization for Standardization (ISO) standards. ISO 26000 (ISO, 2010) added value to existing initiatives for social responsibility by providing harmonized globally relevant guidance based on international consensus among expert representatives of the main stakeholder groups. As a result, ISO 26000 encourages the implementation of best practices in social responsibility worldwide. Zink and Fischer (2013) linked in an interesting and important way human factors and sustainability. Specifically, they noted their aim:

> ... to contribute to a broad discourse about opportunities and risks resulting from current societal "mega-trends" and their impacts on the interactions among humans and other elements of a system, e.g., in work systems. This paper focuses on the underlying key issues: how do the sustainability paradigm and human factors/ergonomics interplay and interact? (p. 348)

Zink's (2014) illustration clarified the equal role of social dimension in sustainability and its links with systemic modeling by WSs (i.e., by the ones that are also a function of time; see Figure 5.2). For instance, a little sign of "common interest of work systems and sustainability" was our team's article about a WS for truck drivers published recently in a journal dealing with sustainability in the transportation industry (Reiman et al., 2013). Management might be enriched by a WS and design science emphasis as a tool for participatory strategic company development (Rajala and Väyrynen, 2011a,b).

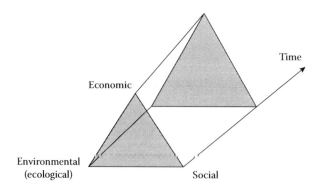

FIGURE 5.2   Sustainability as a triangle-model (cf. Zink, 2014).

Reason (1991) emphasized proactive monitoring of organizational capability by explaining that

> in order to promote proactive accident prevention rather than reactive "local repairs", it is necessary to monitor an organization's "vital signs" on a regular basis. "Vital signs" can be categorised aided by so-called general failure types among the ones related to organizational failures and communication failures are on the top of the listing. (p. 9)

Reason (1991) linked a very practice-related concrete field daily phenomenon—a near miss and related events—with the abstract concept of safety culture. Countries around the world define a near miss in various ways. For example, in the United States, a near miss is an unplanned event that did not result in injury, illness, or damage, but had the potential to do so (NSC, 2013). In contrast, Finland includes in its reporting systems even more static and potential hazard conditions and situations observed and reported by employees that could become an event concluding with a near miss or an accident. These differences make Finland, and our cases, an area of interest for investigating WSs. Maybe Finland is closer to the concept of deviation that, according to Harms-Ringdahl (2013), is an event or a state that diverges from the correct, planned, or usual function in equipment and activities, that is, in WSs and their human, organizational, technological, and environmental elements. Harms-Ringdahl (2013) explained, "Systems do not always function as planned. There are disturbances to production, equipment breaks down, and people make mistakes. There are deviations from the planned and the normal. Deviations can lead to defective products, machine breakdowns, and injuries to people" (p. 116). Further, the list can be continued by interruptions to production, accidents leading to environmental harm, fire, and explosion.

NSC (2013) clearly stated that reporting near-miss incidents can significantly improve employee safety and enhance an organization's safety culture. van der Schaaf (1991) described in many interesting ways how significant can be the links between the abstract concept of culture and concrete near-miss incident reporting of everyday events and states with deviations (Figure 5.3).

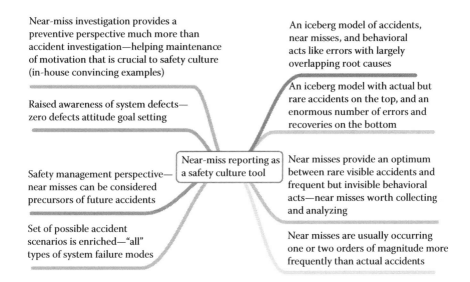

Near-miss investigation provides a preventive perspective much more than accident investigation—helping maintenance of motivation that is crucial to safety culture (in-house convincing examples)

An iceberg model of accidents, near misses, and behavioral acts like errors with largely overlapping root causes

Raised awareness of system defects— zero defects attitude goal setting

An iceberg model with actual but rare accidents on the top, and an enormous number of errors and recoveries on the bottom

Near-miss reporting as a safety culture tool

Safety management perspective— near misses can be considered precursors of future accidents

Near misses provide an optimum between rare visible accidents and frequent but invisible behavioral acts—near misses worth collecting and analyzing

Set of possible accident scenarios is enriched—"all" types of system failure modes

Near misses are usually occurring one or two orders of magnitude more frequently than actual accidents

FIGURE 5.3 Insights to key aspects provided by high-quality near-miss incidents reporting. (Set compiled from the first chapter of the book of van der Schaaf [1991]).

Multiple management systems (MMSs) are related quite closely to IMSs and standards, but in a sense are relative to quality prize models as well. The current MMS standard (ISO, 2011) describes how to "survive" better because in today's business environment, many organizations incorporate a number of management systems, such as quality, environmental, information technology (IT) services, and information security. As a result, these organizations want to harmonize and, where possible, combine the auditing of these systems (like ISO 9001 and ISO 14001, in conformity with EU's Ecomanagement and Audit Scheme (EMAS) on environmental issues, and ISO 22301 plus OHSAS 18001 on occupational hazards). A new standard, "Societal Security—Business Continuity Management Systems—Requirements," was written by leading business continuity experts and provides the best framework for managing business continuity in an organization. Furthermore, the standard can become certified by an accredited certification body (ISO, 2012). The new standard's main assumption is that business continuity is part of overall risk management in a company, with overlapping areas with information security management and IT management.

Quality prize evaluation criteria are modeling holistic excellence management as far as various businesses are concerned. A European excellence model, EFQM, focusing on four stakeholder groups, that is, society, people, customers, and shareholders, is most important in Finland (EFQM, 2013). The U.S. Baldrige national excellence framework criteria for improving an organization's performance are well known around the world and emphasize both work processes and WSs (Baldrige, 2013). The last mentioned WS concept, not included in EFQM, is described as "how the work of your organization is accomplished" (Baldrige, 2013). In general, these criteria examine how an organization plans, implements, manages, and improves its

TABLE 5.1

Some Emerging Issues Related to HSEQ Assessment Promotion

Issue	References
More detailed implementation of WS approach, e.g., according to those for health care	Carayon et al., 2014
Looking for synergies between human factors/ergonomics and sustainability	Dul et al., 2012; Zink and Fischer, 2013; Zink, 2014
Promoting by Lean thinking	Hafey, 2009; Averill, 2011; Raja, 2011; Jadhav et al., 2014
More specific regulations, like those dealing with a contractor's obligations and liability	TEM, 2013
More conscious use and promotion of the concept of safety culture	Eeckelaert et al., 2011; ACSNI, 1993
Extended definition of an accident	Hugnes and Ferrett, 2003; Harms-Ringdahl, 2013

key work processes to deliver purchaser value and achieve organizational success, responsibility, and sustainability.

Table 5.1 lists some more emerging issues related to HSEQ assessment promotion.

5.2 FRAMEWORK FOR APPLYING HSEQ ASSESSMENT PROCEDURE

Based on the above literature review, our chapter aims to examine empirically how case organizations implement, work, and manage the HSEQ assessment approach, as well as to identify its results, especially as far as smaller company developments are concerned (indicators of an HSEQ situation and trends in both purchasing large enterprises and their suppliers). Further, the main emphasis of the empirical review comprises the concepts of HSEQ, WS with outcomes, integrated management, sustainability and responsibility, safety culture, holistic competitiveness, microenterprises, and SMEs.

Microenterprises and SMEs play an important role in the Finnish economy both generally and in terms of supplying services for industrial companies. Many large manufacturers are purchasing a significant share of their human work contribution, resulting in collaborative environments. Many employers with their employees are involved in the purchasing company's value chain of production, often in the same sites, plants, mills, and factories. As pointed out by Porter (1985), value chains consist of a series of activities that create and build value.

The Federation of Finnish Enterprises (2013) described the business environment in the country:

> Finland has a total of 322,183 enterprises (excluding agriculture; data from 2012), of which 99.8% are SMEs employing fewer than 250 people. There are 93.3% Finnish companies that have fewer than ten employees. The role of SMEs in Finnish employment and the economy is quite significant. Of all private-sector employees, as many as 63% work for companies employing fewer than 250 people. These enterprises generate

about 50% of the combined revenue of all Finnish businesses. SMEs are responsible for more than 13% of Finland's export revenue. The proportion of entrepreneurs in Finland is below the EU and the Organisation for Economic Co-operation and Development (OECD) average. The same is also true regarding the number of people who plan to start their own companies. The fact that fewer people in Finland intend to choose entrepreneurship as a career, when compared to other countries, is somewhat surprising as many surveys show entrepreneurship is held in high regard within Finland.

Before the summer of 2011, the Finnish government had a political program called the Political Programme for Employment, Entrepreneurship and Working Life that included elements specific to SMEs. Following the end of this program, the present government (in power since June 2011) chose "Actions against the Grey Economy" as one of its most important projects. This work requires considerable cooperation between many sectors (SBA Fact Sheet, 2012). Both programs are related to SR (social responsibility), sustainability, and HSEQ.

Networking is a typical solution for companies of different sizes to combine and manage their contributions in a competitive way in a contemporary business environment. Employees from several supplying companies or contractors typically work simultaneously for the production in industry companies purchasing contributing services and materials. Recent years have seen an increase in the number of organizations using this type of production and sharing workplaces. The situation mentioned has set up new requirements for managing HSEQ, enabling issues, and achieving desired results within that framework. Those requirements are partly regulation based; however, they are partly voluntary, business driven, to gain more competiveness within this kind of WSs.

Wilpert (1998) gave an important notice of "interorganizational responsibility" to all manufacturing and service industries, not only to high-accident-risk ones. Wilpert stated that

> an example would be an industry with high hazard potential (aviation, nuclear or chemical industry) whose organizations (manufacturers, utilities, regulators, service organizations) share the safety concern of the industry. A first step to change structurally determined defence orientations and fragmented role perceptions within the interorganizational field is to create an awareness that all actors are part of the same system. (p. 236)

Largely according to the same way of thinking, large-scale process and other manufacturing industry companies in Finland have developed during more than 15 years, and started to apply, the HSEQ assessment procedure (AP) for measuring and evaluating their suppliers' capabilities (Väyrynen et al., 2012; Koivupalo et al., 2015). The HSEQ Cluster is a voluntary consortium for supplier holistic quality control according to the HSEQ AP among the suppliers working inside factory areas in shared workplaces. The main objective of the HSEQ AP is to ensure that outside employees in shared workplaces have HSEQ knowledge and skills good enough to operate in the premises of the principal companies (purchasers). The puzzle of managing business and human resources currently involves a large number of pieces in WSs. Optimization of WSs (Figure 5.1) may be evaluated based on measures of three categories: (1) health and well-being, (2) safety, and (3) production performance (the

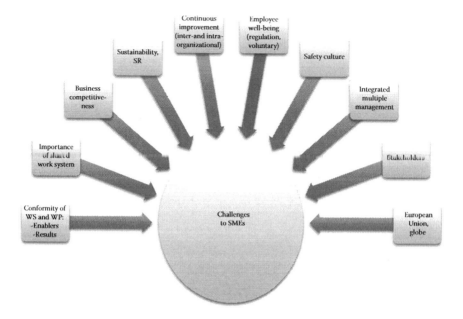

FIGURE 5.4 The HSEQ approach helps in managing regulation-based issues and all contemporary criteria for success in SMEs, e.g., the above. (WP means work processes.)

quantity and quality of production) with minimal nonconformities (International and European Standard, 2004; Väyrynen, 2005; Väyrynen and Nevala, 2010). According to this holistic thinking, occupational risks threaten both factors of well-being and productivity at work. On the contrary, an optimal WS can simultaneously promote workforce well-being and productivity.

Figure 5.4 shows some of the key challenges, especially concerning microenterprises or SMEs servicing high-level purchasing industrial companies.

The HSEQ procedure is an assessment system that was developed through the collaboration of three partners: the process industry companies in northern Finland (the major purchasers in the area), the University of Oulu (a research and development [R&D] organization), and POHTO (a training organization). The purchasing companies have agreed upon the criteria and principles of the HSEQ AP (Figure 5.5). The special auditors are evaluating supplying companies (e.g., contractors) upon basic requirements ("questions based on criteria") (Koivupalo et al., 2015; http://www.HSEQ.fi). The HSEQ AP is predominantly meant to evaluate supplying companies offering services for shared workplaces. Supplying companies as key stakeholders were also involved in developing the HSEQ AP. Each supplier needs only one audit for all purchasing companies. So, both the purchasing and supplying companies benefit in saving resources, money, and time (Figure 5.6). A supplying company can also utilize its own results when selling services to other purchasers. An ideal group of auditors (assessment team) includes persons from at least two different purchasing companies and a

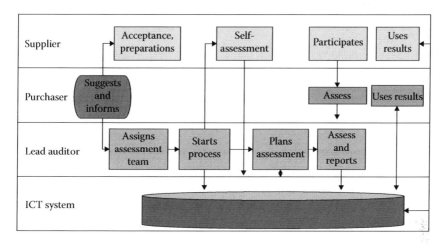

FIGURE 5.5 HSEQ AP and system elements. (ICT means Information and Communications Technology.)

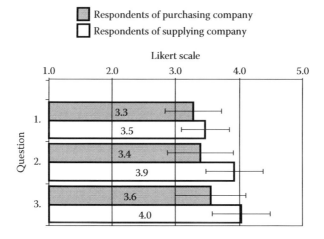

FIGURE 5.6 Both purchasing and supplying companies of the cluster express quite high satisfaction for the HSEQ AP (question 1, "has increased cooperation between purchasing company and supplying company"; question 2, "has been useful"; question 3, "is suitable for my branch") (right).

lead auditor who is a representative of the auditing service provider (the system operator). The group of auditors is chaired by the auditing service provider as moderator. The assessment focuses on the organization's functioning at shared workplaces. During an assessment session, HSEQ capability is scored based on the evidence presented by the supplying company. All the assessments are valid in all participating purchasing companies. The HSEQ AP is controlled by a steering board that includes representatives from each purchasing company, training organization, and R&D organization, and POHTO is responsible for providing the HSEQ AP web-based information system (see http://www.HSEQ.fi; Inspecta,

2013; Väyrynen et al., 2015). The HSEQ AP code of conduct has been accepted. It states, among others, the following:

> The Cluster and its members respect and adhere to the legal obligations of the European Union and any national legislation applying to their business operations; This Code of Conduct is for the purpose of promoting adherence to the legal requirements which relate to competition laws, including those of the European Union.

The HSEQ Cluster acts as an initiator of each assessment. The lead auditor always represents a system operator (for the present, Inspecta, a company for auditing services), and purchasing companies appoint other auditors who have been well trained in HSEQ AP. The registers are maintained by impartial administrators (currently POHTO). The purchasing companies decide in which ways they will utilize the results of assessments. Total scores and HSEQ capabilities and performance profiles are presented anonymously via the Internet, and with identification data in the HSEQ Cluster intranet. Some organizations, particularly R&D organizations, have produced general comparisons (e.g., trend of average accident frequency rate) (see Figures 5.7 and 5.8).

For the history of the HSEQ Cluster, the assessment model was influenced by the EFQM (2013) excellence model. Every question is assessed on a four-step maturity scale. Total scores of HSEQ capabilities and the performance profile by criteria categories aimed to comprise a useful approach for measuring, benchmarking, managing, negotiating, purchasing, controlling, accepting, sharing, agreeing, communicating, and so forth. The first aim of the HSEQ Cluster was to evaluate the needs for the AP. Based on those, we outlined the HSEQ AP on the basis of the ideas

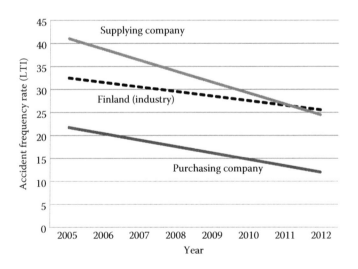

FIGURE 5.7 Accident frequency rates (LTI frequency: accident frequency [≥1-day absence due to injuries] per million work hours) converted to a linear regression trend. Supplying and principal (purchasing) companies represent the HSEQ AP Cluster results, which are compared to the national situation.

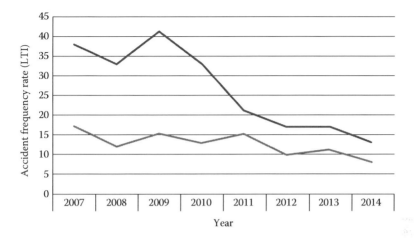

FIGURE 5.8 A case purchasing company and its tens of suppliers, on average, show a quite consistent 8-year downward trend in accident figures (accident frequency rates [LTI frequency: accident frequency {≥1-day absence due to injuries} per million work hours]). The supplying network's accident frequency is coming nearer and nearer to the purchaser (Kurppa, 2015).

and experiences in literature and case company-specific data. Each of the participating company cases had some practices of its own specific to service provider assessment and contracting. So, benchmarking and benchlearning (Freytag and Hollensen, 2001) played an important role during development. We collected both qualitative and quantitative data inside the cases. We primarily based the path toward the goal on so-called design science (Väyrynen et al., 2012). We utilized a constructive design science research approach (Järvinen, 2004) in the cases to formulate proposals for common and specific guidelines. The design science paradigm has its roots in engineering and the sciences of the artificial (Simon, 1996). Design science consists of two basic activities, building and evaluation (March and Smith, 1995). During the development, we combined the design science research approach with the general model of planned change (Cummings and Worley, 2004).

5.3 EMPIRICAL STUDY: METHOD AND MATERIALS

Throughout the HSEQ AP, the auditors "walk through" tangible and intangible enablers within a WS in question and related work processes, management systems, and desired and undesired results within the WS (Figure 5.1). In other words, for each potential supplying company or company unit, the auditors are able to focus on both enablers and results, indicators of both the WS and the culture, as well as the positive and negative outcomes.

We compiled quantitative data from databases on HSEQ AP results from the HSEQ Cluster. We also used safety and other performance HSEQ data that we obtained from some purchasing companies. Those companies represented the ones having the most experience in developing and utilizing the HSEQ AP. The corresponding data of the supplying companies came mainly from the purchasers' statistical records. We used the lost-time injury (LTI) rate as a loss-based safety performance indicator (Kjellén, 2000). We defined the LTI rate as the number of lost-time injuries per 1 million hours of work. An LTI is an injury that causes at least one whole day of absence from work. In addition to injurious incidents (accidents), we gathered noninjurious incident records from the companies as well. The latter consisted of real near-miss incidents, as well as potential hazard conditions and situations observed and reported by employees.

We carried out the empirical section of this study partly through qualitative research and more specifically through interviews used to answer the research questions. The interviewees represented the five case study companies operating for the HSEQ Cluster's purchasing companies. Case studies can be used for the purposes of organizational and management studies. A case study is an in-depth investigation that uses qualitative research methods, such as direct observation and interviews, to gather information for analysis. Findings can then possibly be used when creating a generalization or theories (Yin, 1989). We used semistructured interviews in this study; that is, we chose predominantly qualitative interviews rather than focus group–style discussions (Sinclair, 2005). This method lets the interviewees express their opinions and perspectives (Hirsjärvi et al., 2007; Pope et al., 2000; Sinclair, 2005). Our approach was inductive. Further analysis and interpretation followed according to the principles of respecting the issues raised by interviewees and phenomenography (Rissanen, 2006). We interviewed key persons related to HSEQ from the five case companies that were studied. The first mentioned quantitative figures were available for some of these case companies.

Further, we used formal questionnaires to gather respondents' experiences about HSEQ AP. We primarily used the Likert scale in the construction of the questionnaire. We sent the questionnaires ($N = 73$) by email and received 39 replies. Of the respondents' organizations, the purchasing companies ($N = 15$) were big units of process industry, and the sizes of the supplying companies ($N = 24$) varied between 1–9, 10–39, and 40–239 employees (each size represented one-third of the responding supplying companies).

5.4   RESULTS

Our presentation of the main results is twofold: First, quantitative ones originate from both purchasing companies and their suppliers, typically representing time series, comparisons, or differences caused by variations and changing situations. Mostly, various graphics in figures are utilized. Second, qualitative results describe the five supplying case companies.

5.4.1   QUANTITATIVE INDICATORS OF CULTURAL ENABLERS AND POSITIVE AND NEGATIVE OUTCOMES

Figure 5.6 demonstrates quite positive attitudes toward the HSEQ AP in enterprises actively involved. In all, the representatives of the assessed supplying companies rated the HSEQ AP higher than so-called auditors coming from the purchasing companies. Auditors are educated for being specialists, especially while taking part in assessment sessions.

We used the LTI rate as a main loss-based safety performance indicator of the HSEQ AP Cluster. Figure 5.7 presents regression lines describing the linearized trend of LTI from 2005 to 2012. The decreasing trend of accident frequency rates has been significant, especially in cases in which supplying companies contributed to their purchaser's activities at a shared workplace (Figure 5.8).

Figures 5.9 and 5.10 show separate scorings in the first assessments, whether totally HSEQ scoring or separately concerning HS, E, and Q scorings. Although the trend has been mainly positive, capability improvements are still possible and needed.

Figure 5.11 indicates that the trend based on comparing reassessments with first assessments has been clearly positive; however, capability improvements, that is, higher scoring in assessment, in all supplying case companies are still needed.

Figures 5.7 and 5.8 showed that purchasers themselves are still superior to their suppliers in safety performance. One explanation for this trend may lie in differences in safety cultural heritage and resources. Figure 5.12 shows that a high level of

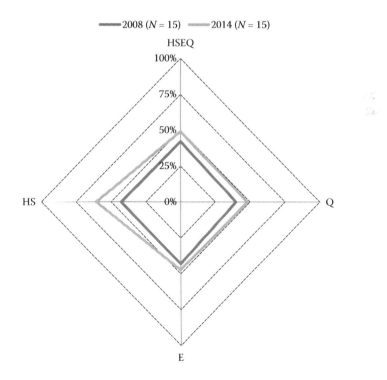

FIGURE 5.9 A profile of HSEQ outcomes shows that HS scoring was at the highest relative level of single factors, and all the separate scorings were increasing during a 6-year-long period, from 2008 (gray) to 2014 (light gray). The relative level indicates the corresponding scores compared with the maximum achievable scores (%).

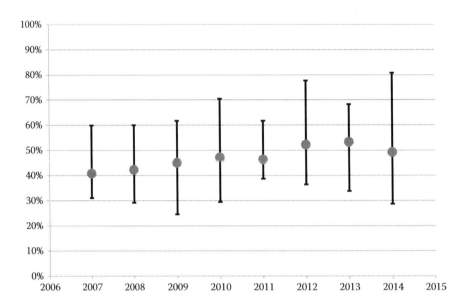

FIGURE 5.10   Eight-year trend of total HSEQ scores compared to the maximum achievable scores (%). This shows the yearly average in first assessments.

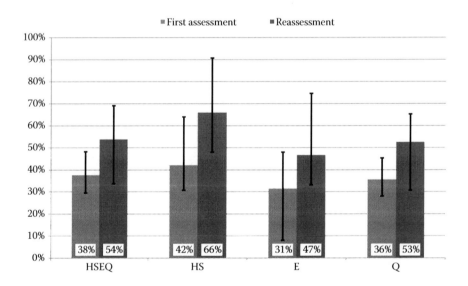

FIGURE 5.11   For total HSEQ scores, as well as the scoring of HS, E, and Q, respectively, there is a clear improvement in all supplying companies; comparison (average, max, and min) with the maximum achievable scores is presented (%). This shows the positive effects of the 3-year time span between first assessments and reassessments of the same companies ($N = 10$).

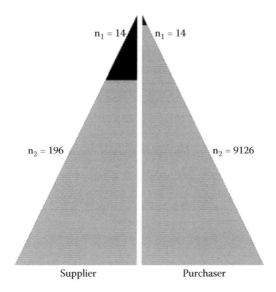

FIGURE 5.12 Comparison of accidents ($\geq$1-day absence due to injuries) (n_1) with near-miss and hazardous situations (n_2) in the same case companies (purchaser on the right, suppliers on the left).

activity in reporting near-miss and hazardous situations could be another explanation, though related to safety culture.

5.4.2 SUPPLIER EXPERIENCES WITH BUSINESS BENEFITS IN HSEQ ASSESSMENT

Business benefits occurred, for example, as increased and improved documentation. The assessment allowed suppliers to consider past events and ponder what types of effects certain selections have had regarding business functions. Furthermore, in the assessments, business functions are vastly processed in different sectors; hence, companies get a clear understanding of what kind of operating model they have and what defects might occur. Thus, companies get a full picture of how they should operate from the purchaser's perspective. In one supplier's case, advanced documentation was seen as improved risk assessment in the company's work sites.

Suppliers use reports and results gained from HSEQ assessment in their annual report and monthly meetings. There, they discuss deviations and nonconformities that were found in assessment, and thus try to generate improvement proposals about how to develop their activities and practices further. Thus, the assessment activity promotes a combined way of thinking and developing business functions, but it is hard to measure these benefits in financial terms. The value can be seen when the company detects nonconformities in activities, ponders jointly with the purchaser ways to improve them, and manages to implement the information within the organization. One of the suppliers concretized the value of the assessment activity by saying that the company has succeeded in creating a coherent way of acting as a result of the assessment activity; furthermore, this new method of acting has resulted in saved resources and expenses in different activities. Therefore, tools and models from assessment activities help organizations to deal with different business activities and develop them in a coherent way.

The greatest assets from the assessment activity can be seen as improvements in safety culture. The purchaser–supplier cooperation enables discussion about safety-related issues in common workplaces, and the organizations can seek solutions for defects through safety observations and incident reports. One of the suppliers organizes monthly safety meetings with the purchaser; as a result of these meetings, the supplier has managed to adapt safety issues and take advantage of ready-made and functioning models from the purchaser. Furthermore, reports of near-miss incidents have been used in the supplier's workplaces in order to anticipate and avoid work-related injuries and absences. Thus, the improvement in information flow has also supported cooperation and resulted in benefits. An interviewee pointed out that this situation was at times reversed; in other words, the purchaser was able to adapt safety procedures from the supplier and hence benefit from it in other areas as well. As a result of the cooperation, both the supplier and the purchaser have seen benefits in anticipating safety culture and risk avoidance.

Accident frequency is one critical way to measure companies' safety level. One of the suppliers highlighted that a strong effort in HSEQ activities has greatly improved the company's work safety. A few years earlier, its LTI, indicator of accident frequency, was approximately 50–60, but the corresponding number was around 10 in the year 2014. The suppliers emphasized that measurement is important in different activities because the organization cannot control what it cannot measure. Furthermore, they added that HSEQ cannot be a separate unit of a company's activities, but it must be present in every process at every level. Another supplier noted that if one pays attention to different aspects of HSEQ assessment, one can prevent work accidents and thus avoid unnecessary expenses. Therefore, it is possible to minimize expenses by reducing safety hazards and avoiding risk to lower deviation and nonconformity costs.

Some suppliers felt that HSEQ assessment has benefitted their relationships with the purchasers. They agreed that today's purchasers are aware of different perspectives of HSEQ, and that suppliers must take some steps before they can begin collaboration. Furthermore, the interviewees concluded that the purchasers appreciate when the supplier is able to present key numbers and statistical information regarding business aspects, and thus prove that the business is organized and managed. In this context, HSEQ assessment has worked well as an image improvement activity for some of the suppliers, and therefore created extra value for the companies.

Further, the suppliers have noticed diverse types of benefits in their activities. One of the suppliers reported a reduction in its quality deviations and reclaims. The supplier also mentioned that these third-party HSEQ auditions now involve many more purchasers; thus, the company may get many new ideas and perspectives about how to increase and improve its business. Another supplier pointed out that the HSEQ assessment has improved its safety and, in that way, created savings. The supplier added that the assessment evaluates daily routines and activities, allowing it to improve storage costs and supervision costs, for example.

5.4.3 THINGS THAT AFFECT SUPPLIER PERFORMANCE

The supplier interviews highlighted that the HSEQ activation has affected daily routines in companies and, in some cases, clearly improved the companies' performance ability. The organizations have taken into consideration the deviations, nonconformities, and remarks that were revealed in assessment; as a result, they have introduced development mechanisms so that their company will continue to improve. Even if some suppliers already had environment, safety, and quality certificates, they understood the need for HSEQ assessment in order to fulfill purchaser expectations regarding produced goods. Thus, the suppliers received concrete information about how to develop their services. They also added that the HSEQ assessment gave support to certified management systems and affected their performance ability in a good way. This transmitted especially when the suppliers noticed that the questions in the HSEQ assessment were very similar to certified system audits.

The suppliers' goal in participating in the audition was mainly to improve collaboration with the purchaser organization. The interviews also revealed a current trend that nowadays, in many cases, the purchaser requires the suppliers to take part in HSEQ assessment in order to renew their current cooperation agreements. Thus, more and more suppliers need to be able to produce results proving that the company's HSEQ section is in order, and this requirement has a positive effect on the performance ability for both suppliers and purchasers. The effects can be seen most easily in improvements in safety knowledge. One interviewed representative of the suppliers noticed that before performing the HSEQ AP, the company had lots of room for improvement in safety issues, and found the HSEQ assessment to be a great tool to change the environment for the better.

The suppliers felt that the assessment was a good way to recognize issues that normally cannot be detected. Thus, the audit done by a third party brings out different visions and proposals that the supplier itself would not notice. Furthermore, the deviations and nonconformities in the assessment can help organizations to realize what development methods the company must take in order to get operations and work procedures to the required level. One of the interviewed suppliers had taken the development activities into consideration by supervising the improvement proposals and then attempted to put them into use to avoid defects. Therefore, it can be noted that the assessment activities partly affected suppliers' management systems, for example, by enhancing organizing documentation.

One supplier saw performance benefits in the development of safety guides and instructions for the company's own employees and for subcontractors. The supplier took advantage of the HSEQ assessment to also evaluate its own suppliers and gather more detailed information about them. Collaboration with the purchaser resulted in improved safety and benefits in the work conditions in both the supplier's WS and the purchaser's WS. Thus, the supplier noticed that the HSEQ assessment resulted in improvements in many safety-related aspects, such as work procedures, the use of protective devices, and observation and information sharing. Furthermore, one of the suppliers highlighted that using alternative or substitute work is directly related to the assessment activities. Alternative or substitute work is applied during the cases of lowered work ability due to injury or illness of an employee. Further, the company has managed to control absence costs and create a more advanced safety-related mind-set among employees.

Environmental impacts can be managed in shared work sites and systems through activities such as effective and successful waste sorting and ensuring that rules and regulations are obeyed. One of the suppliers noticed that it has managed to make even more ecological choices regarding work tools and instruments, for example, by paying attention to tools' lifetime and support compared to regular tools. Thus, the supplier has succeeded in reducing harmful impacts to the environment as well as decreasing work-related environment risks.

5.5   DISCUSSION AND CONCLUSIONS

This chapter describes a proven solution for how companies that utilize various deliveries for enabling their production are now able to assess holistically their supplying companies. Networking for the same business and working often in the same premises at one single site now characterize production in many industrial sectors. This naturally raises a question of equal HSEQ requirements for conditions, quality, and management of all employers. Moreover, the potential achievable advantages touch key stakeholders as well, including employees, shareholders, the regulating society, unions of employees, and confederations of employers—in other words, especially social partners. Diverse driving forces have been present and active in developing the HSEQ AP, that is, the desire to create an equal level of quality and shared requirements of management for all companies operating at the sites and within a network of each business leader manufacturing industry company. The HSEQ AP features key indicators describing both resulting outcomes and enablers of holistic excellence for a total of three dimensions of sustainability: social, economic, and ecological. The HSEQ AP scales the three-dimension whole to fit SMEs, facilitating them to be more updated and contemporarily competitive. The HSEQ AP offers SMEs timescale continuity for their excellence and sustainability efforts, for example, trends of single criterion results, bundles of criteria like HS and E, and of course, the total HSEQ AP scoring for a specific company or company unit. The near future should bring greater popularity for the HSEQ AP because a national standardization body aims to promote the assessing system by publishing standards based on the current HSEQ AP.

The review part of this chapter comprised updated backgrounds from the literature. We emphasize links between sustainability, safety culture, and HSEQ criteria within WSs, with outcomes of various companies contributing to shared workplaces and other premises. The experiences and results of the HSEQ AP have been very promising for many years, but new member companies and developments are still needed. There exists a well-grounded opinion that with the strengthening economy since 2015–2016, the number of participating companies will increase. New companies, in the role of either supplier or purchaser, will become new members of the HSEQ AP Cluster. In terms of suppliers, the role that microenterprises and SMEs play in both businesses and employment is already large and is increasing, so their importance for optimal WSs with outcomes is decisive. Embedded solutions are necessary, and at the same time, they need to be utilizable for fulfilling many synergic goals. The HSEQ AP's measured scorings are clearly dependent on time; in that sense, it also provides a follow-up tool with many key features for managing and developing WSs, human factors, and sustainability, as Zink (2014) presents being needed.

Eeckelaert et al. (2011) concluded:

> As already mentioned before, the safety culture approach and many related diagnostic tools have their origins in high-risk industries, primarily with the aim to prevent orga-nizational accidents and disasters. Looking at OSH (Occupational Safety and Health) from a cultural viewpoint is most suitable and useful when all regulatory aspects and related risks are already addressed by the organization in a systematic way. It is, there-fore, questionable as to whether the concept of safety culture, and the use of safety cul-ture assessment tools, is really relevant for many SMEs—particularly for the smaller and micro enterprises outside the high-risk industries. (p. 48)

Based on its purchasing companies and extensive supplying network with many SMEs, we think that this network utilizing HSEQ AP can at its best be called a cluster of safety culture, or maybe even organizational culture, and sustainability. A key feature of being called a cluster with a safety culture is the broad acceptance and planned continuous use of safety culture assessment tools. We can show many signs of the influence of the specific safety culture indicated by the safety performance. The signs are most significant in the purchasing companies with networks that have a long tradition within the HSEQ AP Cluster. By being members of the supplying network of an HSEQ AP–style purchasing company, SMEs not only have an oppor-tunity to participate in a safety culture, but it is a "must" for them. They have to provide—and utilize—the key issues listed by ACSNI (1993), Wilpert (1998), and Reason (1991): positive safety culture, mutual trust, shared perceptions, confidence in the efficacy of preventive measures, interorganizational responsibility, avoidance of fragmented role perceptions, awareness that all actors are part of the same system, proactive monitoring of organizational capability (i.e., proactive accident preven-tion rather than reactive "local repairs"), and monitoring of vital signs on a regular basis (e.g., general failure types among the ones related to organizational failures and communication failures, and near misses).

The concept of an HSEQ culture might be definable and needed beyond the safety culture to include closely related issues in the HSEQ AP, as well as sustain-able development. This chapter comprises an exercise in how HSEQ AP brings properly scaled practices of economic, social, and environmental sustainability to SMEs.

This study used the concept of HSEQ. However, Harms-Ringdahl (2013) has argued that the expression *safety, health, environment,* production (SHEP), or HSEP, might be equally in use. This study emphasized the addition of positive issues to production while simultaneously preventing undesired consequences. Therefore, Harms-Ringdahl's (2013) definition of P ("to prevent problems with production, quality, etc., and loss of property [p. 67]") is a good one as well. Qualitative descrip-tions of pros and cons are not enough because figures and facts are needed as well for follow-up, benchmarking, and managing in companies. The HSEQ AP is able to provide the SME sector with quantitative indicators that are needed for evidence of guaranteeing sustainable progress. In this chapter, we provided both quantitative and qualitative examples of the progress within industrial networks. The interviews gave this chapter a qualitative "direct voice" from supplier SMEs that was both interesting and useful to practitioners and future researchers alike.

According to the principles and tools presented in this chapter, HSEQ management fulfills both regulatory and business competitiveness requirements while promoting developments toward balanced sustainability. Employers are responsible for playing key roles and fulfilling key requirements to ensure that they are adequately managing a productive WS, health, safety, and the environment in the context of holistic quality management. Many large companies within the Finnish process, energy, and manufacturing industries have begun utilizing the HSEQ AP to encourage prerequisite actions for guaranteeing success in their value networks, which may include SMEs. Both core and assisting partners are contributing and collaborating in modern value networks with a better organizational and safety culture, actually with a sustainable integrated HSEQ culture. On a daily basis, investing in both WS and HSEQ helps supplying companies to meet delivery commitments too.

The quantitative and qualitative results show interesting positive evidence on the HSEQ AP network, both generally and specifically, related to sustainability and responsibility issues, and even total network management. The literature review and our empirical study have also provided new insights on culture, near-miss and hazardous situation reporting, experiences from purchasing and supplying companies, especially SMEs, and communication inter- and intraorganizationally; however, more research is needed. For example, the links between HSEQ AP and sustainability should be clarified in more detailed way, and the stability of performance in deliveries compared to scoring in HSEQ assessment should be followed more precisely. The HSEQ Cluster most probably will also consider the possible needs for continuous development of HSEQ scoring criteria.

REFERENCES

Averill, D. (2011). *Lean Sustainability: Creating Safe, Enduring, and Profitable Operations.* Boca Raton, FL: CRC Press, Taylor & Francis Group.

Baldrige. (2013). Criteria. Gaithersburg, MD: National Institute of Standards and Technology. http://www.nist.gov/baldrige/publications/criteria.cfm (accessed March 3, 2014).

Carayon, P., and Smith, M. (2000). Work organization and ergonomics. *Applied Ergonomics* 31, 649–662.

Carayon, P., Wetterneck, T.B., Rivera-Rodriguez, A.J., Schoofs Hundt, A., Hoonakker, P., Holden, R., and Gurses, A.P. (2014). Human factors systems approach to healthcare quality and patient safety. *Applied Ergonomics* 45, 14–25.

Cecich, T., and Hembarsky, M. (1999). Relating principles to quality management. In Christensen, W., and Manuele, F. (eds.), *Safety through Design: Best Practices.* Itasca, IL: National Safety Council, pp. 67–72.

Cummings, T.G., and Worley, C.G. (2004). *Organizational Development and Change.* Mason, OH: Thomson South Western.

Dul, J., Bruder, R., Buckle, P., Carayon, P., Falzon, P., Marras, WS., Wilson, JR., van der Doelen, B. (2012). A strategy for human factors/ergonomics: Developing the discipline and profession. *Ergonomics* 55(4), 377–395.

Dzissah, J.S., Karwowski, W., and Yang, Y.N. (2000). Integration of quality, ergonomics, and safety management systems. In Karwowski, W. (ed.), *International Encyclopedia of Ergonomics and Human Factors*. Vol. 2. London: Taylor & Francis, pp. 1129–1135.

Eason, K. (2005). Ergonomics interventions in the implementation of new technological systems. In Wilson, J.R., and Corlett, N. (eds.), *Evaluation of Human Work*. 3rd ed. London: Taylor & Francis, pp. 919–932.

Eason, K. (2009). Socio-technical theory and work systems in the information age. In Whitworth, B., and de Moor, A. (eds.), *Handbook of Research on Socio-Technical Design and Social Networking Systems*. Hershey, PA: IGI Global, pp. 65–77.

Eeckelaert, L., Starren, A., van Scheppingen, A., Fox, D., and Brück, C. (2011). *Occupational Safety and Health culture assessment – A review of main approaches and selected tools*. European Agency for Safety and Health at Work (EU-OSHA), Luxembourg.

EFQM. (2013). EFQM excellence model. Helsinki: Excellence Finland. http://www.efqm.org/the-efqm-excellence-model (accessed March 3, 2014).

Federation of Finnish Enterprises. (2013). The small and medium-sized enterprises. Helsinki: Federation of Finnish Enterprises. http://www.yrittajat.fi/en-GB/federation_of_finnish_enterprises/entrepeneurship_in_finland/ (accessed February 27, 2014).

Freytag, P.V., and Hollensen, S. (2001). The process of benchmarking, benchlearning and benchaction. *The TQM Magazine* 13(1), pp. 25–33.

Hafey, R.B. (2009). *Lean Safety: Transforming Your Safety Culture with Lean Management*. London: Taylor & Francis.

Harms-Ringdahl, L. (2013). *Guide to Safety Analysis for Accident Prevention*. Stockholm, Sweden: IRS Riskhantering AB.

Hatch, M.J., and Cunliffe, A.L. (2006). *Organization Theory. Modern, Symbolic and Postmodern Perspectives*. 2nd ed. Oxford: Oxford University Press.

Hignett, S., Carayon, P., Buckle, P., and Catchpole, K. (2013). State of science: Human factors and ergonomics in healthcare. *Ergonomics* 56(10), 1491–1503.

Hirsjärvi, S., Remes, P., and Sajavaara, P. (2007). Tutki ja kirjoita. 13th ed. Helsinki: Tammi.

HSE [Health and Safety Executive]. (2005). *A review of safety culture and safety climate literature for the development of the safety culture inspection toolkit*. Research report 367. HSE Books, 100 p.

Hugnes, P., and Ferrett, E. (2003). *Introduction to Health and Safety at Work*. Oxford: Elsevier Butterworth-Heineman.

Hutchison, D. (1997). Safety, *Health and Environmental Quality Systems Management: Strategies for Cost-Effective Regulatory Compliance*. Sunnyvale, CA: Lanchester Press Inc.

Inspecta. (2013). HSEQ assessment. Helsinki: Inspecta. http://www.inspecta.com/en/Our-Services/Certification/Management-Systems/HSEQ-assessment/#.UiBXgpzqpIQ (accessed September 11, 2013).

International and European Standard. (2004). EN ISO 6385: Ergonomic principles in the design of work systems. Geneva, Switzerland: International Organization for Standardization.

ISO [International Organization for Standardization]. (2010), ISO 26000: Guidance on social responsibility. Geneva, Switzerland: ISO.

ISO [International Organization for Standardization]. (2011). ISO 19011: Guidelines for auditing management systems. Geneva, Switzerland: ISO.

ISO [International Organization for Standardization]. (2012). ISO 22301: Societal security—Business continuity management systems—Requirements. Geneva, Switzerland: ISO.

Jadhav, J.R., Mantha, S.S., and Rane, S.B. (2014). Development of framework for sustainable Lean implementation: An ISM approach. *Journal of Industrial Engineering International* 10, 72.

Järvenpää, E., and Eloranta, E. (2000). Organisational Culture and Development. In Karwowski, W. (ed.), *International Encyclopedia of Ergonomics and Human Factors*. Vol. 2. London: Taylor & Francis, pp. 1267–1270.

Järvinen, P. (2004). *On Research Methods*. Tampere, Finland: Opinpajan Kirja.

Kjellén, U. (2000). *Prevention of Accidents through Experience Feedback*. New York: Taylor & Francis.

Kleiner, B.M., and Hendrick, H.W. (2008). Human factors in organizational design and management of industrial plants. *International Journal of Technology and Human Interaction* 4(1), 114–128.

Koivupalo, M., Sulasalmi, M., Rodrigo, P., and Väyrynen, S. (2015). Health and safety management in a changing organisation. Case study global steel company. *Safety Science* 74, 128–139.

Kurppa, N. (2015). Manuscript of MSc thesis. Oulu, Finland: University of Oulu, Faculty of Technology.

Levä, K. (1998). Pk-yrityksen laatu-, turvallisuus- ja ympäristöjohtaminen: Integroidun laatu-järjestelmämallin kehittäminen. Licentiate's thesis, Tampere University of Technology, Department of Industrial Engineering and Management [Integrated management of quality, safety, and environment issues within SMEs].

March, S.T., and Smith, G.F. (1995). Design and natural science research on information technology. *Decision Support Systems* 15(4), 251–266.

NSC [National Safety Council]. (2013). How do near miss reporting systems prevent future incidents? Itasca, IL: NSC. http://www.nsc.org/WorkplaceTrainingDocuments/Near-Miss-Reporting-Systems.pdf.

Pope, C., Ziebland, S., and Mays, N. (2000). Analysing qualitative data. *BMJ* 320(7227), 114–116.

Porter, M.E. (1985). *Competitive Advantage: Creating and Sustaining Superior Performance*. New York: Free Press.

Raja, M.I. (2011). Lean manufacturing—an integrated socio-technical systems approach to work design. Doctoral thesis, Glemson University.

Rajala, H.-K., and Väyrynen, S. (2011a). Participative design science approach on the optimum work system: An argumentative review-based model with a case. *Theoretical Issues in Ergonomics Science* 12(6), 533–543.

Rajala, H.-K., and Väyrynen, S. (2011b). Participative approach to strategy communication: A case of small- and medium-sized metal enterprises with a review after seven years. *Human Factors and Ergonomics in Manufacturing and Service Industries* 23(4), 346–356.

Reason, J. (1991). Too little and too late: A commentary on accident and incident reporting systems. In van der Schaaf, T., Lucas, D.A., and Hale, A.R. (eds.), *Near Miss Reporting as a Safety Tool*. Oxford: Butterworth-Heinemann, pp. 9–26.

Reiman, A., Pekkala, J., Väyrynen, S., Putkonen, A., Abeysekera, J., and Forsman, M. (2013). Delivery truck drivers' and stakeholders' video-assisted analyses of work outside the truck cabs. *International Journal of Sustainable Transportation* 9(4), 254–265.

Reiman, A., and Väyrynen, S. (2011). Review of regional workplace development cases: A holistic approach and proposals for evaluation and management. *International Journal of Sociotechnology and Knowledge Development* 3(1), 55–70.

Rissanen, R. (2006). Fenomenografia. Luku 5.1. In Saaranen-Kauppinen, A., and Puusniekka, A., KvaliMOTV—Menetelmäopetuksen tietovaranto. Tampere: Yhteiskuntatieteellinen tietoarkisto. http://www.fsd.uta.fi/menetelmaopetus/kvali/index.html (accessed May 20, 2014). (In English: Phenomenography.)

SBA Fact Sheet. (2012). Finland. The Small Business Act (SBA). Enterprise and Industry, European Commission.

Simon, H.A. (1996). *The Sciences of the Artificial*. 3rd ed. Cambridge, MA: MIT Press.

Sinclair, M.A. (2005). Participative assessment. In Wilson, J., and Corlett, N. (eds.), *Evaluation of Human Work*. 3rd ed. London: CRC Press, Taylor & Francis Group, pp. 83–111.

Smith, M., and Carayon, P. (1995). New technology, automation, and work organisation: Stress problems and improved technology implementation strategies. *International Journal of Human Factors in Manufacturing* 5(1), 99–116.

Smith, M., and Carayon, P. (2000). Balance theory of job design. In Karwowski, W. (ed.), *International Encyclopedia of Ergonomics and Human Factors*. Vol. 2. London: Taylor & Francis, pp. 1181–1184.

TEM. (2013). Contractor's obligations and liability. TEM. http://www.tem.fi/en/work/labour_legislation/contractor_s_obligations_and_liability (accessed March 3, 2014).

van der Schaaf, T. (1991). Chapter 1. In van der Schaaf, T., Lucas, D.A., and Hale, A.R. (eds.), *Near Miss Reporting as a Safety Tool*. Oxford: Butterworth-Heinemann, pp. 1–8.

Väyrynen, S. (2003). Vahinkoriskien hallinta, turvallisuuskulttuuri ja johtaminen: Katsaus lähtökohtiin [Accident risk control, safety culture and management: Basic review]. In Sulasalmi, M., and Latva-Ranta, J. (eds.), *Turvallisuusjohtaminen teollisuuden toimittajayrityksissä* [Safety management in industrial supplying companies]. Helsinki: Ministry of Labour, pp. 5–21.

Väyrynen, S. (2005). Review of machinery risk prevention through efforts expended on design, management, quality, ergonomics and usability. Project Reports of the Work Science No. 20. Oulu, Finland: Oulu University Press.

Väyrynen, S., Hoikkala, S., Ketola, L., and Latva-Ranta, J. (2008). Finnish occupational safety card system: Special training intervention and its preliminary effects. *International Journal of Technology and Human Interaction* 4(1), 15–34.

Väyrynen, S., Jounila, H., and Latva-Ranta, J. (2015). ICT as a tool in industrial networks for assessing HSEQ capabilities in a collaborative way. In Khosrow-Pour, M. (ed.), *Encyclopedia of Information Science and Technology*. 3rd ed., vol. 10. Hershey, PA: IGI Global, pp. 787–797.

Väyrynen, S., Koivupalo, M., and Latva-Ranta, J. (2012). A 15-year development path of actions towards an integrated management system: Description, evaluation and safety effects within the process industry network in Finland. *International Journal of Strategic Engineering Asset Management* 1(1), 3–32.

Väyrynen, S., Koutonen, M., Hietala, J., and Kisko, K. (1997). Teräsvalssaamon puhtaanapitopalvelun kehittäminen -hanke: Esittely ja yhteenveto [Development of cleaning services at a steel rolling mill]. Project Reports of Work Science Division No. 2. Oulu, Finland: University of Oulu, pp. 5–10.

Väyrynen, S., and Nevala, N. (2010). VIDAR as a tool in ergonomic development: Double utilisation model and work system cases. In *Proceedings of the 8th International Conference on Occupational Risk Prevention ORP2010*, Valencia, Spain, May 5–7, 2010, pp. 1–10.

Väyrynen, S., Röning, J., and Alakärppä, I. (2006). User-centered development of video telephony for servicing mainly older users: Review and evaluation of an approach applied for 10 years. *Human Technology* 2(1), 8–37.

Wilkinson, G., and Dale, B.G. (2007). Integrated management systems. In Dale, B.G., van der Wiele, T., and Iwaarden, V.V. (eds.), *Managing Quality*. 5th ed. Chichester, UK: Wiley-Blackwell, pp. 310–350.

Wilpert, B. (1998). Conclusions, after the event: What next? In Hale, A., Wilpert, B., and Freitag, M. (eds.), *After the Event: From Accident to Organisational Learning*. Oxford, GB: Pergamon, Elsevier Science, pp. 233–244.

World Steel Association. (2013). Sustainable steel policy and indicators. Brussels, Belgium: World Steel Association.

Yin, R.K. (1989). *Case Study Research: Designs and Methods*. 4th ed. Newbury Park, CA: Sage.

Zink, K. (2000). Quality management, continuous improvement and total quality management. In Karwowski, W. (ed.), *International Encyclopedia of Ergonomics and Human Factors*. Vol. 2. London: Taylor & Francis, pp. 1312–1316.

Zink, K. (2014). Designing sustainable work systems: The need for a systems approach. *Applied Ergonomics* 45, 126–132.

Zink, K.J. (2011). Stakeholder-oriented management concepts as challenge for macro-ergonomics—how can macro-ergonomics successfully contribute to quality and sustainability? In Lindfors, J., Savolainen, M., and Väyrynen S. (eds.), *Proceedings of NES2011 (Wellbeing and Innovations through Ergonomics)*, Oulu, Finland, September 18–21, 2011, pp. 7–14.

Zink, K.J., and Fischer, K. (2013). Do we need sustainability as a new approach in human factors and ergonomics? *Ergonomics* 56(3), 348–356.

Zülch, G., Keller, V., and Rinn, A. (1998). Arbeitsschutz-Managementesysteme—Betriebliche Aufgabe der Zukunft. *Zeitschrift für Arbeitswissenschaft* 2, 66–72.

6 Ergonomics Point of View of Work Accidents in Safety Management Perspective

Mario Cesar R. Vidal, Rodrigo Arcuri Marques Pereira, Renato José Bonfatti, Alessandro Jatobá, and Paulo Victor Rodrigues de Carvalho

CONTENTS

6.1   INTRODUCTION

Working systems rely on their intrinsic variability (Vidal, 1985). For example, in systems like those found in the construction industry, large oil refineries, health care, and aircraft design, variability is the central regency of the process control. Therefore, managing variability is the core of work performance, and is thus directly related to its success. In other words, one should shift the old "keep control" motto to a new "keep under control" motto.

Regarding work accidents, one must understand that they occur due to loss of control of a physical process capable of causing damage to either people or property. An accidental course of events propagated by the activity of people can either trigger an accidental flow of events or divert a normal flow (Rasmussen and Svedung, 2000).

In this chapter, we propose a statement for work accidents under an ergonomics point of view, that is, that work accidents should develop within a culture of variability management rather than under an asymptotic search for stability. Thus, to define the structure of our research problem, we explore the research topic of the formulation of work accidents based on contemporary foundations of ergonomics. We study this topic to understand the diversity of ergonomics approaches of working systems, when confronted with interpretation frameworks of the occurrence of work accidents.

This work has the conceptual significance of presenting a useful and applied contribution for safety management in the search for constantly monitored dynamics of nonevents, in order to manage them well (Hollnagel, 2014, p. 6). Moreover, the convergence of concepts into the ergonomics point of view promotes a useful formulation for safety management, which stems from the conceptual evolution of two disciplines—ergonomics and safety management.

The practical significance of this chapter relies on the fact that operators do not spend part of their time working and the rest managing variability. Thus, from an ergonomics point of view, these parts are transparently merged.

We believe that, in normal circumstances, it is more effective to understand that human factors are not an issue for work performance. Instead, adopting a normalizing strategy and defining rigid procedures for every context might become problematic for workers. When accidents occur, investigators do not understand the changes workers have made in their work processes—especially if they have been successful most of the time—as part of the system. Accident investigators use hindsight and event-driven accident models to search for causes of accidents that will consider responsible all those determined to have committed violations (Carvalho, 2006).

In addition, human factors instrumentation remains the innermost solution in dealing with complications and disturbances and responding to emergencies and urgencies. The ergonomics point of view of work accidents can be helpful to the analysis and enable the improvement of work situations.

Furthermore, we present a debate across the intersection of ergonomics and work accidents for enabling an acceptable framework for safety management.

This chapter is divided into six sections. We begin with an explanation of conceptual evolution. Next, we proceed to the development of the conceptual evolution of our two major concepts: work accident occurrence and the ergonomics point of view. Section 6.5 explores the crossing of the two conceptual frameworks in focus. Finally, we present a discussion and conclusions.

6.2   CONCEPTUAL EVOLUTION AS A SUBJECT

As part of a specific community, one exposes research, theoretical frameworks, and methodological constructs using a set of words or expressions. This communication approach is not unique. In fact, its meaning is variable, even within near communities. This is the matter of intersubjective verification (Hollnagel, 2014); a word or an expression should actually mean the same to two or more people. This is usual in ergonomics. Consequently, our research papers could have, depending on the reader, precise or vague terminology. Sometimes choosing the definition of a word is a challenge. In fact, most communication breakdowns occur due to two people using the same word in different ways, which hampers the relevancy of scientific research.

The essential problem is the process of appropriation of concepts by different communities, especially in the pragmatic (situated) dimension of speech. Hence, we are considering conceptual appropriation as a rich and complex process, having an epistemological equivalence to the concept creation process. That richness, combined with deep complexity, makes conceptual appropriation a challenging operation; at the same time, it provides concrete support for interactions. On the other hand, it can introduce undesirable effects to communication—misunderstanding. The two core concepts of this chapter—work accident and ergonomics point of view—are supposed to produce such an effect. Thus, it is necessary to do a conceptual evolution analysis of them, allowing intersubjective verification.

The appropriation of a notion into a scientific community lies on the basis of this process, but afterwards a debate should enlighten its use in order to avoid pragmatic deviation. A word such as *activity* reveals great vagueness; a current dictionary lists no less than six distinct definitions for it. How do we decide among these definitions? A reader typically decides instantly and without effort which definition of a word is correct. He uses hints deriving from (1) the grammar (is the word a noun, verb, direct object, and so on?); (2) the larger context, the discourse; and (3) the pragmatic situation—about what circumstances the speaker is talking. Therefore, what does he expect us to do in response? These three contextual factors—grammatical settings, discursive pertinence, and situated properties—should be enough to reveal which of several definitions of a concept is being used. In fact, these are the basis of the complementary looks for the conceptual evolution of task analysis. Briefly, the conceptual evolution analysis denotes the debate around core concepts of scientific discourse to avoid pragmatic deviation.

6.2.1   FRAMEWORK FOR CONCEPTUAL EVOLUTION ANALYSIS

The following methodological framework presents a heuristic for conducting conceptual evolution analysis (Figure 6.1). The conceptual evolution observatory targets have a set of diverse data sources, according to our essential keywords (*work*

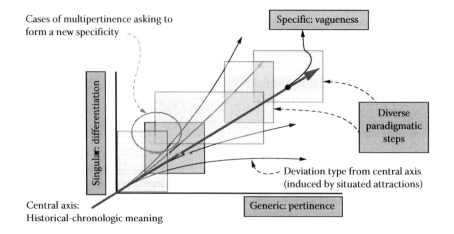

FIGURE 6.1 Schematic framework for conceptual evolution analysis.

accident and *ergonomics point of view*). The initial composition of a data set forms a subset with a list of experts, according to the subject. Those experts should agree on a list of significant official websites, as well as core references related to the subject. Website analysis, the indications of the experts, and adequate criteria to use inference engines produce an initial bibliographic list. The final list of reference issues results from a pertinence analysis of this list.

Analysis consists of examining a core expression in its historic–chronological axis. One forms an initial paradigmatic sequence in order to allow a first partition of the chronological data. Then, we perform an assessment of the contents of gathered data through three or more categorizations. The partition criteria are

- To what extent they meet a general sense, bringing forth the generic quality, the *generality*
- To what extent they exhibit a differentiation, characterizing their *singularity*
- To what extent they combine to form a fuzzy pertinence between generic and singular properties under situated influences, outlining a *specificity*

According to Kuhn (1962), a paradigm evolution can be illustrated as steps of a ladder. Each paradigm is a step, and it is essentially composed of core concepts. A core concept evolves into the paradigm trying to fill the step as a whole. This calls for a definition of wide concepts. A given wide concept is a particular explanation that solves the compromise between the generality and singularity of its objects. Previous studies and applications (Saldanha et al., 1998; Vidal and Bonfatti, 2003; Carvalho et al., 2005) show that specificity more adequately establishes a wider conceptualization by means of pertinent management of the vagueness. However, the first step is clear establishment of generic and singular characteristics of the initial gathered data. It follows that once the initial data are available, they should be organized in appreciation of the completed lectures, the consulted websites, and the testimonials to form the initial paradigmatic class set.

For climbing to the upper steps, extraordinary research should stress the core concepts of the paradigm. This extraordinary research issues a paradigmatic rupture that inaugurates a new paradigm set. Hence, the partition process progresses, identifying the extraordinary concepts' contribution to the general goals of a theoretical framework means testing the structure of the paradigmatic sequence.

6.2.2 Defining the Number of Data Suppliers

The number of data suppliers (collected papers, inquired entities, and the experts involved) was estimated by the mean of a dedicated mathematical model (Arcuri and Saldanha, 2013). We show here a part of this model. Let n be the number of data suppliers. Let m be the total number of potential critical comments induced by the general explanation of the subject (referential speech). If so, it is possible to define the following events:

- A_{ij} = a data supplier i who reports a critical comment j for $i = 1, 2, ..., n$ and $j = 1, 2, ..., m$
- D_{ij} = a critical comment j is reported by at least one of the n data suppliers

Thus,

$$D_j = A_{1j} U A_{2j} U ... U A_{nj}{}_j \tag{6.1}$$

Considering $p_{ij} = P(A_{ij})$, if the events A_{ij} were collectively independent, then the events D_j would be also collectively independent. Hence, we can write

$$P(D) = 1 - (1 - p_{1j}) * (1 - p_{2j}) ... * (1 - p_{nj}) \tag{6.2}$$

Like applying a list of eligibility criteria for supplier set composition, we can assume that all A_{ij} are equally likely. Thus, $p_{ij} = p$ for any i and j. Consequently,

$$P(D_j) = 1 - (1 - p)^n \tag{6.3}$$

Now, let $q_n = P(D_j)$. *Ipso facto*, let us define the following variables:

- Q_n = number of critical comments reported by at least one of the n data suppliers, where $Q_n = 0, 1, 2, ..., m$
- M_n = proportion of critical comments reported by at least one of the n data suppliers, where $M_n = (Q_n/m)$

M_n measures the representativeness of the data supplier set related to critical comment screening. We also observe that Q_n correspond to the number of successes in m Bernoulli trials. In fact, the jth trial is the critical comment j concerning n sources. We consider the success in the jth try as event D_j. Since the events D_j are independent and equally likely, we infer that Q_n has a binomial distribution with parameters m and q_n. That means $Q_n \sim Bi(m, q_n)$.

Thus, concerning Q_n, we have

$$E(Q_n) = m * q_n = m * \left[1 - (1-p)^n\right] \tag{6.4}$$

$$V(q_n) = m * q_n * (1 - q_n) = m * \left[1 - (1-p)^n\right] * (1-p)^n \tag{6.5}$$

Since $M_n = (Q_n/m)$, it follows that $E(m_n) = [E(Q_n)/m]$ and $V(M_n) = [V(Q_n)/m^2]$ Hence,

$$E(M_n) = 1 - (1-p)^n \quad \text{and} \quad V(M_n) = \frac{\left\{\left[1 - (1-p)^n\right] * (1-p)^n\right\}}{m} \tag{6.6}$$

Additionally, if $m > 10$, $m * [1 - (1-p)_n] > 5$, then Q_n has a sufficient adherence to normal distribution, as well as M_n. We can now state that the representativeness of the data supplier set related to critical comment screening depends on the number of data suppliers (n), the probability of report of a pertinent critical comment by at least one data supplier (p), and the total number of critical comments potentially issued from a problem upon analysis (m). So, prior to defining the size of the sample set of data suppliers, one must estimate the values p and m in the research context. Nielsen and Landauer (1993) searched the estimation for interface quality evaluations using think-aloud tests. Table 6.1 shows the empirical results obtained.

Nielsen and Landauer claim that the realization of think-aloud tests has generated adjusted parameters (p) between 0.12 and 0.48. Moreover, in their specific case studies, where it was possible to find the actual values for the m parameter, these ranged

TABLE 6.1
Proportion of Critical Comments during Quality Evaluations

Data Suppliers	Proportion of Critical Comment Reports	Increase Generated by the Last Supplier
1	0.29	0.29
2	0.49	0.20
3	0.63	0.14
4	0.73	0.10
5	0.81	0.07
6	0.86	0.05

Source: Nielsen, J. and Landauer, T.H., in *INTERCHI 93 Conference Proceedings*, Amsterdam, 1993, pp. 206–213.

from $9 < m < 14$. Hence, the calculations show that in our case study, that comprises high value for m, if we assume $p = 0.25$ and $n = 11$ (for example), we can get $E(M_{14}) = 0.96$, a quite robust mapping of critical comments. Finally, the data supplier set should be (1) equilibrated between its subcategories (papers, organization, and experts) and (2) diversified into each subclass, considering the existing scientific and practice communities.

6.2.3 Conceptual Analysis Results

A specific device, the inserting matrix (Vidal and Bonfatti, 2003), continuously organizes the gathered data. Actually, we adopted a two-dimensional matrix, with each row dedicated to a data source (a writing, a website excerpt, or an eyewitness testimony) and each column defined by a specific class in the paradigmatic sequence. Just like a big display, the inserting matrix allows us an agile screening of the data universe, merging bibliography annotations with testimonial transcriptions. Figure 6.2 shows the generic inserting matrix for this study.

Multipertinences are the launchers for the formation of new categories: the more multipertinences that occur, the more data rearrangements there are to be done. Rearranging one adds a new column—what explains the denotation "inserting matrix." Otherwise, one can conduct several rearrangements during the analysis, including testing for an *ad hoc* hypothesis. Thus, it is important to ensure the lowest

	Paradigmatic steps												
	Paradigm$_1$				Paradigm$_2$				...	Paradigm$_N$			
Inquiries	Paradigmic class 1a	Paradigmic class 1b	...	Paradigmic class 1z	Paradigmic class 2a	Paradigmic class 2b	...	Paradigmic class 2z	...	Paradigmic class Na	Paradigmic class Nb	...	Paradigmic class Nz
1	®®	®®®®	...	®®	®	XX	...	®	...	®®®®	®	...	®
2	®®®	®	...	XX	XX	®®	...	®®	...	®®	®®®®	...	®®
...	...	...	...	...	...	...	...	...	...	...	...	...	...
n	®	®®	...	®	®®®®	®®	...	®	...	®®	®®	...	®®®®
$\int_1$	®®®®	®®®®	...	®®®	®®®®	®®®®	...	®®®®	...	®®®®	®®®®	...	®®®
$\int_2$	Thematic evaluation$_1$				Thematic evaluation$_2$					Thematic evaluation$_n$			
$\int_3$	Global assessment												

® = single data register

$\int$ = integration of data in an assessment type

Note: 1, 2, ..., n is a data series that can be fulfilled with elements of different natures.

FIGURE 6.2 Insertion matrix for conceptual evolution data gathering.

number of multipertinences, since we are dealing with crisp analysis within a binary logic.

Besides the establishment of stream within a paradigm, this organized database allows diverse kinds of exploration. For this chapter, it we used the following categories:

- Broad definition: The proposed explanation about the causality of accidents of a stream, which forms its generality
- Baseline: Subdivision of a given stream characterized by clear pertinence to the general definition of the stream, but having some particular contents allowing some specific lectures of the generality
- Contribution: The immediate consequence to the accident prevention field throughout the baseline
- Singularities: The role of contextual variables of the baseline, forming its particular causality
- Notable absences: Matters that a baseline is supposed to fulfill for a convergent proposition

If sometimes there are no concrete conditions for establishing a consistent set of baselines, a less detailed analysis can be made considering only the characteristics of the stream.

6.3   CONCEPTUAL EVOLUTION OF WORK ACCIDENT NOTION

The notion of work accident is a polysemy. In order to characterize such a polysemy, we organized an observatory around the explanation concerning the genesis of work accidents in general, and the mechanism of a given work accident. The composition of a data supplier set for this particular conceptual evolution analysis consists of experts, official websites, and selected papers. Table 6.2 summarizes the data supplier set. We added three additional sources to the already sufficient number of data suppliers.

The application of these criteria outlined nine work accident causality explanations, grouped into three paradigmatic sets: atavist conceptions, transitional views, and contemporary approaches. Table 6.3 shows this arrangement.

6.3.1   ATAVISM: FOCUS ON THE VICTIM

We believe that a work accident derives from human properties (failures or misfits). "There is always a human error at the origin of disasters," is a common slogan. In the first stream, *guiltiness*, an accident is explained because of misconduct or bad work performance. This idea prevailed in labor systems of medieval organizations. However, we have been hearing this comment during ergonomics actions and field research in the construction area (Vidal, 1985), agriculture (Cartaxo, 1987), refineries (Vidal and Duarte, 1992), aviation (Saldanha et al., 1998), and so forth. Even nowadays, some preparations for dangerous tasks are preceded by a quick talk emphasizing the individual's risk and the corresponding self-protection actions to be performed.

TABLE 6.2

Data Supplier Set for Conceptual Analysis of Work Accident Meaning

Data Supplier	No.	Diversity	Year
Experts	1	Tenure professor in ergonomics	2014
	2	Active professional practitioner of ergonomics	2014
	3	Senior researcher in safety	2014
	4	Tenure professor in safety	2014
	5	Active professional in safety	2014
Official sites	6	Occupational Safety and Health Administration (United States)	2014
	7	Health and Safety Executive (United Kingdom)	2014
	8	Institut National de Recherche et Securité (INRS)	2014
Papers/books	9	J.M. Faverge	1980
	10	C. Perrow	1984
	11	J. Reason	1997
	12	B. Pavard	1997
	13	J. Rasmussen and I. Svedung	2000
	14	E. Hollnagel, D. Woods, and N. Levenson	2006

The second one, *accident-proneness* (Lahy and Kongorold, 1936), is where the authors tried to establish a scientific basis for hypothetical accident-based behavior. Scientists formulated the hypothesis that accident-proneness was a human property particular to certain individuals, made evident by specific psychological tests. Some inconsistencies of such research were pointed out by Zurfluh (1957).

In the subsequent stream—*accidentability* (Adler, 1951; Tiffin and McCormmick, 1967)—work accidents happen as a consequence of the unsuitability of a workplace to the characteristics of its occupants. Despite appearances, the preventive attitude was, essentially, to refine selection procedures in order to provide "the right man to the right place."

6.3.2 Transition View: The Victim and His or Her Context

This second paradigmatic group explains causality as occurring in the interaction of the person with the context in which he or she operates. Such a framework allows a novelty: it was, from now on, to deal with the analysis of accidents in which the victim is not an active agent. Its initial stream (*technical factors and human factors approach*) brings a broader conception of causality. Technical factors assemble all the elements contributing to dangerousness, whereas human factors revamp the atavist notions around human errors and misfits. Despite this revamp, this stream contributes to a better understanding of work accidents by means of two definitions: connectivity and multicausality. Connectivity is where work accidents have no direct or indirect link with the victim. Multicausality has a dialectic effect in work accident causality: on the one hand, one given accident can be a result of the concurrence of

TABLE 6.3
Three Visions and the Nine Baselines of Work Accident Causality Explanations

Paradigm	Baselines	Contribution	Singularity	Notable Absences
Atavism: Focus on the victim	Guiltiness	Human failure as a causality	Narrow approach on human behavior	Resources for alternative explanations
	Accident-proneness	Human property as a causality	Questionable explanatory variable	Incapability for nonlinear approach
	Accidentability	Compatibility between job profile and human capability	Interface approach	Joint system approach for technical and human factors
Transition: The victim and his or her context	Technical factors vs. human factors	Broader scenario of the accident	Connectivity and multicausality	Lack of overall systemic approach
	Rasmussen's progressive model	Role of macrostructures in accident enlargement	Contingent aspects	Lack of mid- and high-level managers' activity analysis
	Ignition theory	Macrostructures and microprocesses in the same framework	Search of the accident launcher mechanism	Restorations and regulations are not taken into account
Modern: Focus on the context	Sociotechnical reliability	Course of action in accident expertise	Point of view of the activity	Topography of security levels as a determinant
	Anthropotechnologic causality	Culture and organization aims influencing working process	Cross-cultural aspects as a core concept	Lack of prevalidation of reference situations
	Variability management	Epistemological equivalence between success and failure	Focus on the variability by diverse ways	Dedicated tools to implement concepts of management

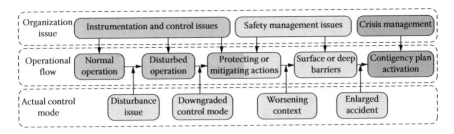

FIGURE 6.3   Rasmussen's progressive accident model.

various processes, but on the other hand, the same causal factor can be at the origin of different accidents.

The next stream is anchored in Jens Rasmussen's *progressive model* (1990). This framework introduces two essential notions: disturbance and accident enlargement. Disturbance is related to accident genesis, referring to the phenomenon that transforms a normal situation into an abnormal one (Figure 6.3). The proposition suggests that several sectors of a company, including neighborhood policies, must be engaged in accident management. Accident enlargement is the complementary concept, which absorbs the involvement process triggered by the occurrence of a work accident.

The transitional paradigm, *ignition theory*, shows the complex relationship between an accident and its context. It is emphasized that the majority of accidents happen in special circumstances whose added durations do not exceed a quarter of the total worked time. Such circumstances have a common starting point, the occurrence of an operating incident. The transition from a normal to an abnormal process derives from the occurrence of one operating incident, but not necessarily a critical one. To bypass the operating obstacle created by the incident, the operator deviates from the usual path, performing an unscheduled, vicarious task. Since it is not scheduled, nothing can be planned, organized, or provisioned for its execution. With the completion of vicarious tasks, the worker can rescue the normality (success) or enter into a succession of incidents (failures). Figure 6.4 schematizes these two possibilities.

This stream promotes the paradigmatic rupture as merging the two precedent ones. It solved the founded deadlocks by the creation of a new concept (the ignition of a work accident). It also leaves us a legacy: *the understanding of a given work situation cannot be restricted to the description in terms of normal situations, but must also take into account situations of recovery*. Contrary to what may appear, these situations are not random or episodic, but are current work situations, and therefore also *normal* (Perrow, 1984).

6.3.3   MODERN VIEWS: FOCUS ON THE CONTEXT

We follow two traces: (1) a dynamic approach and (2) a complexity assessment. The assumption that an accident is an inherent phenomenon in a system's functioning is based on modern views. This should not be taken as its obvious meaning, that a work accident is supposed to happen inside a given work system. This matches

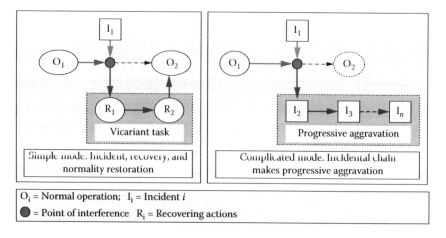

FIGURE 6.4 Incidents: passage from normality to abnormality.

the conception of an accident as a nonevent (Hollnagel, 2014, p. 5). According to this concept, one should search for accident causality in the regular system's nature: its structure, context, and dynamic links. Thus, this mode of examination does not focus on the victim as the previous one.

Three streams compose this group of causalities: (1) sociotechnical reliability, (2) anthropotechnologic causality, and (3) variability management.

The search for sociotechnical reliability, merging the nature and context of a working system, led to the recovery of the sociotechnical system as extraordinary research. It merges the original concepts issued from the Tavistock Institute's formulations for design, and the methodological frameworks of ergonomics work in analysis and macroergonomics. It brings out the triple dimensioning of a working system (people, technology, and organization). In this stream, a work accident is produced by latent failures that are inherent to the systems' structures. Those failures emerge in different modes according to *specific* contexts.

The *anthropotechnologic perspective* introduces cultural materials into the sociotechnical conception. The preceding evolutional stream allowed the formulation of depth defense, a state-of-the-art concept in that time. According to chaos theory, its application was pertinent, until reaching its limits: its deployment in other anthropotechnologic contexts, placed in the boundaries of the original applications. Since sociotechnical systems are frequently designed elsewhere than the production system localization, there exist differences between representations of the designers and the final users. There are two singular cases: the introduction (in a general scale-up of innovations) and the transfers (replication of a sociotechnical system in different emplacement). Those differences grow with the social distance among work system agents in the case of introductions, as well as in cases of transfers.

The *variability management* stream comprises the two latter ones, incorporating the complexity point of view. The opacity of emergent latent failures in the cultural clash context introduces a problem, which has a joint resolution by the complementary approaches of robustness and resilience modes of engineering. According to

Alderson and Doyle (2010), "a *property* of a *system* is **robust** if it is *invariant* with respect to a *set of perturbations*."

Thus, we can define robust engineering as the expertise of designing systems able to resist changes without adapting their initial stable configuration. In the work accident field, this leads to the property of the working system maintaining its structure even when just impacted by incidents. This characteristic requires that the working system be designed considering incidents (typically unforeseen events) and integrate failures as a part of normal functioning—something heretic for classical engineering principles. On the other hand, one does not solve the problem of safety management, but only creates barriers and resistance, prior to coping with the dynamic and complexity of systems. That is the reason for asking for a complementary resource: to allow resilient functionalities for system operationality.

6.4 CONCEPTUAL EVOLUTION OF THE ERGONOMICS POINT OF VIEW

Our second polissemic expression is the ergonomics point of view. Table 6.4 shows the data supplier sets dedicated to this second conceptual evaluation.

The primordial definition of *ergonomics* (Jastrzebowski, 1857) refers to the science of work. A Polish philosopher outlined a scientific program dealing with physical, aesthetic, rational, and moral dimensions of work, that is, work that is kinetic, emotional, intellectual, and spiritual. After this manifestation, there was a large

TABLE 6.4

Composition of the Data Supplier Set for the Ergonomics Point of View

Data Supplier	No.	Diversity	Year
Experts	1	Tenure professor in ergonomics	2014
	2	Senior professional practitioner of ergonomics	2014
Official sites	3	International Ergonomics Association	2014
	4	Human Factors and Ergonomics Association	2014
	5	Societé d'Ergonomia de Langue Française	2014
	6	Brazilian Ergonomics Association	2014
	7	Japan Ergonomics Society	2014
	8	Ergonomics Society (actually Institute of Ergonomics and Human Factors)	2014
Papers/books	9	W. Jastrzebowski	1857
	10	K. Tanaka	1921
	11	E. Grandjean	1974
	12	A. Wisner	1974
	13	H. Hendrick and Benner	1987
	14	K.F.H Murrell	1965

absence. We found only two significant records: (1) *The Science of Labour and Its Organization*, published in 1919 by Józefa Joteyko, a Polish scientist, who dealt in detail with the measurement of occupational fatigue and principles of the scientific management of labor; and (2) *Research of Efficiency: Ergonomics*, published in 1921 in Japan by Kanichi Tanaka.

The discipline of ergonomics reappeared after World War II in 1949 with the establishment of the Ergonomics Research Society, the first national society in our modern times. This triggered an international movement in Europe and North America, which debuted with the creation of the International Ergonomics Association in 1959. At the end of the twentieth century, we saw the emergence of several ergonomics societies and associations. Actually, the intense and extensive connectivity has caused the word *ergonomics* to be heard and spoken everywhere, making real the desideratum of the 9th International Energy Agency (IEA) Congress: "Designing for everybody, everywhere."

In recent last years, we have seen an interesting path of the conceptual evolution of the notion of ergonomics. Data reveal points of view about what the discipline was supposed to be, passing by its general aims and arriving at the presentation of a set of criteria for ergonomics actions assessment. Even considering that the profession really exists, evidenced by research unit activities, available training programs, professional organizations, and certification processes, the misunderstanding has not decreased. Therefore, the word *ergonomics* loses a part of its precise definition, whereas the ergonomics field grows. The word *ergonomics* becomes polysemic. However, our concern is to approach the meaning of the expression "ergonomics point of view." Hence, we must deal with the meaning of a set of words as a whole. Nevertheless, if we want to discuss the meaning of this whole, we must take into account the many ways in which the word set meanings occur in a discourse on ergonomics. We delimited our empirical field to ergonomics authority. It consists of texts and speeches produced by a professional practitioner of ergonomics or a recognized entity (e.g., a society of ergonomics). Thus, gathered data were composed of a set of papers, educational publications, and excerpts of societies' websites, as well as some personal communications. We adopted a historical-chronologic order for data organizations, as exposed above. This issued three current streams on ergonomics practice. The generic and common field is the entourage of human working. All the definitions and testimonies match a common goal described in each aim: to contribute to the improvement of working conditions. We progressed such understanding onto a contemporary aim, the well functioning of the sociotechnical system. This includes people, technology, and organizations in the same reality. We also added correlate subjects, allowing for the fulfillment of the three analytic categories: the contribution of common goals, the singularity of each concern, and its notable absences. Table 6.5 shows this portion of our data set.

6.4.1 Human Factors Formulation

The human factors formulation stream consists of studies and practices aimed at the production of human data (e.g., physical parameters and cognitive capabilities). This group brings the notion of human requirements to the design of work systems. Its

TABLE 6.5

Inserting Matrix of Actual Streams of Ergonomics Practice

Actual Stream		Contributions		
Main Concern	**Correlate Subjects**	**to Common Goals**	**Singularity of Each Concern**	**Notable Absences**
Human data	Human limits and performance ranges	Human data for design applications	Restricted to human variables	Systemic modeling of working
Interface technology	Related objects like software, accessibility, and friendliness	Model of human–systems coupling aimed at the design of working system elements	Confined to focused parts of the system, and so limited to located concerns	Holistic approach of working, including the joint analysis of local coupling and global connections
Activity modeling	Resilience and robustness	Course-of-action modeling	Prevalent to process approach	Issues in terms of charts and screenings

interest lies in its great relevance to the design and design changes of artifacts and "mind facts" in the work situation. It is important to note that ergonomics is, and has been from its beginning, a discipline dedicated to design. Perhaps because of a certain youthful shyness, the human factors field has incurred two structural problems: constraint issues and distancing from its own field of action. There are constraint issues because contributions dealt only with the human variables in systems design. Distancing from its own field of action occurred because human factors specialists were more occupied with producing data for design than approaching the design process of "things" in which human factors are decisively present.

Until now, a classic application of human data has been the integration of anthropometric data in systems design. This is fully right because things have dimensions, and so do people. Hence, integrating them in a man–machine system brings together the mutual adjustment of dimensional data (of things) and anthropometric data (of people using those things). The question is how effective this can be without a systemic modeling of working in which dimensional aspects are not unique or even the most important. Having the basic concern of establishing a pattern language for clerical working, one of our fellows (Cassano, 2014) indicated that design patterns should not be composed only by classical design basis like surfaces texture, shapes, dimensions, and so forth. Instead, design patterns should include a set of prediagnostics of working typologies that will the workplace. In this research, a large set of structural data was organized that allows not only a good anthropometrical relationship, but also some dimensional annotations related to working dynamics. In two applications of the same pattern, the working surface could vary in function for the needs of those who are working. For example, in one working position, the essential action was to consult a set of old reports, which requires more surface for reading. However, if the essential task is to operate software on a computer, the workstation requires two monitors. Both working situations need more space than the space needed for the conventional workstation, but in different ways. Hence, even having a good consideration of human variables for furniture purchase (for reading papers and for operating computers), there still remain significant ergonomics problems. Modeling of working is essential for a good design of work process (technology and organization). It cannot be fully established with only a human variables data set and out of the optics of the design process.

6.4.2 INTERFACE TECHNOLOGY

The *interface* group reads ergonomics as a technology of interfaces in a working system. It operates around the IEA 2000 definition: ergonomics (or human factors) is the scientific discipline concerned with the understanding of interactions among humans and other elements of a system, and the profession that applies theory, principles, data, and other methods to design in order to optimize human well-being and overall system performance. The major concern improves the original idea of ergonomics as a design discipline searching to build models of human–systems coupling aimed at the design of working system elements. The improvement considers the advance obtained in moving from the place of the human data supplier to one of scenario providers.

Models and related scenarios are useful for the development of solutions for system components, as well as protocol configurations. This practice introduces modularity in process design, recovering an old formulation of infinitesimal calculus. Modularity is a needed property for system robustness. Here, we have also two problems, according to the simple or complex nature of the system. In simple systems, modularity requires a decomposability of a whole into parts using a stable criterion. Chaos theory, in turn, dictates that such stability is valuable to the epicenter of a phenomenon; that is, the arising concept names the edge of the chaos. In dealing with complex systems, one of their essential properties is low decomposability, which is especially pregnant in the treatment of sociotechnical systems. This leads to a working concept that overcomes the notion of interaction toward more elaborate topics, such as compositions, cooperation, and interindividual coordination. This creates a paradox: an apparently good system approach has great effectiveness only when confined to well-defined parts of the overall system. The paradox enhances the fundamental problem of this stream: the lack of a holistic approach of the working process.

Very recently, we developed the notion of production logic (Vidal et al., 2012) for educational purposes. The need of such a concept derives from the difficulty of our students in achieving an overall understanding of the specificities of the branch in which ergonomics development is in progress. The central idea was to have a comprehensive framework of the production systems so that a scheme would allow us to interpret the workload of work teams. The textual formula is as follows: *such production logic interferes with the course of action of individual A, producing the following impacts*. For example, in a fuel delivery terminal (Moreira and Vidal, 2013), the precise moment of fuel delivery from the refinery, with a random queue of tank trucks, is an equation to solve. Every day, the facility must get ready by 10:30

a.m., for the dispatching (receiving of fuel). At the same time, it is also is supposed to proceed to the loading of tank trucks and the issue of sales documents (parallel processes taking place in different locations) through the day. Those two processes interlink, of course, but there is a prevalence of the first one, because a fuel sales terminal cannot afford the lack of fuel to sell. On the other hand, a fuel delivery terminal is also supposed to sell and deliver fuel.

The examples above show that even a macroergonomics approach limited to interactions in interfaces does not produce the holistic approach of a working process, especially because it does not include the joint analysis of local coupling and global connections. It is an advance compared to the data supply paradigm, and it brings to the ergonomics field related objects like software and connecting devices, and also introduces concerns such as accessibility and friendliness. However, it is not enough to effectively deal with safety management.

6.4.3 ACTIVITY MODELING

The diversity of fields in which ergonomics is supposed to act does not perfectly cope with basic knowledge applied without some located variables. This assertion leads us to another conception of the discipline of ergonomics. The classical conception of science supposes a field and a corpus. This field of view has become gradually accompanied by another constitution, the object where the term *science* has given way to the notion of *scientific discipline*. A scientific discipline, unlike the philosophical notion of the field to establish a reflection on this body (which this field consists of), is replaced by an epistemological constitution (which reflects what can be done about the methods).

If so, we can make a science of formulation as grounded reflection on fields and methods to propose solutions to relevant problems (e.g., taxonomy, including in kind) or resilience (adaptation of methods and techniques to specific circumstances, limits, or peripheral). The foundation takes place in the vetting productions on a topic, and the form of reflection is the comparison between two or more lots of empirical materials (concrete and relevant situations). This reflection is advised because we all have an understanding from which we examine what we observe in a systematic investigation. It is proper for the scientist, after seeking reasons, to make reflections on his or her methods. With that, he or she enables the addition of content to this "thing" (his or her methods), making it possible to produce "advised comparisons." The philosophical contribution is thus of an epistemological nature.

Safety management requires a framework that gives more than a corpus and always deals with located situations (having particularities and some singularities). One can obviously largely benefit from epistemological contribution. This certainly is the central contribution of the ergonomics point of view to safety management. Indeed, any valuable framework for safety management should not be built on the idea of the dynamics of systems or their related formulation, complex systems.

The activity trend presents not only a large set of concepts, but also a specific method and techniques, revisiting ethnography and psychology. This approach can dialogue with all precedent groups. Its leading concepts are synthetized by the concept of the course of action. A course of action is defined as "the activity of a given actor engaged in a given physical and social environment belonging to a given culture, where the activity is meaningful for the actor; that is, he [*sic*] can show it, tell it, and comment upon it to an observer-listener at any instant during its unfolding" (Theureau and Jeffroy, 1994; Mohamed et al., 2015). Hence, the current work analysis by means of course-of-action modeling fulfills our need for a scientific approach to ergonomics—a scientific approach that matches contemporary needs of safety management.

Activity ergonomics has one positive characteristic: it is fully compatible with a process approach. The actual questions are issues in terms of charts and screenings. This is important for the current reports, as well as for realistic simulation purposes. Reports for data diffusion and realistic simulation are actual and typical tools for safety management. They can be fully enriched with ergonomics contributions. We can show this in two different circumstances. The first one is a safety mitigation experience in offshore diving for oil pump maintenance (Figueiredo et al., 2000). Course-of-action modeling was employed to produce a report of cognitive expertise in order to understand the accident production, as well as the evaluation of the rescue structure of the diving hardware. The second was the design of an air safety training program within a safety philosophy called line-oriented flying training (LOFT) (Saldanha et al., 2008). Its goal was to propose to a crew the experience of a downgrading situation, potentially changing from a normal flight condition to a near-accident situation. Activity modeling was in the center of these two emblematic situations. These two questions strongly evidenced the pertinence of activity ergonomics, the activity modeling of the dynamics of a sociotechnical system, using the course-of-action principles, for evaluation and design.

6.5   ERGONOMICS POINT OF VIEW OF SAFETY MANAGEMENT

The analysis of the conceptual evolution of work accidents have shown that the work accident views changed from human to technology and ultimately to the work context approach. Within this latest trend, three explanation groups were underlined, namely the sociotechnical reliability, the anthropotechnological causality, and the variability management. Table 6.6 shows the relationship between the work accidents modern view and the main axes of the ergonomics point of view (Human Data Supply, Interface Technology and Activity Modeling). This operation outlined a research and development program for the contribution of Ergonomics to safety management.

The exploration of this framework is made by outlining, for each safety management stream, a graphical layout followed by a textual formula. The graphical layout summarizes the modeling, while the textual formula finds the essential descriptive variables, articulating them in a redaction syntax.

6.5.1   ERGONOMICS POINT OF VIEW OF SOCIOTECHNICAL RELIABILITY

Sociotechnical reliability is built in the meaning space defined by the basic concepts' latent failures, incidents, particular circumstances, and restoration actions. How can the different streams of safety management contribute to recognizing these four components of accident production?

This answer should observe that we are dealing with models. Models produce data that refer to preceding situations with very relative likelihood to current situations.

TABLE 6.6
Ergonomics Point of View in Fulfilling the Notable Lack of Safety Management Modern Streams

Work Accident Streams	Ergonomics Point of View		
	Human Data Supply	Interface Technology	Activity Modeling
Sociotechnical reliability	Human data for design of warnings, alarms, and wearable technology	Physical and logical environments and settings for supporting complex activities	Built of emergent constraint scenarios, aimed at safety training
Anthropotechnological causality	Localized requirements concerning human data	Standards or pattern languages adopted in reference situation	Avoidance of problems in transfers and transfer of problems
Variability management	—	Specific software for diagnosis of some process variables	Development of variability monitoring and prevention issues

The same configuration touches the observer's point of view, on the borderline of pertinence. Thus, the action of matching two simplifications can produce a lack of accuracy. Because we need conclusive certainty to apply data into the domain of a given observatory, it will be done at the core of the domain, in which the essential shafts of the model are strongly manifest. In sociotechnical reliability issues, it focuses on incident propagation in the working system's internal logic. Data application, in such a context, was thus fully possible in material signs of a system while progressing toward a downgraded mode. Hence, a contribution of basic human factors data targeted human data for the design of warnings, alarms, and the like (Fitts and Jones, 1957). This is one of the points of debate regarding safety I versus safety II, as proposed by Hollnagel (2014): Should we develop more and more barriers to avoid and/or contain the consequences of accidents rather than attack the nucleus of the problems? In the current state of discipline, ergonomics deals with both trends, even the underlying peripheral level of results issued.

This conjuncture was a step forward. It was the formulation of foundations for physical and logical environments and settings for supporting complex activities. This interesting formulation required the ergonomics discipline to develop an approach of complex systems, until this point just mentioned (but not really used): a conceptual evolution from an approach of a deviation phenomenon (the incident) toward a state change of the systems (as entering in a complex mode). It operates two deepening paths of development: (1) from the incident, taken as evidence impacts that have latent failures as root causes, and (2) from particular circumstances to an activated short-term state in which some specific, eventful, pertinent actions are to be done, aimed at restoring a normal condition. This also rebounds on the other sociotechnical interfaces. It launches a technological and managerial development program: to prepare technology and organizations to cope with complexity. This was the old proposition from J. Rasmussen, now shaped by the construction of the resilient approach to engineering (Hollnagel et al., 2006).

An idea was born and a proposal launched to deal with specific tools to instrument such a modality of engineering. The current task lies over simulation issues. This brings out the possibility for the integration of alarms and warnings from an instantaneous occurrence toward a contemporary process vision. Simulation, especially virtual augmented realities, is an effective tool to explore the dynamics of systems, because all their engines are driven by time functions and time recurrence functionalities. But simulation can be built over a distorted vision of the reality of the working process. This is a classical problem of ergonomics design, also appearing in simulation design for ergonomics purposes. As exposed above, it only can be solved by activity modeling by means of the concepts of regulation and course of action. In turn, ergonomics activity modeling supplies enough material to build emergent constraint scenarios, aimed at safety training and surpassing this subject of distorted representation. One emblematic development was the line-oriented flying training for a Brazilian air company (Saldanha, 2005). In this development, large field research, about the course of the aircraft piloting actions, produced a register of a set of pertinent constraints. After that, it combined the regular flight simulation with the insertion of constraint scenarios built over that constraint list. As a result, it gave rise to a set of unprecedented experiences to the crew in near-accident conditions.

Figure 6.5 presents the first graphic layout. Its textual formula is as follows: the *latent failures* produce *incidents* that arise into a *particular circumstance* in which there are not enough resources to proceed to *recovering actions*. Its prevention process lies over deep defenses, dealing with technological and organizational elements.

6.5.2 Anthropotechnologic Perspective

This sociotechnical reliability stream allowed the formulation of depth defense, a state-of-the-art concept in that time. According to chaos theory, its application was pertinent until it reached its limits: its deployment in other anthropotechnologic

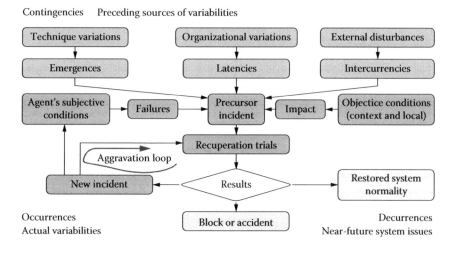

FIGURE 6.5   Sociotechnical reliability model (Vidal, 1984, 1994, 2001, 2013).

contexts, placed in the boundaries of the original applications. This becomes clearer when studying the design of a technological transfer.

The utility, pregnancy, and effectivity of simulations for safety management in relationship are the realm level of the simulation. The specific kind of realm for these purposes is essentially conceptual. Simulation can be built over a distorted vision of the reality of a working process or by taking references from bad references; these premises quickly become a source of high costs for safety management. These costs are primarily to have people trained for useless problems, hence, not actually benefitting from the training process. Ergonomics has, historically, met such a context in its epistemology:

- Data, in general, contrast with localized requirements of human data.
- Local production logic considerably stresses the reference models.
- Match (and mismatch) of a model of current work in a reference situation and the local one implies a specific design difficulty.

These aspects produced singular contributions to safety management. Contingencies and downgraded modes should be assessed and treated by the continuous improvement of contingency attenuation and the current fulfillment of the lack of recovery condition. These ways of forwarding the consequences evoke different procedures according to the stream in focus.

The launcher approach of human factors, human data supplying, met the diversity context because data confronted a list of localized requirements induced by demographic contents, as well as by cultural ones. Yes, people have different characteristics, but they also behave differently, which intensifies this variation data. For instance, a comparison between repetition standards in the assembly lines of sites in Brazil and France indicates different profiles (J.C. Bispo, personal communication). In terms of interface technologies, these variations may strongly impact some design options since they challenge the validity of their references. Meckassoua (1986) reports an effective control mode in a brewery plant in central Africa, where the controller spends much more time overseeing the production area rather than the control center dashboards. The number of disruptive incidents (package stop, cram in line feeding, and the like) are lower than in the referential brewery plant in Europe. However, the control room design placed the control workstation at a distance from the window for the production area. This obliged the controller to repetitive displacement during the day.

In that sense, the ergonomics point of view in the anthropotechnologic perspective enhances the activity modeling as an essential process. This modeling should combine microsituated approaches with macroergonomics conceptual models. Hence, the anthropotechnologic perspective enlarges the concepts and introduces the macroergonomics perspective as a part of the ergonomics point of view for safety management issues. However, it does not fulfill the essential indication concerning practices.

Introductions and transfers have two major dimensions: technical and organizational. This approach brings out the core concepts of variability, contingency, and downgraded mode in order to connote the diverse impacts from introductions and

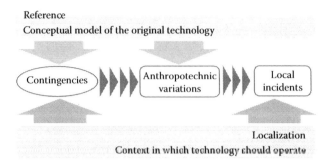

Reference
Conceptual model of the original technology

Localization
Context in which technology should operate

FIGURE 6.6 Anthropotechnologic perspective of work accidents.

transfers. The safety level depends on the extension and depth of the cultural clash issued from a technological introduction or transfer. Consequently, the four-pillar foundation of the sociotechnical reliability approach needs to be progressed to an enlarged formulation. In this sense, the root causes of accident components progress from latent failures to production contingencies; incidents that provoke the ignition of an accident become part of the variations that induce it. The particular circumstance in which an accident occurs is now an event in the downgraded mode to which it belongs. Finally, isolated restoration actions have less sense than their occurrence in a wise work accident chain (Figure 6.6).

Hence, its textual formula becomes the following: *the contingencies* produce *variations* in the work process, which turns into a *downgraded mode*, the actual context for *the work accident chain*. Its prevention process lies over the management of contingencies and downgraded modes by the continuous improvement of contingency process attenuation and the current fulfillment of the lack of recovery conditions in downgraded mode chains.

6.5.3 VARIABILITY MANAGEMENT

Simulation can be synthetized as the performance of screenplays within built scenarios. The two precedent ergonomics contributions had developed a set of considerations based on scenarios. Concerning screenplays, the basic model* (Figure 6.7) should be explored in terms of its two principal directions (resilient condition maintenance and robustness assessment). The screenplays are special objects of design, fully inserted into the dynamics of the working system and examined under the complexity point of view. The core point is to insert agents in the precursory space of a microincident. Its time function begins with a normal context, knowing that normality is only in appearance, because combined (internal and external) perturbations are continuously dealt with by type I regulations (little recoveries and mitigations). Observed activity is not entirely procedural, is absolutely not improvised, and is much closer to the actual situation in which some occurrences are taken into account for operating decision making.

---

* This framework is based on propositions by Pavard et al. (1991, 1997), with further development aggregated by our research team for complex systems.

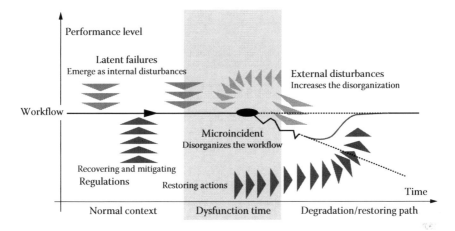

FIGURE 6.7   Core domain of variability management.

They are not entirely procedural because procedures do not cover all the little events and their combination in a just moment. They are not improvised because an operational experience exists, even in a new device (which is partially based on precedent technical bases). They are much closer to the current situation because this is where microincidents occur. This is also where one can identify the precursor's signs.

The solution seems to be a paradox introduced by Hollnagel et al. (2011), with the proposition of laying down the classical antagonism between success and failure. *Ipso facto*, it formulated the core foundations of the *resilience engineering* framework as follows:

1. Incidents do not necessarily bring up perturbation, but emergent variations in a process.
2. Success and failures are epistemological equivalents as process issues; they have the same origin, the same process, the same context, and the same crew.
3. Success and failures occur by a particular and instantaneous combination of variabilities.

Considering these assumptions, it can be emphasized that keeping the work accident chain under control involves coping with contingent nature impacts for robustness reinforcement, combined with resource availability for variability management, which requires resilient contents in technology and organization design. The two precedent approaches supply enough material for coping with the contingencies in their principles (sociotechnical reliability) and in their enlargement (anthropotechnologic perspective). Hence, the variability management settings turn to the resources for doing it. Here, we will introduce (or rather, we will revisit), the core concepts of regulation (Faverge, 1992; Leplat, 2006), emergence (Holland, 1998), and course of action (Theureau and Pinsky, 1979; Theureau, 1992).

The specific textual formula is as follows: *external disturbances* combined with latent failures, which emerge as *internal disturbances* that impact normal workflow, are continuously counterbalanced with the *regulations* performed by the agents, which *recover and mitigate* such impacts. This works in a resilient way until the emergence of a particular *microincident* that overcomes the system robustness. This special kind of impact disorganizes the *workflow*, sending the system onto an incidental path, requiring *recovery of the system*, and thus sending the workflow back to the normal path.

6.6 DISCUSSION

The most general aspects of the expression "ergonomics point of view upon work accidents" establishes a relationship between safety management and human factors. It is common sense that an accident is an undesirable system output, a deviation from its major objectives. However, there are differences if we consider work accidents as intrinsic, extrinsic, or emergent phenomena related to work functioning.

In singular terms, it is important to underline the seminal (but pessimist) view of Perrow (1984) that in all complex systems, accidents are part of normality. However, some organizations are more efficient than others, which has inspired research on high-reliability organizations (Weick and Roberts, 1993). This approach exhibits a considerable lack: it does not take into account the Emergence, an essential property of complex systems. It simply does not take into account the occurrence of nondesired issues (accidents) and their necessary explanation as an overflow of the variability management capability of a system.

Despite the complexity theory statement that it is not always possible to clearly match causes and their effects, it is necessary to consider the different process failures in human, technical, and organizational parts of a sociotechnical system. Hollnagel et al. (2011) assimilates success and failures as *epistemic* equivalents. He also infers that the modes and the context in which actions are performed, as well as the principles under which they should be understood, are the same, independent of the positive or negative results. However, the differences between system components are not strongly emphasized. This entails a resurgence of the atavist notion of human error.

Their positive contribution is the enunciation: *a particular accident is a specific instantaneous overlap of diverse variabilities*. Hence, variability management appears as the logical issue in work accident prevention, the core concern of the relationship between safety management subjects and the human factors field. However, managing variability is not only registering variabilities and following up on them. This means active interaction as the path correction of a spatial vessel in the cosmos.

In this sense, we outline a set of devices aimed at improving prevention in sociotechnical systems. First, sociotechnical systems should have the availability of dedicated devices to provide this action field (device 1). The idea of resilience completes robustness engineering by producing structures that will resist up to a certain degree of variability (device 2). This also requires the flexibility of resources, which could reinsert in a nonaccidental path (device 3). These robust flexibility settings for sociotechnical systems have conditions for effective prevention: to manage variability effects in technical and organizational components, rather than asking to the human dimension of the system to perform it. In summary, an ergonomics point of view, as exposed, has the following consequences in theoretical aspects.

In factual terms, *a work accident is not only the damage to a worker.* This means that a work accident cannot be reduced to a moment or place. Damages are just the results of a progressive process. In the same way, it is not enough to classify areas by any kind of danger level. Examining the structure of the progressive process that tragically ends by injury (damage) to workers in a given area produces better ways to assess the danger. In examining a potential process, work accidents can be studied without the occurrence of tragedies. This opens a forensic perspective for work accident studies aimed at consolidating the safety management subject.

In processing terms, *a work accident implies a mechanism involving a combination of diverse factors, not necessarily close to the accident's place or moment.* This introduces the notion of the systemic distance between a concurrent factor and the accident as a whole. An accident combines various factors having different distances of the final events. Those systemic distances are related to time, for example, combining an event that occurred the day before and something that happened a few minutes before the crash. They are also related to physical distance, for example, combining elements in place of the tragedy and other elements taken away. This reinforces the low utility of the contextual analysis of accidents limited to the circumstantial approach. However, this consequence places emphasis on an enlarged causality component set, in temporal terms, as well as on topological ones.

Concerning its genesis, *a work accident is launched by mechanisms arising from the work process.* In an accident genesis, the workers were surely doing the same things that they usually do. This means that success and failures have no genetic differences. Nevertheless, some variations produce the inflexion toward an accidental issue. Finding these factors and their specific combination is a matter of having a process view of work: a systemic view, not the static one. Moreover, the model of accident investigation should place the event in chronological and topological terms. It is the juxtaposition of a set of variabilities in normal processes that are in the genesis of accidents.

In order to be established under a variability management conception, the accident—as a phenomenon seen by a systemic and dynamic approach—should be explained in sociotechnical terms in a plan defined by the combination of brittleness–robustness and rigidity–resilience.

6.7 CONCLUSIONS

The core finding of our study is that an ergonomics point of view of work accidents can be useful for safety management. It provides accident expertise and supports the elaboration of an outline of prevention issues, as it contributes to

- Inserting human data into the specification of requirements, with knowledge of the scenarios built by course-of-action analysis
- Enabling the design of working elements (tools, instruments, furniture, procedures, and information technology) and their interfaces, improving the monitoring of variability

- Driving the analysis to a lengthwise and comprehensive mode—from macrostructure to activities flow

Thus, the ergonomics point of view brings scientific and methodological contributions to variability management, to the extent that the sociotechnical reliability of a working system should be ensured by its capacity to manage variability. It enables the analysis of work accidents at the operational and technological levels, highlighting the lack of robustness and resilience for each level of analysis.

In this chapter, we presented an explanation of conceptual evolution, applied to a safety management approach, and ergonomics, in order to provide a suitable framework for an ergonomics point of view of work accidents. We believe that the organizational and technological design will take advantage of the ergonomics point of view to improve variability management, resilience, and robustness in keeping work accidents under control.

The ergonomics point of view allows us to understand the distinction between the noted variation in work accidents and its properties as a systemic dysfunction, as well as its causal link with external factors. Moreover, no system is so perfectly designed that it avoids any kind of dysfunction; therefore, it is impossible to expect an uneventful working system.

ACKNOWLEDGMENTS

We thank the Brazilian National Council of Technological and Scientific Development and The Fundação de Amparo à Pesquisa do Rio de Janeiro, for financial support.

We also thank to the Applied Human Factors and Ergonomics 2014 organizers for proposing the scientific challenge in the genesis of this chapter.

REFERENCES

Adler, A. (1951). Some analytical concepts related to occupational and personal accidents. *Journal of Applied Psychology*, 4(3), 27–33.

Alderson, D., and Doyle, J.C. (2010). Contrasting views of complexity and their implications for network-centric infrastructures. *IEEE Transactions on Systems, Man, and Cybernetics—Part A: Systems and Humans*, 40(4).

Arcuri, R., and Saldanha, J.M.B. (2013). *Usability assesment of the SIGA platform: An exploratory study*. Graduation report, Federal University of Rio de Janeiro, Brazil.

Cartaxo, C.J. (1987). The battle of Josefa Moenda against Biu Metedor: Work conditions in sugar cane plantations at Brazilian nord-east region [in Portuguese]. MSc dissertation, Production Engineering Program, Federal University of Paraiba, Brazil.

Carvalho, P.V., Vidal, M.C., and Luquetti, I.S. (2005). Nuclear power plant shift supervisor's decision-making during micro incidents. *International Journal of Industrial Ergonomics*, 35, 619–644.

Carvalho, P.V.R. (2006). Ergonomic field studies in a nuclear power plant control room. *Progress in Nuclear Energy*, 48, 51–69.

Faverge, J.M. (1980). Working as recuperation activity. *Bulletin de Psychologie*, 33(314).

Faverge, J.M. (1992). *Work analysis within a regulation approach*. In Leplat, J. (coordinateur), *L'analyse du travail en psychologie ergonomique*. Toulouse, France: Éditions Octarés, pp. 61–68.

Figueiredo, M.G., Vidal, M.C., Marchand, T., and Pavard, F. (2001). Cooperation and safety in complex systems: Case study of deep dive in offshore oil facilities in the Campos Basin. Presented at Proceedings of the ABERGO 2001, Brazilian Congress of Ergonomics, Gramado, Brazil.

Fitts, P., and Jones, R. (1957). Analysis of factors contributing to 460 pilot error experiences in operating aircraft controls. Reprinted in: Sinaiko, H.W. (ed.), Selected papers on human factors in the design and use of control systems, 1961. New York: Dover Books, pp. 332–358.

Grandjean, E. (1974). *Fitting the Task to the Man*. London: Taylor and Francis.

Hendrick, K., and Benner, L. (1987). *Investigating Accidents with STEP*. New York: Marcel Dekker.

Holland, J.H. (1998). *Emergence: From Chaos to Order*. Redwood City, CA: Addison-Wesley.

Hollnagel, E. (2014). *Safety-I and Safety-II, the Past and the Future of Safety Management*. Farnham, UK: Ashgate.

Hollnagel, E., Pariès, J., Woods, D.D., and Wreathall, J. (eds). (2011). *Resilience Engineering Perspectives*, vol. 3: *Resilience Engineering in Practice*. Farnham, UK: Ashgate.

Hollnagel, E., Woods, D., and Levenson, N. (2006). *Resilience Engineering*. Farnham, UK, Ashgate.

Jastrzebowski, W. (1857). Rys Ergonomji, czyli Nauki o Pracy. *Ergonomia*, 1979(2), 13–29. Special translation for commemorating International Energy Agency 2000 edition.

Kuhn, T. (1962). *The Structure of Scientific Revolutions*. Chicago: University of Chicago Press.

Lahy, J.M., and Kongorold, S. (1936). Experimental research upon psychological causality of work accidents. *Le Travail Humain*, 11, 1–64.

Leplat, J. (2006). The notion of regulation in activity analysis [in French]. *Pistes*, 8(1).

Meckassoua, K. (1986). Étude comparée des activités de régulation dans le cadre d'un transfert de technologie: une approche antropotechnologique. Doctoral thesis, CNAM/Paris (Thesis advisor: A. Wisner).

Mohamed, S., Favrod, V., Philippe, R.A., and Hauw, D. (2015). The situated management of safety during risky sport: Learning from skydivers' courses of experience. *Journal of Sports Science and Medicine*, 14(2), 340–346.

Murrel, K.F.H. (1965). *Ergonomics: Man and His Environment*. London, UK: Chapman & Hall.

Nielsen, J., and Landauer, T.H. (1993). A mathematical model of the finding of usability problems. In *INTERCHI 93 Conference Proceedings*, Amsterdam, pp. 206–213.

Pavard, B., and Decortis, F. (1994). Communication et Coopération: de la théorie des actes de langage à l'approche ethnometodologique. Em: Pavard B. (Org.) Systèmes coopératifs: de la modelisation à la conception. Octarès Edition, Toulouse.

Pavard, B., and Dugdale, J. (1997). The Contribution of Complexity Theory to the Study of Socio-Technical Cooperative Systems. Toulouse: ARAMIS-IRIT.

Pavard, B., and Vidal, M.C. (1997). Plongee Off-Shore: Rapport d'expertise cognitive, Accident SC-300. GRIC-Arhamiihs, Université Paul-Sabatier, Toulouse.

Perrow, C. (1984). *Normal Accidents: Living with High Risk Technologies*. New York: Basic Books.

Rasmussen, J. (1990). Mental models and the control of action in complex environments. In Ackermann, D., and Tauber, M.J. (eds.), *Mental Models and Human-Computer Interaction 1*. Amsterdam: Elsevier Science Publishers, pp. 41–46.

Rasmussen, J., and Svedung, I. (2000). *Proactive Risk Management in a Dynamic Society*. Swedish Rescue Services Agency.

Reason, J. (1997). *Managing the Risks of Organizational Accidents*. London, UK: Ashgate.

Saldanha, M.C.W. (2004). Ergonomic Design of a platform Line Oriented Flight Training (LOFT) in a Brazilian air company, Rio de Janeiro, COPPE/UFRJ, D.Sc. Thesis, Production Engineering, (Thesis Advisor: M. C. Vidal).

Saldanha, M.C.W., Carvalho, R.J.M., and Vidal, M.C. (2008). PRO-LOFT: Line oriented flying training experimental program development in a Brazilian air company. Presented at Proceedings of the 9th International Symposium on Human Factors in Organizational Design and Management (ODAM), Guaruja, Brazil.

Tanaka, K. (1921). *Research of Efficiency: Ergonomics.* Tokyo: JES.

Theureau, J. (1992). *Le cours d'action: analyse sémio-logique. Essai d'une anthropologie cognitive située.* Berne, Switzerland: Peter Lang.

Theureau, J., and Jeffroy. (1994). *Ergonomie des situations informatisées: la conception centrée sur le cours d'action des utilisateurs.* Octarès Editions, Toulouse.

Theureau, J., and Pinsky, L. (1979). *Analyse du travail en saisie chiffrement.* Research report. Paris: LENET/CNAM.

Tiffin, J., and McCormmick, E.J. (1967). *Psychologie Industrielle.* Paris: PUF.

Vidal, M.C. (1984). The conceptual evolution of work accidents causality [in Portuguese]. Presented at Proceedings of the IV National Meeting of Undergraduated Training of Production Engineering, Piracicaba, Brazil.

Vidal, M.C. (1985). *Mason's working in France and in Brazil: Sources and management of differences and variations* [in French]. DrIng thesis, Conservatoire National des Ats et Métiers, Paris.

Vidal, M.C. (1994). Sur l'interaction en Analyse du Travail: La conversation dans une analyse collective en situation. Proceedings of the 12th Triennal congress of the International Ergonomics Association, 353–356. Reprint em Feitosa, V. and Vidal, M. (1996) Linguagem e trabalho, GENTE/ COPPE.

Vidal, M.C. (2001). *Ergonomics in the firm: useful, practical and applied techniques.* Rio de Janeiro, Brazil: EVC.

Vidal, M.C. (2003). *Guidelines for Ergonomic Work Analysis in Workplaces: A Realistic, Ordered and Systematic Methodology* [in Portuguese]. Rio de Janeiro, Brazil: EVC.

Vidal, M.C. (2013). Socio-technical system safety and corporate sustainability: A complexity approach [in Portuguese]. Presented at Opening Conference, IV ERGONODIA, Natal, Brazil.

Vidal, M.C., Bonfatti, R.J. and Mafra, J.R.D. (2012). Guidelines for the ergonomic work analysis for beginner in Ergonomics. Keynotes of the MA in ergonomics, Federal University of Rio de Janeiro, Brazil.

Vidal, M.C., and Duarte, F.J. (1992). The introduction of digital systems in the control rooms of oil refineries in Brazil. In Shwarazan, K. (ed.), *Advances in Industrial Ergonomics and Safety IV.* London: Taylor & Francis, pp. 1447–1453.

Vidal, M.C., and Moreira, L.R. (2013). *Ergonomics Appreciation of a Fuel Distribution Terminal,* Consulting reporting, GENTE/COPPE/UFRJ, Rio de Janeiro, Brazil.

Vidal, M.C., Moreira, L.R., Bonfatti, R. J., Ricart, S.S.I. and Mafra, J.R.D. (2015). EAMETA: a friendly method of participatory analysis of work situations. 6th International Conference on Applied Human Factors and Ergonomics (AHFE 2015) and the Affiliated Conferences, AHFE 2015.

Vidal, M.C., and Nunes, A.M. (1994). L'organisation du travail et l'interaction dans l'analyse collective en situation. Presented at Proceedings of the XIIth IEA Congress, Toronto.

Vidal, M.C.R., and Bonfatti, R.J. (2003). Conversational action: An ergonomic approach to interaction. In Grant, P. (ed.), *Rethinking Communicative Interaction.* Amsterdam: John Benjamin Publishing Company, pp. 108–120.

Weick, K.E., and Roberts, K.H. (1993). Collective mind in organizations: Heedful interrelating on flight decks. *Administrative Science Quarterly,* 38, 357–381.

Zurfluh, A. (1957). *Work Accidents and Safety Training* [in French]. Paris: Dunod.

7 S-MIS
Identifying, Monitoring, and Interpreting Proactive Safety Indicators

Toni Waefler, Simon Binz, and Katrin Fischer

CONTENTS

7.1   S-MIS BACKGROUND

Safety management systems (SMSs) provide integrated organizational mechanisms designed to control safety risks, as well as current and future safety performance (Cooper, 1998). Part of an SMS consists of instruments focusing on assessing both the safety level of an organization and the effectiveness of measures an organization takes in order to improve its safety. However, in practice, the information basis regarding these two aspects is often very vague. This vagueness is due to a multiplicity of factors contributing to safety, as well as a lack of knowledge regarding their complex interrelations. Furthermore, there is also a lack of methods for objective, reliable, and valid measurement of these factors (Waefler et al., 2008). In order to proactively mitigate risks or provide resilient processes able to cope with safety threats, a more reliable information basis would be very helpful.

Against this background, the project "Safety Management Information System" (S-MIS) aimed to develop an information system that supports decisions in safety management. The development of the S-MIS process was based on state-of-the-art insights regarding safety indicators and safety assessment, as well as on case studies analyzing respective processes in practice. In pilot projects, together with industrial partners operating in high-risk industries such as aviation and nuclear power, the S-MIS process was tested and fine-tuned.

7.2   INDICATORS FOR SAFETY ASSESSMENT

Generally, for assessing safety, both reactive and proactive indicators can be applied (Choudhry et al., 2006). Reactive indicators focus on safety outcomes. Thus, these indicators are based on retrospective data. Respective indicators are also referred to as downstream indicators or lagging indicators (Hinze, 2005; Mohamed, 2002). Representatives of this kind of indicator are, for example, the number of safety-related occurrences, such as incidents and accidents and follow-up costs of accidents. This implicates that safety can be measured on the basis of (near) system failures. However, the suitability of reactive indicators for assessing safety is very restricted, as the absence of unwanted occurrences does not necessarily mean that operations are safe. This is because unsafe operations do not compulsorily create adverse events. Hence, the absence of occurrences in the past does not guarantee an absence of occurrences in the future.

Due to the limited value of reactive indicators, a shift to proactive indicators (or upstream indicators, leading indicators) is advocated, referring to aspects such as safety climate (Flin et al., 2000; Mohamed, 2002), safety culture (Choudhry and Fang, 2005; Cooper, 2000), or observable safe behavior (Cooper and Phillips, 2004; Strickoff, 2000). This proactive approach focuses on enablers of safe operations rather than on system failures in the past. According to Grabowski et al. (2007), proactive indicators are suitable to identify hazards, assess risks, and consequently mitigate risks.

Other authors, such as Amalberti (2013) and Hollnagel (2014), go further. They emphasize the importance of resilience for safety. Resilience is defined as the ability of a system to keep control under changing conditions. This broadens the focus of safety management. Traditionally, safety management is rather deficiency oriented, aiming at identifying and mitigating risks. The aim is to find defects in the system (e.g., potential human error) and fix them by barriers, which either correct the defect or contain negative consequences of the defect. In contrast, the concept of resilience suggests that unsafe situations do not only occur due to defects in the system. They are also a result of the system's inability to adapt to dynamic working conditions. Hence, safety promotion requires not only the fixing of defects by reactive barriers. It also requires the ability to actively sustain safety by proactive adaptive processes. Whereas the former aims at avoiding weaknesses, the latter aims at promoting strengths. For S-MIS, this means that safety indicators should not refer to defects and weaknesses only, but also to safety enablers in terms of strengths.

7.3   APPLICATION OF SAFETY INDICATORS IN PRACTICE

Two in-depth case studies in the aviation and nuclear power industries revealed the following critical aspects of safety indicator application in practice (Waefler et al., 2008):
   Indicators and measurement

- Indicators applied in practice are rather reactive than proactive, and hence represent safety outcomes rather than safety enablers.
- Indicators not referring to hard, countable facts are measured in an *ad hoc* manner rather than by applying state-of-the-art methods for data collection and analysis.

Aggregation of data, interpretation, and decision making

- Even though respective processes are well systematized and described in a comprehensible way, and responsibilities are assigned clearly, aggregation of data, interpretation of information, and decision making are driven by

distributed subjective assumptions of the safety experts regarding cause–
effect relations and interactions of different indicators. These subjective
assumptions incorporate a lot of tacit knowledge, the quality of which is
extremely difficult to assess.

- Decision making based on tacit knowledge incorporates the risk of cognitive biases (cf. excursus below). In the case studies, no measures have been
found that systematically avoid these biases. Rather, the decision making
depends on the decision makers' (managers and safety experts) personal
preferences.

7.4   CONSEQUENCES FOR THE S-MIS PROCESS

The aim of S-MIS is to provide a structured process for identifying and applying
clearly defined safety indicators. The conceptual considerations as described above,
as well as the critical aspects of indicator application in practice as revealed in the
case studies, led to the following consequences for S-MIS.

First, the indicators to be considered should be proactive and refer not to weaknesses only, but also strengths. To achieve this objective, a hierarchical indicator
model is envisioned. At the top of the hierarchical model stands safe acting, which
defines the workers' safe acting as a main proactive indicator of safety. The assumption behind this is that safety is, to a large extent, brought about by a workforce
that works safely. Having safe acting at the top of the hierarchical indicator model,
S-MIS provides a process for identifying underlying levels of indicators, which refer
to enablers of safe acting. These enablers are considered to be proactive, and respective indicators can concern weaknesses as well as strengths.

Second, S-MIS suggests a participatory process of indicator identification and
application. This is mainly due to two reasons. On the one hand, reliable scientific knowledge regarding factors contributing to safety is far too scarce to allow
for deriving relevant safety indicators and their interrelations. On the other hand,
experienced safety practitioners show an extensive tacit knowledge regarding factors
having an impact on safety. A participatory process combining the practitioners'
tacit knowledge with latest insights from safety science is considered to be most
promising.

Third, S-MIS needs to ensure process rationality (cf. excursus below). As participatory, tacit knowledge–based processes of indicator selection, safety assessment,
data interpretation, and decision making are vulnerable to cognitive biases, they
need to be designed systematically, promoting a continuous and critical reflection of
the subjective assumptions that are driving information search, information assessment, and decision making. Ensuring process rationality aims at making the processes as robust, transparent, and rational as possible.

Considering these three major requirements, S-MIS supports the identification
and application of clearly defined safety indicators. However, one true safety model
to be programmed into an automated decision engine is not envisioned, but rather a
continuous process that allows for both (1) making explicit individual mental models regarding causes and effects of safety-related factors and enablers and, on that
basis, (2) the joint elaboration of a safety indicator model shared by the participating

safety experts. The shared model itself is subject to continuous critical reflection, and hence continuous reconstruction—if required. Since this continuous modeling is a social and collective process, it is also a systematic exchange of (tacit) safety knowledge. It thus supports knowledge management, making the organization less dependent on experiences distributed throughout the organization and hidden in individual minds.

7.5 EXCURSUS: COGNITIVE BIASES AND PROCESS RATIONALITY

As S-MIS relies on a participatory process, it is vulnerable to cognitive biases. This excursus discusses such biases and the consequences for S-MIS. *Cognitive bias* is a broad term for all distortions of the human mind that are hard to avoid and that lead to judgments that deviate systematically, involuntarily, and rather distinctly from reality (Pohl, 2005). Cognitive biases can occur in individual information processing as well as in group decision making. They may affect rating of frequencies, probabilities, or other numerical variables in an uncertain environment. They affect memory retrieval processes, and they occur in thinking, reasoning, and decision making. Most of these biases are associated with simplifying heuristics people use rather than extensive algorithmic processing. Heuristics are cognitive rules of thumb that help people to make difficult and important judgments in an uncertain world. Mostly, these heuristics are simple and efficient. They save time and effort, and they typically yield rather accurate judgments, but they can also give rise to systematic error. Failures occur unconsciously and are difficult to detect. Furthermore, decisions based on heuristics "feel good." They are intuitively satisfying regardless of their (in-)correctness (Gilovich et al., 2002).

7.5.1 COGNITIVE BIASES ON INDIVIDUAL LEVEL

The following heuristics underlying many intuitive judgments under uncertainty have been identified (Gigerenzer et al., 1999; Gilovich et al., 2002; Kahneman et al., 1992; Russo and Shoemaker, 1989):

> **Availability:** This is the ease with which a given instance or scenario comes to mind, and typically salient memories override normative reasoning. Many people overestimate the likelihood to die in a plane accident compared to the likelihood to die in a car accident, although most of them more often drive by car than fly by plane. The real base rate, the statistical background information about the frequency of given events, is often underweighted or ignored entirely in intuitive human reasoning, especially when events are salient, vivid, or dramatic.
>
> **Representativeness:** This relates to people's tendency to classify objects based on surface features. When asking people which sequence of randomly played dice is more probable, sequence A [1, 2, 3, 4, 5, 6] or sequence B [3, 5, 2, 6, 4, 1], people judge B as the more probable although the probability of both is the same. Sequence B appears to be more representative for a random die.

Anchoring and adjustment: Due to this, people adjust insufficiently for the implications of incoming information. When estimating the product of the two series [$10 \times 9 \times 8 \times 7 \times 6 \times 5 \times 4 \times 3 \times 2 \times 1$] and [$1 \times 2 \times 3 \times 4 \times 5 \times 6 \times 7 \times 8 \times 9 \times 10$], people's estimates are strongly biased by the figures at the beginning of the series. In studies requiring participants to give their answers as quickly as possible, the average estimate for the first series is about 900, and for the second series only around 150. The correct answer is 3,628,800. Using the anchoring and adjustment heuristic means that people form judgments around an anchor, and additional incoming data must fight against the inertia of the anchor, even when the anchor is objectively irrelevant to the judgment at hand.

Confirmation bias: This refers to the unwitting selectivity in the acquisition and use of evidence. People tend to be willing to gather facts that support a certain conclusion, but disregard other facts that support different conclusions. Decision making always takes place within a specific context that provides the decision maker with information about the issue. This information may be interpreted by the decision maker beforehand, eventually on the basis of earlier decisions made in the same or in a similar context. Consequently, the decision maker may develop preliminary beliefs regarding the issue. People tend to seek evidence confirming their beliefs and ignore the need to look for additional information necessary to test their beliefs.

Repetition bias: Decision making also depends on the amount and quality of information taken into account. Whether selected information is perceived to be especially significant or persuasive is strongly influenced by its source. In this respect, people tend to a repetition bias, which refers to the willingness to believe what they have been told most often and by the largest number of different sources.

7.5.2 Cognitive Biases on Group Level

Looking at decision making within groups, there are further specific factors affecting information processing, judgments, and decision making caused by social interaction.

Groupthink: This is a type of thought exhibited by group members trying to minimize conflicts and reach consensus without critically testing, analyzing, and evaluating ideas. The desire for conformity in the group results in an irrational or dysfunctional decision making. The following factors contribute to groupthink (Janis, 1982):

- Structural deficiencies in the organization, such as insulation of the group, lack of tradition of impartial leadership, lack of norms requiring methodological procedures, a high homogeneity of group members' social background, and ideology
- Provocative situational contexts with high stress from external threats, recent failures, excessive difficulties on the decision-making task, or moral dilemmas

Risky shift: This is the tendency for decisions made by a group after discussion to be less conservative and riskier than decisions made by individuals acting alone and prior to any discussion (Six, 1981). It is the difference between the average risk taken by individuals and the risk taken by a group. There are a number of reasons for that: diffusion of responsibility in group decision making (Wallach et al., 1964), a higher confidence and persuasiveness of people who take greater risks (Collins and Guetzkow, 1964), and higher social status of risk-prone group members (Brown, 1965).

7.5.3 PROCESS RATIONALITY IN S-MIS

In safety management, information interpretation, probability estimations, frequency rating, or weighting processes are complex, and mainly done under uncertainty. Respective judgment and decision making is not immune to cognitive biases. Hence, process rationality is required. It creates conditions that prevent cognitive biases and make the decision processes as robust, transparent, and rational as possible. Thus, it is not only necessary to evaluate data as an input for safety-related decisions but also to analyze and design the decision-making process itself.

To ensure process rationality in safety-related decision making, it is necessary to separate the different stages of the decision process and carefully go through them. These stages are (1) determining and selecting relevant safety indicators, (2) weighting the indicators, (3) collecting data, (4) aggregating data, (5) interpreting the data, and finally, (6) making the decision.

When selecting relevant safety indicators, it is important to guarantee their content validity. The indicators should be accepted, exhaustive, and nonredundant. The clear distinction of stages 2 and 3 should prevent the decision maker from confounding weight and measurement, and should minimize biases like a *post hoc* underweighting of indicators with negative outcomes or an overweighting of those with positive values. Another method to overcome the constraint of confounding weight and value of a safety indicator is to set up of two distinct groups of experts: one group weighting the indicators with respect to their importance for safety and the other group measuring the indicators. However, aggregation of data should follow previously defined aggregation rules, and decisions derived from the data should be documented in a comprehensible and appraisable manner. Going through these stages does not guarantee perfect decisions, but ensures the highest possible process rationality and makes the decision process robust, transparent, and comprehensible.

As discussed above, human decision making is not completely objective and rational. In safety management, aggregation, judgment, and decision making are driven by distributed subjective assumptions of the safety experts involved. These assumptions include subjective views on the safety-related importance of measured indicators, on reasonable ways to interpret information, and on cause–effect relations and interactions of different indicators. Decision makers seem to apply tacit knowledge gained during years of safety management practice and remarkable insider knowledge of a plant's functioning. In decision making, this knowledge very often comes into play as affectively mediated "gut feeling"; that is, underlying knowledge and assumptions can only partly be explained. However, such implicit assumptions and knowledge may also lead to biased judgments and decisions. For these reasons, special attention should be turned to the cognitive biases that could influence the individual and collective safety-related decision-making processes.

On the individual level, the availability bias, for example, may lead to restricted selection of data sources and safety indicators. Safety experts involved in updating the applied list of safety indicators have only little experience with soft factors and qualitative data, which may lead to a concentration on hard facts and a disregard of social science methods.

Furthermore, safety experts may be prone to a confirmation bias since they "know" the results of measured and interpreted indicators before they record additional qualitative data. Thus, information search could be limited to data confirming those conclusions that have been primed by the previously found hard facts.

As safety controlling reverts to several sources of information, there is also the risk of a repetition bias. The more often certain information has been reported by different sources, the higher is the tendency to believe that this information is particularly relevant for the plant's safety status.

On the group level, data interpretation and the preparation of decisions regarding safety-related measures are usually made by groups working under high time pressure. Furthermore, high group cohesion can be assumed, since the groups' composition remains stable over time and members are exclusively recruited from top management and safety staff. Thus, the groups may also be relatively isolated. Additionally, the self-perception of high expertise, only partly established clear procedures for decision making, and the explicit aim to reach consensus in group discussions are further risk factors for groupthink processes. Over the last decades, several techniques have been developed to prevent groupthink and increase the chance of a high-quality decision, including the devil's advocacy decision-making technique, rotating in new group members, inviting attendance by outsiders, and allowing a temporary delay before the final decision is made (Corey, 2011; Lunenburg, 2012).

Since many decisions in safety management are made by groups, one should be aware of the risky shift phenomenon. However, there are several methods and techniques to overcome this constraint: avoiding diffusion of responsibility, enhancing the accountability of each person within the group, letting the group members make the decision on their own before sharing it, and encouraging everyone to be a critical evaluator of the decision process and the decision outcome.

It is very clear that cognitive biases cannot be fully avoided, on either the individual or the group level. Nevertheless, a number of precautionary measures have been taken to ensure process rationality for S-MIS as much as possible. This mainly includes a clear distinction of process steps and their respective objectives, a clear distinction of different roles the various participants have throughout the whole process, and a methodological sound approach in all steps for information gathering, processing, and interpretation. However, as the most problematic aspect is that there is only little reflection of cognitive biases, a continuous awareness needs to be fostered.

7.6   SEVEN STEPS OF THE S-MIS PROCESS

In detail, the following aims are set for the S-MIS process:

- To identify proactive safety indicators considering an organization's characteristics
- To periodically assess these proactive safety indicators
- To aggregate the assessed proactive safety indicators to an appropriate level
- To interpret the results of the periodically assessed proactive safety indicators
- To enable well-founded, safety-related decisions

To achieve these aims, seven process steps are required (cf. Figure 7.1). Three distinguishable groups with clearly different roles participate in these steps.

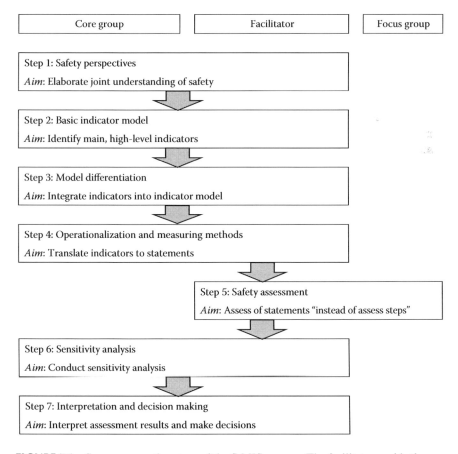

FIGURE 7.1   Seven consecutive steps of the S-MIS process. The facilitators guide the core group through Steps 1–4 and 6–7. The focus group participates in Step 5.

Core group: Group of experienced safety experts representing the organization in question. The main role of the core group is to develop the indicator model as well as the instruments for indicator measurement, the interpretation of measurement data, and the preparation of safety management decisions.

Focus group: Representatives of the organization or the organizational unit in which the safety indicators are applied. The main role of the focus group is to apply the instruments for indicator measurement developed by the core group. So, the focus group provides quantitative as well as qualitative data regarding the indicators.

Facilitators: The role of the facilitators is to facilitate the S-MIS process by guiding the core group as well as the focus group through their respective process steps. To do so, the facilitators provide methodological know-how required in the different steps, as well as extensive knowledge regarding the state of the art in safety sciences.

Steps 1–4 of the S-MIS process aim to define and operationalize safety indicators on the basis of the core group's expertise. In Step 5, the safety state of the organization is assessed by the focus group. Finally, in Steps 6 and 7, the core group interprets the results of the assessment and prepares safety-related decisions. The seven steps are described below in more detail. Results from pilot studies are presented for illustrative purpose.

7.6.1 Step 1: Safety Perspectives

7.6.1.1 Aim
In the first step, the core group elaborates in a participatory process a shared understanding of safety.

7.6.1.2 Method
The facilitators present different concepts from safety science (e.g., Amalberti, 2013; Dekker, 2006; Hollnagel, 2004; Perrow, 1992; Reason, 1990) to the core group. In a guided discussion, the core group's members connect these concepts with their own organizational experiences. Based on their discussions, the core group members determine requirements for an appropriate model of safety indicators.

7.6.1.3 Example
The facilitators introduce, for example, the different types of unsafe acts according to Reason (1990) to the core group. Reason differentiates unintended from intended unsafe acts. Unintended unsafe acts are caused by either a lack of attention (slip) or a memory flaw (lapse). Intended unsafe acts occur because of either errors in planning or violations. Errors in the planning of a task are more difficult to detect than slips or lapses. Most likely, an error in the planning of a task is not detected before the objective of an action is obviously missed (Hofinger, 2008).

The core group discusses situations in which unsafe acts happened. Doing so, a core group member remembers a situation in which the workload was very high. As

TABLE 7.1

Examples of Requirements for an Appropriate Model of Safety Indicators as Derived by the Core Group

Requirement for an Appropriate Model of Safety Indicators	Rationale of Requirement
The context of unsafe acts needs to be considered.	If an unsafe act happens, not only the unsafe act and its consequences need to be considered, but also the context of the situation in which the unsafe act happened.
Attitudes need to be represented.	An occurrence investigation can be biased by unconscious attitudes. Hence, attitudes need to be critically reflected.
The concept of resilience needs to be represented.	A safety management system should include resilience as a trait. In case of variation and disturbances, the organization should be able to adapt and sustain normal functioning.

a consequence, he had to cut some formal safety rules. According to Reason (1990), the intended skipping of formal procedures is a violation. However, there are reasons for such violations. In order to understand why violations happen and what measures are suitable to prevent violations, the context of the violation needs to be taken into account. This insight is generalized into a requirement for an appropriate model of safety indicators: the context of unsafe acts needs to be considered by the model. Similarly, further requirements were derived from the discussion of the theoretical safety concepts (cf. Table 7.1).

7.6.2   Step 2: Basic Indicator Model

7.6.2.1   Aim

The aim of the second step is to identify main, high-level indicators on the basis of the core group members' tacit knowledge regarding enablers of safe acting.

7.6.2.2   Method

Step 2 is completed in three substeps. In substep 2.1, the tacit knowledge of each core group member regarding enablers of safe acting is elicited. To do so, the Model Inspection Trace of Concepts and Relations (MITOCAR) method is applied (Pimay-Dummer, 2006). Each core group member separately completes sentences, for example,

- Safe acting is enabled by … if ….
- Safe acting is hindered by … if ….
- Safe acting is dependent on … because ….

In substep 2.2, the facilitators apply a qualitative content analysis. Diverse types of enablers of safe acting (e.g., influencing factors, preconditions, conditions, and

TABLE 7.2

Examples of Results of Steps 2.1 and 2.2: Elicitation of Tacit Knowledge Regarding Enablers of Safe Acting

Step 2.1: Data Collection Examples of Sentences Completed Individually by the Members of the Core Group	Step 2.2: Extraction of Concepts Examples of Concepts Regarding Enablers of Safe Acting Extracted by Facilitators
Safe acting is enabled by "education, instruction, and training" if "the focus is on development potentials"	• Education, instruction, and training • Development potentials
Safe acting is enabled by "proactive thinking" if "supervisors allow flexibility for employees"	• Proactive thinking • Flexibility for employees
Safe acting is enabled by "just attention" if "avoidance of errors and insecure acting are supported"	• Just attention • Avoidance of errors and insecure acting
Safe acting is hindered by "mutual recriminations" if "incentive systems suppress the communication of errors and insecure acting"	• Mutual recriminations • Incentive systems • Communication of errors and insecure acting
Safe acting is hindered by "a loss of commitment and self-initiative"	• Commitment and self-initiative
Safe acting is dependent "on situation awareness" because "it leads to desired behavior of individuals"	• Situation awareness • Behavior of individuals

Note: Phrases in quotation marks are added by core group members.

barriers) result from the analysis. Finally, in substep 2.3, the core group clusters the enablers of safe acting. When all enablers are assigned to one cluster, the clusters are labeled. After the labeling, it is verified whether all enablers of a cluster really match the label. If not, the enablers are assigned to a more appropriate cluster. At the end of this process, each cluster represents a high-level indicator regarding enablers for safe acting.

7.6.2.3 Example

Table 7.2 illustrates the results of Steps 2.1 and 2.2. The left column shows sentences the members of the core group had to complete in order to make explicit their tacit knowledge. The concepts extracted from these sentences by the means of qualitative content analysis are presented in the right column.

In the substep 2.3, the core group members cluster the concepts and label them as shown in Table 7.3.

As all core group members agree on the assignment of the safety concepts to an indicator, Step 2 is completed.

TABLE 7.3

Examples of Results of Step 2.3: Four Clusters of Concepts and Corresponding Indicators

Indicator	Concepts
Label of Indicator after Clustering of Concepts	**Cluster of Concepts Extracted in Step 2.2**
Error culture	• Mutual recriminations • Avoidance of errors and insecure acting • Communication of errors and insecure acting
Motivation	• Just attention • Commitment and self-initiative • Incentive systems
Attitude	• Behavior of individuals • Proactive thinking • Situation awareness
Ability	• Education, instruction, and training • Development potentialities • Flexibility for employees

7.6.3 STEP 3: MODEL DIFFERENTIATION

7.6.3.1 Aim

The aims of Step 3 are (1) to integrate the indicators identified in Step 2 into a multilevel, hierarchical model, (2) to define the indicators, and (3) to check the requirements set in Step 1.

7.6.3.2 Method

To create a multilevel, hierarchical indicator model, the indicators are differentiated by the core group. If reasoned overlaps between indicators are detected, indicators are merged together to one basic indicator. Clusters within these basic indicators are differentiated into subindicators. The resulting indicator model consists of safe acting on the top, with basic indicators and subindicators on underlying hierarchical levels. In a next step, for each basic indicator and subindicator a definition is elaborated.

The resulting multilevel, hierarchical model is checked on the basis of the requirements set in Step 1. In an iterative process, Steps 2 and 3 of the S-MIS process are repeated until the core group considers the indicator model to be compliant with the requirements.

7.6.3.3 Example

In Figure 7.2, an example of a multilevel, hierarchical indicator model is presented. On the top of the indicator model is safe acting, followed by the basic indicators on the first level (e.g., "framework conditions," "culture," and "competencies") and the subindicators on the second level (e.g., "ability," "motivation," and "cooperation").

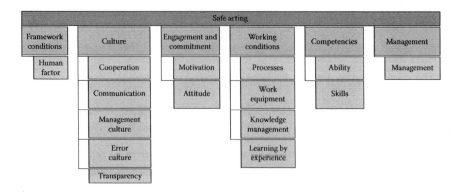

FIGURE 7.2 Example of a multilevel, hierarchical indicator model representing enablers of safe acting.

TABLE 7.4

Examples of Definitions for Basic Indicator "Engagement and Commitment" and Its Subindicators "Motivation" and "Attitude"

Basic Indicator	Definition of Basic Indicator
Engagement and commitment	• Motivation and the attitudes of employees

Subindicators	Definition of Subindicators
Motivation	• Appreciation of effort and the engagement of employees
Attitude	• Dealing with uncertainties, sources of error, error prevention strategies, and the attitude of employees toward their work

For illustrative purposes, the elaborated definitions of the basic indicator "engagement and commitment" and its affiliated subindicators "motivation" and "attitude" are presented in Table 7.4.

7.6.4 Step 4: Operationalization and Measuring Methods

7.6.4.1 Aim

The aim of Step 4 is to elaborate an instrument for measuring the indicators identified in the previous steps.

7.6.4.2 Method

To enable the assessment of the subindicators, the core group operationalizes each subindicator by translating it into several statements. These statements need to represent observable characteristics of the organization. For methodological reasons, the statements need to be free of negations, double negations, or ambiguous expressions.

The statements of each subindicator are integrated into a questionnaire. In this questionnaire, each statement is assessed on a scale from 1 (strongly disagree) to 6 (strongly agree). In order to check the questionnaire's quality, it is tested with a sufficient sample of employees of the organization concerned. On that basis, an analysis of the statements, as well as of the scales (Cronbach's alpha), is conducted. The final questionnaire contains approximately 80 statements, measuring all subindicators reliably (i.e., Cronbach's alpha at least 0.700).

7.6.4.3   Example

The statements assigned to the subindicator "motivation" are presented in Table 7.5.

7.6.5   Step 5: Safety Assessment

7.6.5.1   Aim

The aim of Step 5 is to gather quantitative and qualitative data regarding the indicators identified and operationalized in the previous steps.

7.6.5.2   Method

In order to assess the safety, a focus group is established. This focus group consists of representatives of the organization or the organizational unit to be assessed. To gain quantitative data, each member of the focus group completes the questionnaire developed in Step 4. To be able to monitor trends, the questionnaire needs to be completed periodically (e.g., quarterly). For obtaining qualitative data, there are two sources. On the one hand, when completing the questionnaire, there is a possibility to add qualitative comments to every question. On the other hand, periodically

---

TABLE 7.5

Examples of Statements of Subindicator "Motivation" with Rating Scale as Depicted in the Questionnaire

	Strongly Disagree					Strongly Agree
	1	2	3	4	5	6
Employees are rewarded for good work.	☐	☐	☐	☐	☐	☐
Recognition for work is arbitrary.	☐	☐	☐	☐	☐	☐
Employees are dedicated to sophisticated solutions.	☐	☐	☐	☐	☐	☐
Winning employees over for work teams is difficult even when resources are present.	☐	☐	☐	☐	☐	☐
Administered tasks are rejected by excuses.	☐	☐	☐	☐	☐	☐
Recognition of employees includes team behavior.	☐	☐	☐	☐	☐	☐

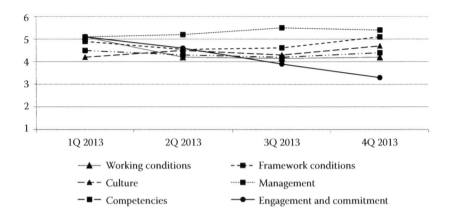

FIGURE 7.3 Example of survey assessment results for the four quarters of 2013. For each basic indicator, a timeline is presented.

Note: The scale ranges from 1 (strongly disagree) to 6 (strongly agree). Only focus group members are included in the results; they participated in each quarter of the corresponding assessment period. For reasons of confidentiality, the data depicted are fictional.

(e.g., yearly) a workshop is organized, in which the focus group discusses the quantitative results from the questionnaire. The focus of these discussions is to gather qualitative arguments for both, for the value of each indicator and for the trend of each indicator over the past period of quantitative measurement (e.g., over the past four quarters).

7.6.5.3 Example

In Figure 7.3, quantitative survey results for basic indicators are presented. The figure shows timelines for six indicators, covering a period of four measurements (first through fourth quarters of 2013). A negative trend can be observed for the indicator "engagement and commitment." Based on the collected data, differences between the means of each indicator can be calculated with dependent *t*-tests. If one or more indicators are not normally distributed, Wilcoxon tests are calculated. Multiple *t*-test or Wilcoxon tests are corrected according to Bonferroni.

In Table 7.6, qualitative arguments referring to means of the survey assessment are presented.

In Table 7.7, qualitative arguments referring to the trends of the survey assessment are presented.

7.6.6   STEP 6: SENSITIVITY ANALYSIS

7.6.6.1   Aim

The sensitivity analysis allows for comparing the relative impact each basic indicator has on the main proactive indicator at the top level of the hierarchical indicator model, that is, on safe acting. The aim of Step 6 is to prepare and perform such a sensitivity analysis.

TABLE 7.6

Examples of Statements and Qualitative Arguments for the Means of the Subindicator "Attitude" Generated by the Focus Group

Statement	Qualitative Argument
Statements out of the questionnaire	Qualitative arguments generated by the focus group reasoning the means of the subindicator "attitude"
Peers or superiors are asked for help in case of uncertainty.	• There was not enough time to ask peers or superiors for help during the first quarter of 2013. • Asking peers and superiors enables the exchange of experience.
Employees work according to the STAR principle: stop, think, act, and review.	• Time pressure has been too high since the third quarter of 2013 to work according to the STAR error prevention procedure. • Enough time has to be reserved for the briefing of tasks.
Employees carry out their tasks efficiently.	• The efficient work program of the summer of 2013 leads to a better balance between private life and professional life. • It could be improved if employees obtained more responsibility.

Note: For reasons of confidentiality, the data depicted are fictional.

TABLE 7.7

Examples of Statements and Qualitative Arguments for the Trends of the Subindicator "Motivation" Generated by the Focus Group

Statement	Qualitative Argument
Statements out of the questionnaire	Qualitative arguments generated by the focus group reasoning the trend of the subindicator "motivation"
Administered tasks are rejected by excuses.	• Due to increasing time pressure since the first quarter of 2013, the increasing value of the statement can be reasoned.
Recognition for work is arbitrary.	• Due to job cuts on the level of superiors between the second and third quarters of 2013, superiors do not have enough time to give elaborate feedback.
Employees are dedicated to sophisticated solutions.	• Sophisticated solutions at work are the key to smooth work procedures.

Note: For reasons of confidentiality, the data depicted are fictional.

7.6.6.2 Method

The core of the sensitivity analysis is a simplified mathematical formula considering weights and interrelations of the basic indicators. This allows for simulating each basic indicator's impact on safe acting, considering its relative weight as well as its indirect impact mediated by interrelations with all other basic indicators. However, it is important to note that this simulation does not at all claim to compute an exact representation of reality. Even more, it is not based on facts, but aims at visualizing assumptions of the core group's members. Against this background, it only gives hints regarding the sensitivity of each basic indicator compared with the other basic indicators. Hence, it indicates which basic indicators have a high or a low impact on safe acting. On the basis of this information, the core group can deploy the organization's resources more efficiently by concentrating on indicators with high impact.

Two constituents are necessary for the sensitivity analysis: (1) each basic indicator's relative weight and (2) the interrelations between the basic indicators. For the relative weight, the core group assigns each basic indicator a value considering the relative impact of the basic indicator to safe acting. The total of all weights equals 100. For the interrelation, the core group discusses each basic indicator's vectored impact on the other basic indicators. For that matter, the following three dimensions of an interrelation are discussed.

- Orientation of interrelation: positive (1), negative (–1), none (0)
- Linearity of interrelation: proportional (1), amplifying (a_x), restraining $(a^{1/}_x)$
- Time lag of interrelation: short-term effect (1), middle-term effect (2), long-term effect (3)

Weights and interrelations between the basic indicators are integrated into a simplified formula, which enables one to run a sensitivity analysis.

7.6.6.3 Example

Figure 7.4 shows the impact of each basic indicator on safe acting if each basic indicator is changed by the value of +1. It can be seen that the basic indicator "management" has a relative higher middle-term and long-term impact on safe acting than the other basic indicators. Coincident changes of multiple basic indicators are not considered in Figure 7.4.

7.6.7 STEP 7: INTERPRETATION AND DECISION MAKING

7.6.7.1 Aim

The aim of Step 7 is to interpret the results of the safety assessment (Step 5) and sensitivity analysis (Step 6) and prepare safety management decisions.

7.6.7.2 Method

In a discussion, the core group develops possible measures suited to sustain or even improve safety. The measures should consider mainly two aspects. On the one hand, the need for action is determined. A need for action is given, for example, when an

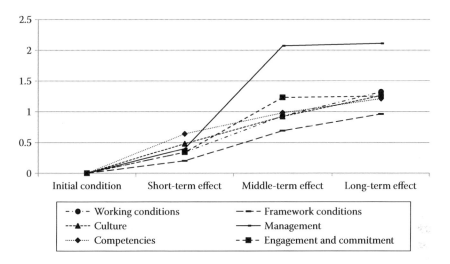

FIGURE 7.4   Examples of results of the sensitivity analysis.

indicator shows a low level in the safety assessment (Step 5) or when a negative trend is detected over the course of several safety assessments. On the other hand, the core group considers the results of the sensitivity analysis (Step 6). The aim is to identify those indicators that have the most impact on safe acting. This allows for selectively developing measures with a leveraging effect. For concretizing the measures, the core group refers to the qualitative data the focus group collected in the safety assessment (Step 5). These qualitative survey results comprise concrete descriptions of safety-related strengths and weaknesses.

7.6.7.3   Example

Figure 7.3 depicts that the basic indicator "engagement and commitment" shows a clear negative trend over four assessments. The sensitivity analysis (cf. Figure 7.4) implies that the basic indicator "engagement and commitment" has the second strongest middle-term effect on safe acting. As a consequence, the core group decides to propose measures addressing this basic indicator. On the basis of the qualitative survey results (Tables 7.6 and 7.7), it is decided on the content of the measure (e.g., reduce time pressure to allow working according to the error prevention procedure STAR [i.e., stop, think, act, review]).

7.7   EXCURSUS: SUBJECTIVE ASSESSMENT VERSUS QUANTITATIVE CALCULATION

S-MIS is designed as a participatory process. This is because of both, there is no sufficient reliable scientific knowledge available regarding factors contributing to safety, and practitioners show an extensive tacit knowledge regarding such factors. In the participatory S-MIS process, the tacit knowledge of these experts is made available for safety assessment and for the generation of safety-promoting measures. However, by doing so, S-MIS relies on subjective assumptions and interpretations.

This also applies for the determination of indicator interrelations required for the sensitivity analysis (Step 6). In one of our case studies, there are enough quantitative data available to calculate the interrelations between the indicators. This offers the opportunity to compare qualitative judgments with quantitative calculations. In this excursus, the comparison of these two sources for determining indicator interrelations is briefly described. A detailed description is published in Binz and Waefler (2014).

7.7.1 Quantitative Calculation of Indicator Interrelations

Sixty-five employees of the company concerned completed the S-MIS questionnaire developed in this case study. Based on the survey results, a value was calculated for each basic indicator. As a next step, correlations between these values were calculated (cf. Table 7.8). As the values of some basic indicators are not normally distributed, Spearman's rho was calculated. No correction for alpha-failure inflation is implemented, since a hypothesis is available for each tested correlation, which is based on the core group's qualitative judgment (cf. Step 6). Because the hypotheses derived from the core group ratings are directional, one-tailed correlations were tested. All correlations are significant at the level of $p < 0.001$.

To make possible a comparison of these quantitative calculations with the qualitative judgments on an aggregated level, for every basic indicator, a mean of its correlations with all other basic indicators was calculated. As the correlation coefficients are neither normally distributed nor interval scaled, Fisher z-transformation was applied (Bortz, 2005). Table 7.8 presents the means in its last column, labeled M.

7.7.2 Qualitative Judgment of Indicator Interrelations

The qualitative estimation of the basic indicators' interrelations is based on the core group's consensus regarding the orientation of the relation (cf. Step 6). As with the

TABLE 7.8
Correlations between Basic Indicators for Quantitative Data

	1	2	3	4	5	6	M
1. Working conditions	—	0.764*	0.775*	0.827*	0.823*	0.619*	0.770
2. Framework conditions	0.764*	—	0.680*	0.771*	0.658*	0.582*	0.695
3. Culture	0.775*	0.680*	—	0.897*	0.885*	0.781*	0.785
4. Management	0.827*	0.771*	0.897*	—	0.881*	0.724*	0.830
5. Competencies	0.823*	0.658*	0.885*	0.881*	—	0.694*	0.805
6. Engagement and commitment	0.619*	0.582*	0.781*	0.724*	0.694*	—	0.685

Note: Spearman correlations between basic indicators. M equals the mean across all correlation coefficients in relation to one indicator, according to Bortz (2005). $n = 65$.
*$p < 0.001$, one-tailed.

TABLE 7.9

Correlations between Basic Indicators for Qualitative Data

	1	2	3	4	5	6	M
1. Working conditions	—	0.200	1.000	0.800	0.200	1.000	0.640
2. Framework conditions	1.000	—	1.000	1.000	0.800	1.000	0.800
3. Culture	0.800	0.000	—	1.000	0.600	1.000	0.680
4. Management	1.000	0.200	1.000	—	1.000	1.000	0.840
5. Competencies	0.600	0.400	1.000	0.800	—	0.800	0.720
6. Engagement and commitment	0.800	0.200	1.000	1.000	0.800	—	0.760

Note: M equals the mean across all correlation coefficients in relation to one basic indicator. Participants are subject matter experts of the corresponding organization ($n = 5$).

quantitative data, the means across all correlation coefficients concerning one basic indicator were calculated. Because the core group members rated the correlations between the basic indicators on a scale within the range of $[-1, 1]$, an interval scale was assumed. Hence, no Fisher z-transformation was conducted concerning the calculation of the means for the qualitative data. Table 7.9 presents these means in column M.

7.7.3 COMPARISON OF QUANTITATIVE CALCULATION AND QUALITATIVE JUDGMENT

In order to compare the quantitative calculation and the qualitative judgment regarding indicator interrelations, the correlation coefficients based on the quantitative data (cf. Table 7.8) and the correlation coefficients based on the qualitative data (cf. Table 7.9) were categorized (for details, see Binz and Waefler, 2014). The results of the comparison are presented in Table 7.10. Twenty-three (76.7%) categorizations are congruent, while seven (23.3%) categorizations are incongruent. The means of the quantitative and qualitative correlation coefficients were all categorized congruently (cf. column M in Table 7.10).

The aim of this excursus was to compare relations between basic indicators based on a quantitative and a qualitative data source. The results reveal largely comparable basic indicator relations for both sources. In particular, all the means of the quantitative and qualitative correlation coefficients are comparable. This implies that survey assessments with relatively high personal investments may be substituted by the rating of a few experts.

It is notable that six out of seven incongruent categorizations refer to the basic indicator "framework conditions." The core group's qualitative estimation of the interrelations between "framework conditions" and the other basic indicators is consistently lower than the survey-based calculation of the interrelations. The experts of the core group seem to perceive the framework conditions as not controllable. In

TABLE 7.10

Comparison of the Categorization of the Quantitative and Qualitative Correlations

	1	2	3	4	5	6	*M*
1. Working conditions	—	i	c	c	i	c	c
2. Framework conditions	c	—	c	c	c	i	c
3. Culture	c	i	—	c	c	c	c
4. Management	c	i	c	—	c	c	c
5. Competencies	c	i	c	c	—	c	c
6. Engagement and commitment	c	i	c	c	c	—	c

Note: *M* equals the mean across all correlation coefficients in relation to one basic indicator. Congruent categorizations are labeled with a c and incongruent categorizations with an i.

contrast, the questionnaire-based evidence suggests that framework conditions can be influenced more than expected. However, for all other basic indicators, experts' judgments are comparable with survey results (for more details as well as for limitations of this comparison, cf. Binz and Waefler, 2014).

7.8   DISCUSSION AND OUTLOOK

The S-MIS project aimed to providing industry with a process suitable for both (1) a reliable, quantitative, and qualitative safety assessment referring especially to proactive indicators and, on that basis, (2) a well-founded support for decision making in safety management.

There are a number of challenges S-MIS faces. These are mainly the following: (1) science does not provide sufficient insights into safety enablers and their complex interrelations, (2) safety experts in industrial practice seem to have an extensive tacit knowledge in this regard, (3) this tacit knowledge is distributed among different safety experts, and hence a process of consensus building is required, (4) a large number of relevant indicators—especially proactive indicators—refer to human and organizational factors and hence are not quantifiable directly, but require operationalization, and (5) the sheer amount of interrelated indicators makes it very difficult for safety management decision makers to keep an overview and make well-founded decisions.

In the pilot projects, the seven steps of the S-MIS process proved to be suitable to tackle these challenges and hence provide decision makers in safety management with a substantially better quantitative information base, which is enriched with detailed qualitative reasoning. However, the pilot projects also showed that accomplishing the S-MIS process as a whole is quite laborious and time-consuming. The process therefore still needs to be optimized.

REFERENCES

Amalberti, R. (2013). *Navigating Safety. Necessary Compromises and Trade-Offs—Theory and Practice*. Dordecht: Springer.

Binz, S. and Waefler, T. (2014). Exploring Trade-Offs Between Proactive Safety Indicators. In *Proceedings of AHFE 2014*, Kraków, Poland (19–23 July).

Bortz, J. (2005). *Statistik für Human- und Sozialwissenschaftler*. Heidelberg: Springer Verlag.

Brown, R. (1965). *Social Psychology*. New York: Free Press.

Choudhry, R.M., and Fang, D.P. (2005). The nature of safety culture: A survey of the state-of-the-art and improving a positive safety culture. In *Proceedings of the 1st International Conference on Construction Engineering and Management* (pp. 480–485). Seoul, Korea, October.

Choudhry, R.M., Fang, D., and Mohamed, S. (2006). The nature of safety culture: A survey of the state-of-the-art. *Safety Science*, 45, 993–1012.

Collins, B.E., and Guetzkow, H. (1964). *A Social Psychology of Group Processes for Decision-Making*. New York: Wiley.

Cooper, D. (1998). *Improving Safety Culture. A Practical Guide*. Chichester, UK: Wiley.

Cooper, M.D. (2000). Towards a model of safety culture. *Safety Science*, 36, 111–136.

Cooper, M.D., and Phillips, R.A. (2004). Exploratory analysis of the safety climate and safety behavior relationship. *Journal of Safety Research*, 35, 497–512.

Corey, G. (2011). *Group Techniques*. Belmont, CA: Brooks/Cole.

Dekker, S. (2006). *The Field Guide to Understanding Human Error*. Hampshire, UK: Ashgate.

Flin, R., Mearns, K., O'Connor, P., and Bryden, R. (2000). Measuring safety climate: Identifying the common features. *Safety Science*, 34, 177–192.

Gigerenzer, G., Todd, P.M., and the ABC Research Group (eds.). (1999). *Simple Heuristics That Make Us Smart*. New York: Oxford University Press.

Gilovich, T., Griffin, D., and Kahneman D. (eds.). (2002). *Heuristics and Biases. The Psychology of Intuitive Judgment*. Cambridge: Cambridge University Press.

Grabowski, M., Ayyalasomayajuha, P., Merrick, J., Harrald, J.R., and Roberts, K. (2007). Leading indicators of safety in virtual organizations. *Safety Sciences*, 45, 1013–1043.

Hinze, J.W. (2005). A paradigm shift: Leading to safety. In T.C. Haupt and J. Smallwood (eds.), *Rethinking and Revitalizing Construction Safety, Health, Environment and Quality* (pp. 1–11). Port Elizabeth, South Africa: Construction Research Education and Training Enterprises.

Hofinger, G. (2008). Fehler und Unfälle. In P. Badke-Schaub, G. Hofinger, and K. Lauche (eds.), *Human Factors: Psychologisch sicheres Handeln in Risikobranchen* (pp. 36–55). Heidelberg: Springer.

Hollnagel, E. (2004). *Barriers and Accident Prevention*. Surrey, UK: Ashgate.

Hollnagel, E. (2014). *Safety I and Safety II. The Past and the Future of Safety Management*. Farnham, UK: Ashgate.

Janis, I. (1982). *Groupthink* (2nd ed.). Boston: Houghton-Mifflin.

Kahneman, D., Slovic, P., and Tversky, A. (eds.). (1992). *Judgment under Uncertainty: Heuristics and Biases*. Cambridge: Cambridge University Press.

Lunenburg, F.C. (2012). Devil's advocacy and dialectical inquiry: Antidotes to groupthink. *International Journal of Scholarly Academic Intellectual Diversity*, 14, 1–9.

Mohamed, S. (2002). Safety climate in construction site environments. *Journal of Construction Engineering and Management*, 128, 375–384.

Perrow, C. (1992). *Normale Katastrophen. Die unvermeidbaren Risiken der Grosstechnik*. Frankfurt: Campus Verlag.

Pimay-Dummer, P. (2006). Expertise und Modellbildung: MITOCAR. Doctoral dissertation: Universität Freiburg, Germany. http://www.freidok.uni-freiburg.de/volltexte/2806/pdf/pirnay_dummer_dissertation.pdf.

Pohl, R. (ed.). (2005). *Cognitive Illusions. A Handbook on Fallacies and Biases in Thinking, Judgment and Memory.* New York: Psychology Press.

Reason, J. (1990). *Human Error.* Cambridge: Cambridge University Press.

Russo, J.E., and Shoemaker, P.H.J. (1989). *Decision Traps.* New York: Doubleday.

Six, U. (1981). *Sind Gruppen radikaler als Einzelpersonen? Ein Beitrag zum Risikoschub-Phänomen. Praxis der Sozialpsychologie.* Darmstadt, Germany: Steinkopff.

Strickoff, R.S. (2000). Safety performance measurement: Identifying prospective indicators with high validity. *Professional Safety*, 45.

Waefler, T., Ritz, F. Gaertner, K., and Fischer, K. (2008). Decision-making in safety management. In *Proceedings of AHFEI 2008* Las Vegas, NV, July.

Wallach, M.A., Kogan, N., and Bem, D.J. (1964). Diffusion of responsibility and level of risk taking in groups. *Journal of Abnormal and Social Psychology*, 68, 263–274.

Section II

Safety and Human Factors
in Training and Simulation

8 Abilities and Cognitive Task Analysis in an Electric System Control Room for Developing a Training Simulator*

Regina Heloisa Maciel, Rosemary Cavalcante Gonçalves, Luciana Maria Maia, Klendson Marques Canuto, and Vamberto Lima Cabral

CONTENTS

* This study was funded by Companhia Energética do Ceará (COELCE) and research and development program resources of the Agência Nacional de Energia Elétrica (ANEEL).

8.1   INTRODUCTION

Despite technological advances and increasing automation of many of the functions previously performed by people in complex environments, the role of operators is still significant. People can better detect critical incidents, interpret signs, and make essential decisions for the equilibrium of a system's performance (Donald, 2001; De Keyser, 2001). Therefore, understanding the various operators' cognitive activities in modern control rooms becomes a challenge. Control rooms are places where people carry out control and supervision activities of complex systems. Operators are away from the real environment and have to monitor the system through displays, sensors, and communication channels. Aspects of the task involve dealing with system disturbances, which requires a number of cognitive processes, such as perception, planning, decision making, and action control. Operators need to acquire these skills in order to become proficient (Shepherd, 2004).

In this sense, task analysis is a useful approach to identify the tasks performed and their main properties and can serve as a tool for knowledge acquisition and modeling of the skills used by operators in performing their activities (Paternò, 2000). Task analysis methods facilitate the understanding of activities when the intention is to build training programs or simulate activities in a training context. These methods are important to define operational objectives and identify the actions and decisions necessary to meet them, clarifying the activities that require greater expertise and the context in which the work is performed (Shepherd, 2004). Identifying the elements of the task and its goals then becomes an essential step for the examination of the skills needed to perform the job (Annett and Stanton, 2000). Besides that, in the field of assessment and reduction of human errors, task analysis is a useful approach to help identify and eliminate conditions that give rise to errors. This analysis can either assist in the design phase of a new system or suggest changes in an existing system (Embrey, 2000).

However, task analysis is not restricted to observable behavior. To deepen the examination and decomposition of activities, it is necessary to examine cognitive aspects (Hoffman and Militello, 2008). Cognitive elements of the tasks can be inferred as characteristic of a particular transaction, even when not evidenced by observations or in the operator's discourse. It is task analysis that establishes the interaction between cognition and action, and cognitive skills can be extracted from the cognitive elements identified in the analysis. However, it is important to examine the interaction of the various operations performed in order to perceive the complete cognitive skill (Shepherd, 2004).

8.1.1 BACKGROUND

Restructuring of the electricity sector has been occurring in many countries. The privatization and implementation of regulatory mechanisms for electricity distribution and transmission systems have improved the quality of services. However, there remains a concern related to network reliability due to power outages or blackouts when supply failures occur (Joskow, 2008).

These aspects make transactions in electricity networks increasingly complex and sophisticated. Since 1998, electricity distribution in Brazil has been reorganized and transformed. This includes management changes in companies still owned or controlled by state governments and in privatized companies. The direct consequences of these modifications were a significant outsourcing process, reduction of the workforce, insertion of personnel without enough qualifications, and transformation of processes and equipment to make them more agile and less expensive and labor-intensive (DSST/MTE, 2002). The changes also resulted in the creation of government regulations to ensure that distribution systems are safe, efficient, and reliable (ANEEL, 2012). In this context, one of the main challenges of electricity utilities is the provision of quality services, characterized by continuity of supply and sustainability of the voltage levels delivered to customers.

Technological advances in electrical system control rooms have brought growing amounts of information with an excessive number of screens, maps, and alarms, increasing the cognitive demands due to a greater work complexity (Almeida et al., 2007; Francisco and Rodrigues, 2006; Ferreira, 2012). Vitório et al. (2012) carried out a field study to evaluate operators' mental load in an electric power control center. They identified the following as demands with higher weights in the resulting mental workload: (1) mental requirements (such as complex tasks that require a lot of mental effort) and (2) temporal requirements (such as fast-paced and intense work, with a lot of pressure to end problems). Guttromson et al. (2007), in the context of operations in electric networks, discussed the need for a shift from traditional studies on human factors and aspects of vision in human–computer interaction to a focus on questions about situational awareness and shared knowledge. No doubt, the increased complexity of energy industries requires a new look at the human factor. Operators' performance in control rooms is crucial to reduce the consequences of incidents, insofar as they are the ones who make decisions on the functioning of the system, becoming the final link in the processes chain (Faria et al., 2009).

The main operational activities of an electricity distribution company are (1) construction of networks, (2) maintenance to eliminate supply interruptions, (3) emergency services to restore supply after a failure, (4) customers' connections to the distribution system, (5) customers' disconnections, and (6) rewiring of customers' units (Melo et al., 2003).

8.1.2 THE COMPANY

Figure 8.1 shows the basic procedures for the operation of the electrical system in the company studied.

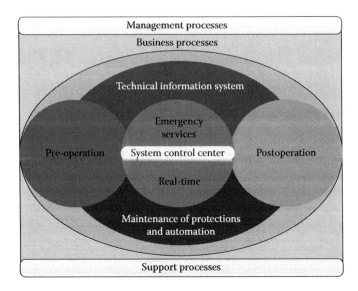

FIGURE 8.1 Map of the electrical system technical operation processes. (From COELCE [Companhia Energética do Ceará], *Manual de gestão da qualidade da operação técnica— MGQ-001/2012 R-09*, Ceará, Brazil: COELCE, 2012.)

Operators of three levels of system control were studied: high- (HV), medium- (MV), and low-voltage (LV) operators. HV operators deal with occurrences involving interruption of energy distribution for a large number of customers, in addition to dealing with network maintenance. MV operators deal with outages and maintenance in a more restricted area. LV operators handle occurrences and emergencies involving only one or a few consumers.

The essence of the work is the management of interruptions in real-time distribution in their respective areas of action. Most of the time, the HV and MV operators manage occurrences through the automated system and sometimes by means of field teams. On the other hand, LV operators manage field teams of up to three electricians, allocating teams to resolve problems and monitoring their work.

For the technical skills that already exist, a simulator (called OTS) assists in the training and evaluation of MV and HV operators. Through this equipment, it is possible to perform training simulations, called drills, in which operators in the control room work with the system as if in real time.

Thus, the workers analyzed in this study are the operators of the operational control center (OCC). They carry out their activities in a special room, with controlled light and temperature, by monitoring computer screens, with specific activities, systems, and communication systems such as radios, tablets, and phones. Emergencies are characterized as being unscheduled and unexpected and can occur at any time or place and require the reestablishment of services in the shortest time possible (COELCE, 2012). In addition, the operators work with scheduled maintenance and network renewal, although these services pertain more to the HV and MV operators than the LV ones.

TABLE 8.1

Operator's Professional and Social Characteristics

		Operators				
		HV	**MV**	**LV**	**Supervisors**	**Total**
N		5	4	17	5	31
Gender	Male	5	4	16	3	28
	Female	0	0	1	2	3
Age (years)		48.4	33.7	39.6	41.2	40.7
Time with company (years)		26.3	10.2	16.2	16.0	17.2
Time in OCC (years)		19.6	4.7	9.2	10.2	11.0
Function	Technician	5	3	17	2	27
	Analyst	0	1	0	3	4

This system is staffed by 31 people: 26 operators and 5 supervisors. Some of the workers' professional and social characteristics are presented in Table 8.1. The work is performed in four uninterrupted shifts of 6 hours, and there are four LV operators, one of MV and one of HV, and a supervisor during each day shift (morning and afternoon). For night shifts, there are only two LV operators. The shifts are rotating 4×2 days (meaning the operators work for 4 days in four different schedules, following the order of morning, afternoon, evening, and night, and then have 2 days off). Despite this rotating schedule, there is freedom for informal exchanges, which occur regularly according to the personal needs of each operator and agreements of colleagues on other schedules. This flexibility is appreciated by the operators.

The work presented here is part of a broader study that aims to describe scenarios for the creation of a training simulator and evaluation of abilities for the control center operators of the electrical distribution system. For the design of this simulator, the devised steps include the identification of the components of the activity for the generation of training interventions aimed at the acquisition of skills and knowledge for beginners in the system or for those less skilled operators, quickly and efficiently. Thus, the research described here corresponds to the initial phase of this project and aims to understand the tasks performed by operators who deal with occurrences of the electric distribution system.

8.2 METHODS

8.2.1 Hierarchical Task Analysis (HTA)

Hierarchical task analysis (HTA) produces a hierarchy of plans and operations that the operator (or a team of operators) must perform in order to meet the goals of a system. HTA starts with the establishment of the objective to be achieved in the activity. Then, the tasks are described in a set of suboperations and arranged in levels, which can be more or less detailed according to the analyst's purposes. Task analysis generally results in diagrams or tables with the collected data. It is a useful method that can be applied together with other task analysis techniques, preferably in an early stage, which provides the context for using other approaches (Kirwan and Ainsworth, 1992; Annett, 2004).

In this study, the task analysis of the operators was accomplished through interviews, analysis of the procedural manuals, and observations in the workplace. A task overview in the form of a diagram was developed using the data collected. A table with specific information, including tasks, subtasks, and plans for each step of the activity, was also prepared. The tasks were numbered according to the order of operations, and the subtasks were numbered progressively, to facilitate the identification of operations and subsequent analysis.

8.2.2 SYSTEMATIC HUMAN ERROR REDUCTION AND PREDICTION APPROACH (SHERPA)

Developed by Embrey (1986, cited in Stanton, 2005) as a technique for predicting human error, the systematic human error reduction and prediction approach (SHERPA) uses behavior classification related to a taxonomy of error modes. While HTA serves as a basis for identifying possible failures in the performance of tasks, in SHERPA analysis, each operation is classified according to the following categories: action, retrieval, verification, and information communication. After that, for each specific activity or operation, associated possible error modes are considered. The possible error modes are shown in Table 8.2. The next step is to identify if there is a later stage of the task where the error can be corrected. The probability of error occurrence, based on historical data or the judgment of an expert on the subject, is then computed. Next, an analysis of the criticality of the error is made to indicate the severity in terms of damage or loss. The last phase involves proposals for corrective measures (Stanton, 2005).

For the purposes of SHERPA, interviews with an experienced operator and a supervisor were conducted. Subsequently, the supervisor validated the results.

Figure 8.2 shows the flow diagram of the method employed in all its stages. It is important to note that the reliability of the results obtained depends on the active participation of supervisors in the steps of validation and consolidation of data, which is not always easy to obtain.

8.3 RESULTS AND DISCUSSION

8.3.1 HTA OF HIGH-VOLTAGE OPERATIONS

HV operations comprise the monitoring and verification of the network maintenance tasks in the operators' respective area of expertise (called planned activities). In the case of HV operators, the main activity is the management of unscheduled occurrences in the HV network (blackouts or variations in voltage level) and solution of the problem as quickly as possible to restore power supply. Thus, activities can be divided into three main operations: management of scheduled incidences (planned activities), performance of network monitoring, and management of occurrences (Figure 8.3).

TABLE 8.2
SHERPA Error Modes

Error Classification	Code	Error Mode
Action errors	A1	Operation too long/short
	A2	Operation mistimed
	A3	Operation in wrong direction
	A4	Operation too little/much
	A5	Misalignment
	A6	Right operation on wrong object
	A7	Wrong operation on right object
	A8	Operation omitted
	A9	Operation incomplete
	A10	Wrong operation on wrong object
Checking errors	C1	Check omitted
	C2	Check incomplete
	C3	Right check on wrong object
	C4	Wrong check on right object
	C5	Check mistimed
	C6	Wrong check on wrong object
Retrieval errors	R1	Information not obtained
	R2	Wrong information obtained
	R3	Information retrieval incomplete
Communication errors	I1	Information not communicated
	I2	Wrong information communicated
	I3	Incomplete information communication
Selection errors	S1	Selection omitted
	S2	Wrong selection made

Source: Stanton, N.A., in *Handbook of Human Factors and Ergonomics Methods*, Stanton, N., et al. (eds.), Boca Raton, FL: CRC Press, 2005.

As shown in Figure 8.3, the management of planned occurrences involves the authorization of requests to release lines and facilities, procedures for releasing the equipment or line where the maintenance work is to be done, monitoring of the service, and standardizing procedures to normalize lines and facilities. HV network monitoring involves monitoring the voltage and electrical current levels of facilities and lines and, if applicable, the insertion or withdrawal of capacitors, as well as the monitoring of installations that are not in ideal conditions of operation. The management of unscheduled occurrences encompasses the verification of incidents, assessment of the criticality of the situation, maneuvers to reset the system, monitoring the service, and forwarding demands generated during the service. For other situations, the operator has to open a normal or urgent anomaly record.

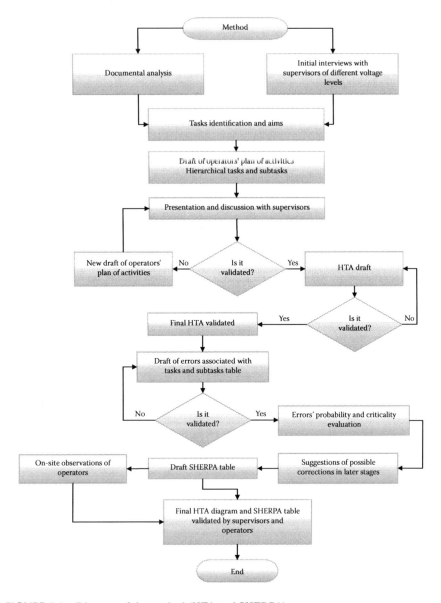

FIGURE 8.2 Diagram of the method (HTA and SHERPA).

8.3.1.1 Management of Scheduled Occurrences

In order to manage the scheduled occurrences, the operator receives a request (by radio or phone) from the person responsible for the scheduled service to authorize the execution of the job. Then the operator checks whether the service request has been previously forwarded by the preoperation sector through the specific system, as well as whether the request is in accordance with the work order (WO) sent by the preoperation sector. Then he verifies whether the electric current and installation

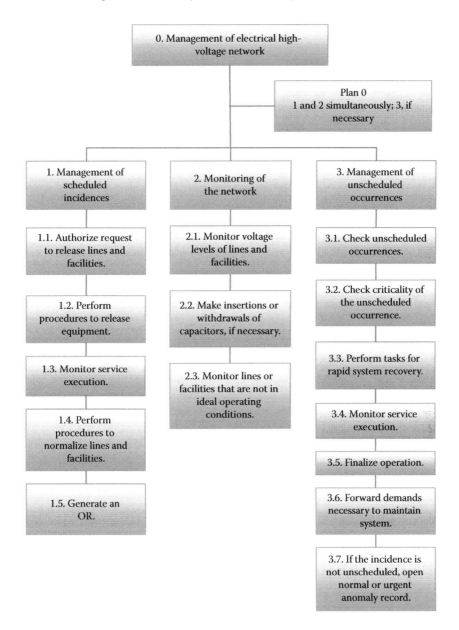

FIGURE 8.3   HTA results of HV operations.

conditions of the network are in accordance with the instructions provided by the preoperation sector. If the request complies with the necessary requirements, the operator authorizes the execution of the service. Sometimes the requisition is incomplete, without the network condition analysis by the preoperation sector. When this happens, the task of authorizing the service gets more complex, because now it involves analysis of these conditions in a relatively short time and, finally, authorization of the service or not. If the request does not comply with require-ments, the operator evaluates the feasibility of implementing the service along with the HV supervisor. If the request is not permitted, the operator informs the per-son responsible for the scheduled service of the impossibility of execution. If the operator authorizes the service without checking the conditions, based only on the preoperation sector analysis, he can risk the occurrence of a major accident. This means the error involved in this operation can bring major consequences. The prob-ability of this happening was considered medium by the supervisor interviewed. This checking requires the operator's attention and technical knowledge based on his experience.

When the service is completed, the operator records the occurrence in the cus-tomer service system (CSS) to generate an occurrence report (OR).

In relation procedures for releasing equipment and lines (1.2 in Figure 8.3), the operator checks the standard procedures to be carried out to enable the implementa-tion of service for field staff; performs remote operations to open, close, or block equipment; performs remote operations to change scenarios (modify parameters); and finally, authorizes the substation operator to carry out the tasks required to release the equipment. Time pressure to free equipment in order for the field team to start operations can lead to judgment errors, including violations of standard pro-cedures. Operations here involve communication between substation operators and field teams. Not checking if the information about the tasks is properly understood by the substation operator, according to the operational standards (POP-12—Verbal Communication and POP-13—Operational Communication), can lead to malfunc-tions in the system, such as power losses and accidents.

In the process of monitoring the execution of the field service, the operator moni-tors the network to check for abnormalities during execution. If an abnormality is detected in the system concerning the area where the team is working, the operator must instruct the team about the procedure to be carried out. In order to do this, the operator contacts, by radio or telephone, the electrician responsible for gathering the necessary information about the situation in the field; analyzes the situation, seeking information on the system or with the protection and maintenance work sectors; and orders the team to suspend or continue the service. One of the main difficulties of this process is the communication with the field team, mainly because in some more distant parts of the distribution area there are no communication services, making obtaining information about what is happening in the field impos-sible. Another difficulty is the need to pay continuous attention to the information system to check for abnormalities. As the operator is monitoring several field teams at the same time, each one working in different areas, he has to observe distur-bances in the network using different systems' screens at the same time. For this reason, the probability of error occurrence was considered medium by the supervi-sor interviewed.

To perform the procedures for the normalization (line/equipment), the operator receives the field team information (by radio or telephone) about the conclusion of the service, carries out remote operation tasks, and authorizes the substation opera-tor to perform the tasks necessary to normalize the system. On finalizing the occur-rence, the operator fills out the OR in the CSS.

8.3.1.2 Monitoring the Network

In order to monitor voltage and electric current levels, the operator observes on the "Synthetic Views" screen the graphs and colors that indicate voltage levels by criticality, as well as the level of transformers and substation lines through the CSS. If necessary, the operator makes the insertion or withdrawal of capacitors to reduce the effect of inducing power. To do so, he checks whether the insertion or removal of a capacitor, made in automatic mode, is performing according to the preestablished levels and conducts the operation of inserting or removing the capacitor if necessary.

In the case of the monitoring installations that are not in ideal operating conditions, the operator must pay attention to the changes in the network until the correction is made. After the correction, the operator carries out normalization of facility procedures.

These two procedures occur simultaneously, so that monitoring the network is done while the maintenance services are being carried out.

8.3.1.3 Management of Unscheduled Occurrences

In order to manage unscheduled occurrences, which is the most important task of HV operators, the operator checks these incidents through alarms (audible and visual) in the CSS, through remote warning systems at substations (via messages sent automatically by radio), or through communication from other internal sectors or sources in the field (consumers, police, etc.).

As soon as an unscheduled occurrence is perceived, the operator makes an analysis of the criticality, involving checking (1) messages on the CSS screen concerning protection systems, (2) the electric map on the CSS screen, (3) the status of the substation's equipment on the CSS screen, (4) whether any device is open or has changed its status due to an anomaly in the system, (5) the current and voltage values, (6) whether there has been a power outage, (7) whether there has been an instant operation by automatic reclosing equipment (relays), and (8) historical records of installation (equipment and line) performance.

Errors in criticality analysis can cause loss of load, system constraints, or delays in normalization of the system. This operation requires the ability to seek information and interpret the data obtained for the decision-making process. In general, this analysis is done together with the supervisors, based on their combined expertise. The complexity of the task consists of the fact that these occurrences are unpredictable and can be highly complex, involving several contributing factors (e.g., blackouts).

To perform the tasks for system normalization, the operator must define the procedures to be performed for system recovery. If necessary, he consults the supervisor or the professional responsible for the sector to establish the most appropriate procedures. After these checks, the operator performs automatic tasks, closing the device that opened or restoring the facilities using other resources. When necessary, the operator asks the maintenance sector to inspect or repair lines and equipment. Then he tells the substation operator to carry out the inspection or repair of the faulty line or equipment. The monitoring of the service occurs through information from the field team. The finalization of the operation of unscheduled occurrences involves completion of the OR in in the CSS.

The operator can make an incorrect decision with respect to responses, such as closing the wrong device or performing the operation at an inappropriate time, which may cause delays in normalizing the system, an even greater drop in loads, damage to installations and equipment, and accidents. If, on the contrary, after the analysis of criticality the operator decides that the incident is not critical, he must report this occurrence, through the system, as an anomaly.

In forwarding demands, the operator can (1) request, by email or telephone, that the maintenance sector carry out an inspection to verify the reliability of the system; (2) register information in the maintenance operation management system (MOMS) about the abnormality (normal, urgent, or emergent); or (3) forward, through the MOMS, a request to ensure that the maintenance sector performs the necessary procedures. When necessary, the operator can open more than one anomaly record, forwarding reports to other departments (e.g., line, protection, and information technology). The operator can also check the return of these maintenance requests and, if necessary, take further steps to solve the problem.

The operator of AT works under time pressure for the quick normalization of load levels, isolation of the defect, or shifting of the load from an affected substation to another one. If the interruption equals or exceeds 3 minutes, the interruption is computed in the continuity indicator, supervised by ANEEL. Even in planned activities, there is pressure not to exceed the time for system normalization, since the scenario foreseen during execution can change, depending on the time of day and other emerging situations. Of course, time pressure ends up being a destabilizing factor in the cognitive dynamics of the operator. In addition, there is no uniformity of equipment in all regions, with some more automated than others, which means that checks and tasks have to be done differently.

8.3.2 HTA OF MEDIUM-VOLTAGE OPERATIONS

In the case of MV operators, the work, as reported by the supervisor, can also be divided into three main operations, as shown in Figure 8.4.

As illustrated in Figure 8.4, to manage the MV electrical system, the operator identifies incidents and carries out corrective and planned actions to reset the system. The MV operator also verifies pending operations that can generate actions, and receives and responds to both LV incidents and occurrences originating from other sources.

At the MV level, the activity of operators begins with the verification of occurrences not finalized in the previous shift, which can be done with the aid of the tools available in the system or verbal information from the colleague from the earlier shift. The identification of pending operations can generate immediate actions to repair the system. Briefing from the previous shift is mandatory within the real-time operational procedures. However, failures can occur, like when the operator for some reason fails to act in time or does so in an unsatisfactory manner. In these cases, it is up to the leader to evaluate, together with other people involved (operators and supervisors), the reason for the failure and determine the measures to solve the problem.

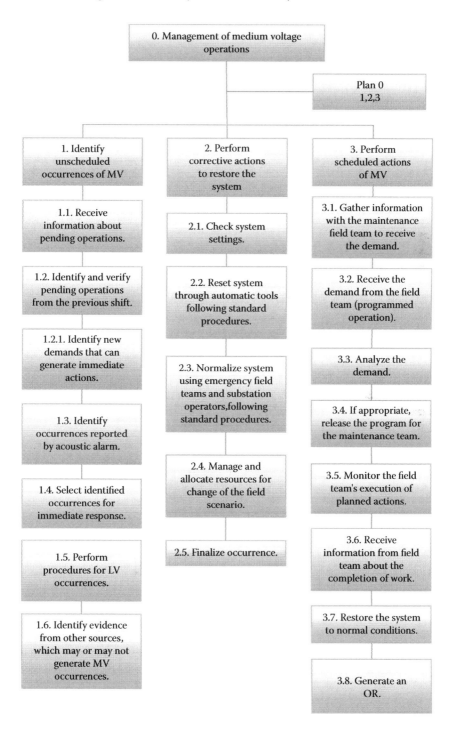

FIGURE 8.4   HTA results of MV operations.

These activities require attention and expertise to anticipate undue occurrences. Furthermore, the MV operator must pay attention to possible repetitive LV incidents that can lead to an MV occurrence.

Corrective actions to restore the system can be done through automatic equipment, emergency field teams, and substation operators. To meet the goals, the MV operator manages and allocates the necessary resources, which can change according to scenario changes. In these cases, the speed of operations is essential. The MV operator must provide a solution to restore the network in 3 minutes, by isolating the location or malfunctioning switches.

In relation to planned actions, the operator receives the demand from the maintenance field team, analyzes it, and if appropriate, releases the action plan to the maintenance team.

The probability of errors in this phase is low, but they are critical.

8.3.2.1 Identify Unscheduled Occurrences

As already mentioned, the grouping of several LV calls to the call center, reporting the same incident, can create a new MV occurrence. This kind of occurrence is then sent to the MV display. New incidents can also be generated by the system itself, indicating that a feeder, or some of its parts, is out of order, identifying the affected area and how many customers are without power. The information can also come via CCS, which sends voltage and electric current values and, when necessary, triggers an audible and visual alarm. The incident can also be detected by parsing information from feeders, received by remote control via General Packet Radio Service (GPRS). In the latter case, the incident is not created automatically, so the MV operator has to detect it from the system. The alarm raised by the system does not always indicate an occurrence, but it can indicate a system oscillation, the need for monitoring, or a change in the voltage or current. Again, the occurrence of errors in this phase has a low probability, but they are critical because they can leave a large number of customers without power.

8.3.2.2 Corrective Actions to Restore the System

Following the identification of an unscheduled occurrence, when necessary, the operator allocates field teams to the sites where corrective action is required. The operator plans the actions to be performed by the field team in order to normalize the system. He can also call substation operators and ask them to perform actions, including, if applicable, changing feeders. The operator also follows the field team in resetting electric switches. In addition, the operator has to share the resources available if there is another urgent demand. If necessary, MV operators can request emergency teams, in both LV operations (from the LV sector) and maintenance operations. When the operator's request is not enough, it is the responsibility of the supervisor to assist the MV operator, or if necessary, the engineer on standby can do the same. However, in the allocation of resources to solve the problem, the operator, given the time pressure, might allocate more resources than necessary to solve the problem, which leaves other operations understaffed.

In order to meet the demand, the operator can perform various actions or ask field teams to do so. Thus, the operator may have to handle multiple simultaneous telephone and radio calls.

8.3.2.3   Scheduled Actions

To perform scheduled actions, the MV operator requests the substation operator to carry out the recovery plan established for the emergency team, reporting the switches that must be reset. In addition, the operator performs the operating instructions (OIs), preestablished actions for system recovery.

The recovery plan involves having to reset multiple switches at various points of the network. The restoration of the system to normal operation rarely occurs with resetting a single switch, since it is not known immediately where the problem is.

Errors in this phase can cause system overload and damage equipment. Therefore, although the probability is low, criticality is high.

8.3.3   HTA OF LOW-VOLTAGE OPERATIONS

The job of LV operators is to manage unscheduled occurrences to restore the electricity supply. Operators receive communication of occurrences and determine steps to be followed by field teams, composed of electricians. Each operator controls a number of field teams and is responsible for a specific area.

The HTA results of the LV operators are presented in Figure 8.5. The figure shows tasks subdivided into levels of operations. Due to the dynamic characteristic of the activities and the large number of occurrences, in practice it is not possible to follow the flow presented in the diagram exactly. That is, managing emergencies, managing teams, and finalizing emergencies occur almost simultaneously.

8.3.3.1   Occurrence Management

LV operators' work shifts begin with a verification of the occurrences list in the CSS, which is forwarded by the relationships center (RC). The operator identifies the occurrence and checks its priority level previously established by RC operators. Occurrence response must be made according to its priority level. Highest-priority occurrences are the ones that involve life-threatening situations. When there is more than one incidence with the same priority, the operator considers the number of affected customers to define the order of response. RC is an outsourced service; that is, RC operators are not always sure about correct priority because they change frequently.

To manage occurrences, constant system updating is necessary in order to verify the input of new data, because the procedure is not yet automated. In addition to the CSS, it is also necessary to make queries to other support system programs for more information about customers' complaints. As the number of occurrences with high priority tends to be high, the ordering of emergencies is a task that requires good analytical skills for the right decisions. For example, the operator may have to decide between first solving the problem of a broken conductor in a public highway and then solving a complaint from a customer who is "electrically dependent" (who needs electrical equipment to survive). Both cases are life threatening.

At this stage, the operation requires continuous attention to the information that appears on the screens and to system data search to aid in decision making. Potential errors mainly involve checking failures or identifying information that results in not

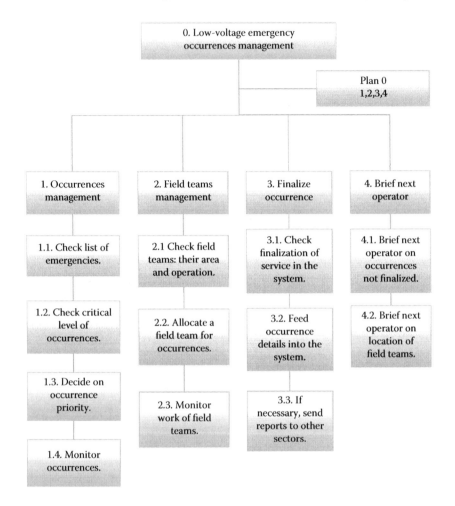

FIGURE 8.5   HTA results of LV operations.

meeting priorities, causing greater impact or delays in the service. The construction of a unique system that allows integration of the necessary information for the development of the activities is a proposal that would facilitate access to data, reducing task overload due to the current need to consult different support programs.

8.3.3.2   Field Team Management

After setting priorities for customer emergencies, the operator passes on to the next step, which is to send a field team to the emergency site to reestablish energy supply. Each operator is responsible for a region of the city and coordinates the work of 7–11 field teams composed of electricians. The number of teams varies according to the shifts and depends on the operator's decision to maintain teams in operation during his shift. Thus, the operator defines and authorizes the teams that will operate in his area during that shift. The teams are classified according to the type of vehicle: (1) truck for heavy operations; (2) truck for highline services, fitted with a ladder with an air basket; (3) smaller vehicle for general services; and (4) motorbike for smaller services in customers' homes. All teams are composed of two electricians, with the exception of the motorbike unit.

The operator initially analyzes the characteristics of emergency to define the type of team (vehicle) that he will send to the site. For example, the fall of a pole requires a truck with a heavy mechanical arm. After that, he checks which field teams are available and their location. This is done to predict the teams' travel time to the site, considering, in addition to the distance, the hour of the day and traffic flow. The main purpose of selecting a field team is to give agility to customer service, so the operator has to make use of his analytical skills and knowledge of the city and field team to make a decision. He has to decide which team will be able to reach the destination in the shortest time and their ability to solve the problem. When no team is available, the operator needs to negotiate with other operators responsible for other city regions that may have staff free to perform the service. During the service's execution, the operator monitors the location of vehicles, the time of arrival at the location, the time to do the job, and any difficulties.

The communication tools used are tablets, radios, and telephones. Currently, most teams use tablets, which facilitates the flow of communication with less waste of time. However, when there are difficulties to be solved, the radio is used because it enables dialogue between operator and electrician. In this way, the team's supervision requires the concomitant use of different media. Besides tablet, radio, and telephone communications with the various field teams in service, operators also have to communicate with other operators in the control room and other sectors within the company with another phone.

At this stage, the operator manages multiple demands that require attention, communication, and coordination of field teams. Demands call on his memory skills, especially short-term memory, in order to retain information about the processes in progress, and on his supervisory skills to continuously detect and evaluate the flow of information. Furthermore, he plans and assesses situations and makes decisions to better meet the needs for completion of services. Some of the possible errors that can occur are failures in the definition of team schedules, inadequate selection of the necessary team type for the service, failure in estimating time to arrive at the location, and insufficient monitoring of the services provided. The main consequence is inappropriate delay of service.

These operations are frequently performed under strong time pressure from managers and customers, or from the perception of risks to the people involved. An example of pressure from risk perception is a delay in the response to broken cable on the ground, with the risk of shock to passersby. Another example is when a power outage occurs in an area of heavy traffic, affecting the operation of traffic lights. In this case, there is the risk of causing traffic accidents. Pressure from managers is related to time to resolve the emergency. Both the company and operators are rated according to the time they take to solve a problem. These statistics are used to calculate salary bonuses, in the case of operators, and rank the services of the company by ANEEL.

During the interviews, it was suggested that one possible remedial measure would be allocation of one inspector in the field to monitor the services, to facilitate the process of team monitoring. Another suggestion was the implementation of a rotation system for field teams that could facilitate acquisition of spatial knowledge about all city regions, avoiding the preference of electricians to work in certain areas in which they are most familiar.

8.3.3.3   Finalizing Occurrences

Upon completion of the service, the next step is to finalize the occurrence in the incident management system (SGI). To this end, the team passes the information on the conclusion of the service by either tablet, radio, or phone. If the information is delivered by tablet, it goes directly to the computer and the operator has only to translate the codes to the correct description of the emergency (according to a system of codes), time of onset and end of service, and other details related to the occurrence. If the information is delivered by radio or phone, the operator has to type all the information in the system. There is a special system screen for doing that. The operator needs to check if the data are correct and fill in the blank fields with instance-specific data. When necessary, he forwards a request to the service center to make a definitive repair.

The task requires the operator's attention to details to fill in the managerial report correctly. Although it is a relatively simple task, many errors can occur at this stage due to failures in checking the information submitted by the teams, incomplete or incorrect data recording, and others. Frequently, the volume of cases combined with the urgency to answer calls reporting further emergencies causes the operator not to record the completion of service in real time. Operators delay this stage and try to finalize services (complete the form) at the end of the shift, or when there are fewer emergencies on the list. Thus, forgetfulness and mistakes can happen due to time pressure rather than lack of skill.

The preparation of managerial reports with inaccurate information may affect analysis for system improvements. The corrective measure suggested by the professionals interviewed was the establishment of a standard procedure where the registration of service completion has to be done immediately. However, the proposed measure might not be effective if the operator does not have the means to properly juggle the tasks required and service demands.

8.3.3.4   Briefing of the Next Operator

As already pointed out, operators work in shifts of 6 hours with 2 days off. There are four shifts: morning, afternoon, evening, and night. They work with frequent shift rotation, which means operators move at regular intervals from the morning shift to the afternoon shift, evening shift, and night shift.

As the activity of the OCC is continuous, team change from one shift to the another is an important step because it is the time when one operator transmits to the next the information about incomplete occurrences. He also informs the next operator about field teams and location of vehicles that are acting in the region. The task requires attention and memory to convey important details of the occurrences. The most common error is failure to transfer pending cases that appear in the system as finalized. That is, the operator sees on the CCS screen (list of occurrences) that the field team has finalized a particular service, but the occurrence actually has not been effectively completed, because it depends on additional support to be completed. As a result, the operator can have difficulties recovering information, which delays conclusion of the service.

The interviewees suggested the implementation of a fixed procedure for shift change. However, it important to take into account the time available for performing this task. Fatigue is also a factor that needs to be considered, because this has a detrimental effect on the transmission of information at the end of the shift.

8.4   SHERPA ANALYSIS

Table 8.3 presents the classification of types of errors as HV, MV, and LV operations and their respective percentages, following the first step of the SHERPA analysis.

As can be observed from the data in Table 8.3, at all three levels, action errors (EA) and verification errors (EV) are more likely to occur, especially for HV and LV operators. Information retrieval errors (ER) seem to be characteristic of MV operations. On the other hand, communication (EC) and selection (ES) errors are more frequent, respectively, for HV and MV operators. Communication in HL operations is important because the operator must be permanently in contact with the substation operators to perform the scheduled tasks. MV operators have to retrieve system information for checks.

With respect to the likelihood of errors and their criticality, due to the type of operations performed, the likelihood of errors in HV and MV operations is generally low (10%). However, their criticality is almost always considered high (86%). With respect to errors in LV operations, the probability is higher (40%), but the criticality is between medium and low, depending on the type of operation and the number of customers affected.

8.5   CONCLUSIONS

The results showed cognitive elements of operators' tasks of an electric distribution system, which are relevant to simulator-based training design. From the HTA, the task could be broken down, and cognitive processes underlying the operations

TABLE 8.3
Percentages of Error in Each Voltage Level
Classified according to the SHERPA Analysis

Error Type	Code	HV	MV	LV
Action errors	EA	35.19	40.00	43.33
Checking errors	EV	37.04	25.00	33.33
Retrieval errors	ER	5.56	20.00	13.33
Communication errors	EC	22.22	10.00	3.33
Selection errors	ES	0.00	5.00	6.67
Total		100	100	100

performed could be inferred, such as attention, memory, vigilance, planning, evaluation, and decision making. The SHERPA served to identify potential critical operational errors, pointing to aspects that need to be better addressed in both training and possible work redesign. The observations and interviews with operators when developing the task and error analysis enabled the observation of relevant aspects in the operators' job context, such as resource constraints and time pressure.

From the analyses, it can be concluded that electrical system operators are involved in planning actions and decision making whose errors can result in serious consequences for the system (such as life-threatening situations, accidents, and interruptions in supply). The more frequent errors are related to omission of actions or verifications. This may be due to the nature of the work, which involves a large number of procedures and diversity of systems and control devices. It is possible that the large amount of information to be analyzed in real-time decision making, in the emergencies operators normally deal with, has a significant influence on operators' errors.

It is important to note that in this work, the notion of human error is related to deviation from what has been established and expected (Leplat, 2011), having no negative connotation of an action that necessarily leads to undesirable outcomes. Errors are considered starting points for performance analysis, which should take into account the complexity of work, system vulnerabilities, and strategies used by operators to deal with these circumstances (Woods et al., 2010).

An interesting point to note here is that the HTA proved to be slightly different from the procedures listed in the operational manuals. This is mainly due to the change of focus proposed by this kind of analysis: the goal is not to verify the procedures necessary for the proper functioning of the system, but to identify the cognitive aspects involved in the tasks in order to deduce the mental operations and operating strategies of the workers in the system.

A drawback of the method used here is that it did not enable a deeper analysis of the work context and its implications for operators' actions. Future research should include natural environment observations for cognitive task analysis. The intention is to clarify the strategies and resources used by operators to solve problems and deal with the work constraints and limitations, in order to build training scenarios with the highest degree of fidelity possible, reflecting the complexity of the operating environment of this company.

The design of a unique system that allows integration of the information necessary for the development of the activity is a proposal that could facilitate access to data, reducing the task overload of bringing together the involvement of different support programs.

Thus, the main corrective measures to be adopted to improve processes, according to the operators and supervisors interviewed, are (1) construction of a unique system to integrate information from the various support programs, (2) introduction of supplementary information in the system, (3) improvements in the communication flow of operators with the various sectors with which they interact, (4) creation of standard procedures for routine tasks, and (5) better monitoring of field services and teams.

REFERENCES

Almeida, F.R., Kappel, G.B., and Gomes, J.O. (2007). Análise ergonômica do trabalho cognitivo dos operadores da sala de controle do COSR-SE. In *XXVII Encontro Nacional de Engenharia de Produção*, Foz do Iguaçu, PR, Associação Brasileira de Engenharia de Produção.

ANEEL [Agência Nacional de Energia Elétrica]. (2012). Procedimentos de distribuição de energia elétrica no sistema elétrico nacional—PRODIST. Módulo 1—Introdução. http://www.aneel.gov.br/arquivos/PDF/ Modulo1_Revisao_6.pdf.

Annett, J. (2004). Hierarchical task analysis. In *The Handbook of Task Analysis for Human-Computer Interaction*, Diaper, D., and Stanton, N. (eds.). London: Lawrence Erlbaum Associates, pp. 67–82.

Annett, J., and Stanton, N.A. (2000). Research and developments in task analysis. In *Task Analysis*, Annett, J., and Stanton, N.A. (eds.). London: Taylor & Francis, pp. 1–8.

COELCE [Companhia Energética do Ceará]. (2012). *Manual de gestão da qualidade da operação técnica—MGQ-001/2012 R-09*. Ceará, Brazil: COELCE.

De Keyser, C. (2001). Evolution of ideas and actors of change. In *Error Prevention and Well-Being at Work in Western Europe and Russia: Psychological Traditions and New Trends*, De Keyser, C., and Leonova, A.B. (eds.). Dordrecht, the Netherlands: Kluwer Academic Publishers, pp. 3–23.

Donald, C. (2001). Vigilance. In *People in Control: Human Factors in Control Room Design*, Noyes, J., and Bransby, M. (eds.). London: Institution of Engineering and Technology, pp. 35–35.

DSST/MTE [Departamento de Segurança e Saúde no Trabalho, Ministério do Trabalho e Emprego]. (2002). *Manual setor elétrico e telefonia*. Brasília: Ministério do Trabalho e Emprego.

Embrey, D. (2000). Task analysis techniques. Lancashire, UK: Human Reliability Associates Ltd. http://www.humanreliability.com/articles/Task%20Analysis%20Techniques.pdf.

Ericsson, K.A., and Simon, H.A. (1993). Protocol Analysis: Verbal Reports as Data (rev. ed.). Cambridge, MA: Bradford Books/MIT Press.

Faria, L., Silva, A., Vale, Z., and Marques, A. (2009). Training control centers' operators in incident diagnosis and power restoration using intelligent tutoring systems. *IEEE Transactions on Learning Technologies*, 2(2), 135–147.

Ferreira, E. (2012). Relatório das condições de trabalho dos operadores do COSE e subestações, assistentes de operação, técnicos de proteção e controle da Eletrosul. Relatório de pesquisa da Intersindical dos Eletricitários do Sul do Brasil—INTERSUL. http://www.intersul.org.br/lista.php?type= Assuntos% 20diversos (accessed July 10, 2013).

Francisco, L.G., and Rodrigues, P.H. (2006). Análise cognitiva do trabalho: estudo de caso com operadores do sistema de baixa tensão da Light S.A. In *XXVI ENEGEP*, Fortaleza, CE, Associação Brasileira de Engenharia de Produção.

Guttromson, R.T., Greitzer, F.L., Paget, M.L., and Schur, A. (2007). Human factors for situation assessment in power grid operations. Report PNNL-167803. Richland, WA: Pacific Northwest National Laboratory.

Hoffman, R.R., and Militello, L.G. (2008). *Perspectives on Cognitive Task Analysis: Historical Origins and Modern Communities of Practice*. Boca Raton, FL: Taylor & Francis.

Joskow, P.L. (2008). Lessons learned from electricity market liberalization. *The Energy Journal*, special issue, 9–42.

Kirwan, B., and Ainsworth, L.K. (1992). *A Guide to Task Analysis*. London: Taylor & Francis.

Klein, G.A., Calderwood, R., and MacGregor, D. (1989). Critical decision method for eliciting knowledge. *IEEE Transactions on Systems, Man, and Cybernetics*, 19, 462–472.

Leplat, J. (2011). *Mélanges ergonomiques: Activité, compétence, erreur.* Toulouse, France: Octarès Éditions.

Paternò, F. (2000). *Model-Based Design and Evaluation of Interactive Applications.* London: Springer-Verlag.

Shepherd, A. (2004). HTA as a framework for task analysis. In *Task Analysis*, Annett, J., and Stanton, N.A. (eds.). London: Taylor & Francis, pp. 9–24.

Stanton, N.A. (2005). Systematic human error reduction and prediction approach (SHERPA). In *Handbook of Human Factors and Ergonomics Methods*, Stanton, N., Hedge, A., Brookhuis, K., Salas, E., and Hendrick, H. (eds.). Boca Raton, FL: CRC Press.

Vitório, D.M., Masculo, F.S., and Melo, M.O.B.C. (2012). Analysis of mental workload of electrical power plant operators of control and operation centers. *Work*, 41, 2831–2839.

Woods, D.D., Dekker, S., Cook, R., Johannesen, L., and Sarter, N. (2010). *Behind Human Error.* Farnham, UK: Ashgate.

9 Immersive Virtual Environment or Conventional Training? Assessing the Effectiveness of Different Training Methods on the Performance of Industrial Operators in an Accident Scenario

Salman Nazir, Alberto Gallace, Davide Manca, and Kjell Ivar Øvergård

CONTENTS

9.1   INTRODUCTION

Chemical processes are flooded with complex human–machine interfaces that are generally coupled to automated systems involving data that have varying levels of reliability (Burns et al., 2008). Industrial processes are large, complex, distributed, and dynamic. They are also socially dependent in that several plant operations call for a teamwork culture (Kluge and Frank, 2014; Nazir et al., 2012a, 2014b). In fact, a number of field activities are managed by crews, shifts, and varied functional groups (Nazir et al., 2013; Wickens, 2000). Although the study of human factors, human–machine interfaces, and the cognitive issues related to these topics has always been more focused on aviation, the armed forces, and command and control, nonetheless, since the last decade, the role and significance of human errors in the process industry have started to gain some attention (Antonovsky et al., 2013).

It is widely accepted that the engagement of different cognitive mechanisms, such as those related to the deployment of focused, divided, spatial, and selective attention, is required by industrial operators during normal operations (Nazir et al., 2013; Wickens and McCarley, 2007). According to Gopher (1996, p. 28), "the operator would gain most if he or she could fully attend to all elements, at all times. However, such full attention is not possible. Hence, some priorities and tradeoffs must be established along with attention allocation strategies." In other words, industrial operators must learn to allocate their attention strategically according to the given scenario and situation. Abnormal situations present operators with several challenging issues, such as unexpected events, unknown operating conditions, alarms, and decisions to be made in short times and under stress and increased mental load (Colombo et al., 2012; Kletz, 1999; Naderpour et al., 2015). As soon as an abnormal situation is involved, a problem to be solved is formulated and the need for replanning the operation becomes essential. In order to effectively respond to an abnormal situation, operators should have problem detection skills and rapid replanning strategies, and thus weigh the situation by considering all the required parameters that are necessary to reach a decision that can solve the abnormality. Under these conditions, mental workload and attention are mutually interlinked. Figure 9.1 shows conceptually the necessity of operating the plant within the suggested and optimal operating region. Failing to operate it in that region may put the whole process in danger.

Continuous production, for which a safe operating zone is a prerequisite, is only feasible when various parameters are monitored, controlled, and if possible, optimized. The range of such parameters is well defined during the conceptualization and erection of a plant, even if external disturbances, uncertainties, and fluctuations play an upsetting role for plant operation. Incidental deviations can introduce off-specification production with an increase in impact that may reach the catastrophic failure of equipment. Human errors, such as pressing a wrong button or switch, operating a wrong valve, misinterpreting the data, making slips (or lapses), and miscommunicating, may result in accidents characterized by significant financial losses and causalities (Antonovsky et al., 2013; Plant and Stanton, 2012; Wickens, 2000).

Training of operators is an essential element of process industry, and finding methods that can reduce the possibilities of human errors or the reaction time in responding to abnormal situations is extremely important (Nazir and Manca, 2014).

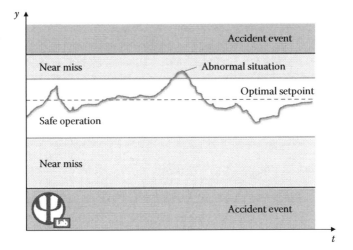

FIGURE 9.1 Qualitative trend of a process variable, y, as a function of time, t. Since the process is intrinsically nonlinear, the horizontal bands for safe operation, near miss, and accident event are usually not symmetric.

Training methods can improve the skills of human beings up to a certain level (Kluge et al., 2014). According to Salas et al. (2012), these skills are mainly understanding, comprehension, responsiveness, attention allocation, mental modeling, productivity, reliability, safety, and commitment among the trainees. Unfortunately, the growth in industrial safety has not been focused mainly on training and assessment methods. In fact, the process industry has seen an increase in automation, emergency shut-down procedures, alarms, and multilevel control loops in recent decades, which has made the role and job of operators more challenging and equally more important (Nazir et al., 2014a).

Conventional training methods applied to commercial operator training simulators (OTSs) are widely used for the training of industrial operators. At the same time, the number of accidents that take place every year in the process industry is alarming and often not well understood (Burns et al., 2008; Wood et al., 2008). The Major Accident Reporting System (MARS) of the European Commission database shows that the number of accidents per year is growing in Europe despite efforts to control the major hazards since 1982, with the first European Community "Seveso" directive (Wood et al., 2008).

The available training methods lack team training and integration of theories from human factors (Kluge et al., 2011; Salas et al., 2010). Moreover, according to a number of authors, lack of realism and fidelity keep existing training methods from achieving the best training and understanding of the operators (Dalgarno and Lee, 2010; Nazir et al., 2012b). Conversely, the training of operators in immersive virtual environments (IVEs) may result in better understanding and performance of industrial operators. Recently, Manca et al. (2013a) and Nazir et al. (2012b) discussed in detail the concepts and theoretical benefits of employing IVE to train the industrial operators to deal with the aforementioned challenges. They highlighted the depth of understanding associated with IVE training and compared the synoptic displays of a distributed control system (DCS) with the three-dimensional (3D) model of a specific section of a chemical plant to demonstrate the usefulness of the IVE training method. The current study is a step forward from theoretical benefits to an experimental assessment. A detailed review by Dalgarno and Lee (2010) summarizes various studies to show the benefits associated with learning and training in immersive environments. However, the current scientific literature lacks experimental studies emphasizing the impact of training in immersive environments applied to the process industry and specifically in case of accident scenarios. Much of the literature signifying the improvements in the performance attributed to training in IVE has been focused on sectors like medical sciences and surgical training (Aggarwal et al., 2006), mechanical systems (Leu et al., 2013), aviation (Rupasinghe et al., 2011), driver training (Damm et al., 2011), and military (Gallace and Spence, 2014; Lele, 2013). Hence, applications of IVE to the process industry and their possible benefits are topics that should be explored further by researchers. In fact, none of the above-mentioned contexts require either exactly the same human cognitive resources or the same pattern or time of response to face abnormal situations compared to those involved in normal operating conditions. For example, a pool of flammable material increases rapidly its probability of ignition (with all the deleterious consequences related to this) as a function of time passed from the initial spillage or leak, of the external temperature, pressure, thermal fluxes, and so forth. Similarly, once ignited, the higher the fire generated by the liquid, the more serious can be the consequences of such an accident and, most likely, the slower the response of an operator (consider the emotional engagement involved in facing a fire in a usually congested area). Therefore, finding a method that is capable of assessing exactly the speed of response of an operator to such an event under those environmental conditions and an effective training method that can speed up the required amount of time of that response might be of substantial importance in this field. To our knowledge, no study has ever addressed in a comprehensive way this relevant topic of the process industry.

Burkolter and Kluge (2012) and Kluge and Frank (2014) demonstrated the impact of training on the performance of operators in a simulated two-dimensional (2D) process control task, and this positive correlation was also highlighted by Bouloiz et al. (2010). However, these studies were focused on the operator activities performed only in control rooms, whereas generally, industrial operators are distributed between the control room and the equipment, called "in the field" (for further details, see Nazir et al., 2012a).

Besides the lack of thorough training methods, there is also a lack of assessment methods based on exhaustive task, person, and performance indicator analyses. The assessment of operators is equally important and is linked directly to their job allocation (Burkolter et al., 2009). If the operator's assessment is not meant to evaluate the real understanding and skills improvement achieved by the training session, the benefits of training are kept from achieving their potential. The most common procedure adopted by several organizations is the conventional approach to assessment based on the direct contribution of the trainer. Under these circumstances, the trainer evaluates the trainees' performance by means of some tests, observations, questionnaires, and the like. In fact, human judgment can be rather weak since it is based on subjective impressions (Saaty and Shih, 2009). Then, it would be desirable to ground the operators' assessment on a reliable, repeatable tool capable of being intrinsically and completely neutral, thus facilitating the assessments of the trainer and the decision makers. In order to reach and satisfy as much as possible the neutrality feature, the assessment procedure should be automated and avoid only the use of questionnaires and the consequent analysis from an examiner. Questionnaires and posttests normally used in conventional training methods produce a final result only after the whole set of actions have been performed. Conversely, the evaluation of performance during the course of achieving a goal provides a comprehensive analysis of the performance at each step or action. As far as human factors are concerned, it is relevant to note that several psychological mechanisms act at an implicit or even automatic level of information processing (e.g., Fitzsimons et al., 2002). This makes the researcher wonder whether responses to questionnaires are necessarily representative of people's actual behavior and performance (for a discussion on this point, see Gallace and Spence, 2014). Moreover, *post hoc* questionnaire procedures have also shown to suffer from important limitations related to memory failure or memory distortions that often affect human participants. These are even more severe following emotionally relevant events (Christianson and Loftus, 1991). Advanced assessment tools call for an automated procedure to evaluate dynamically, objectively, and in real time the performance of operators. It is then advisable that a computer program perform the assessment automatically by not only evaluating the performance marks, but also registering, storing, and analyzing the actions and decisions made by the operators during the training session. Under this perspective, the assessment procedure should become an algorithm to be implemented in a computer program by means of an automated procedure. By using this procedure, it is possible to analyze exactly where and when errors occur, and thus determine how to reduce reaction times to potentially relevant events.

Given this context, the goal of this study is to face the above-mentioned challenges in training and assessment by proposing a new software tool called plant simulator (PS), where the operator's training is done in a 3D IVE (further details of the PS are covered in Section 9.2). Based on the availability of the PS, two mutually excluding hypotheses are formulated:

1. Trainees perform better within a simulated accident scenario after a training session in the 3D IVE.
2. Trainees perform better within a simulated accident scenario after a training session with a conventional training method.

In particular, the fact that the 3D IVE training session involves the presentation of dynamic 3D images reasonably helps the participants to build more complex and detailed spatial representations of the environment compared to participants exposed to static images (see also Williams et al., 2007). It is also relevant to mention that Oulasvirta et al. (2009) showed that significant cognitive loads are necessary to build a connection between bidimensional and real-world environments. That is, being trained in a 3D environment might not require the same deployment of cognitive resources required by a 2D presentation in order for the participants to build a complex representation of the actual work environment.

For the sake of correctness, it is also advisable to cite a different literature opinion or perspective about 2D and 3D environments to train industrial operators. Despite the potential benefits that can be obtained by using the PS to train the operators, there are also possible concerns about using this training procedure compared to a conventional one. That is, based on the recent literature on human interactions with 3D interfaces, one might expect better performances for the participants trained with the 2D conventional presentation than for those trained with the PS. In fact, Cockburn and McKenzie (2001) showed that performance in a spatial memory task deteriorates when using 3D compared to 2D physical or virtual interfaces. That is, the PS training method might require more cognitive resources (or drive the participant's attention away from the information to be learned) than those required by a conventional training method presentation, thus resulting into a possible worsening of a participant's performance. The experiment discussed in this chapter was also aimed at addressing this point in a more ecologically valid situation and possibly answering this critical point.

The chapter presents the design and achieved results of the experiment aimed at understanding which of the two mutually excluding hypotheses is true.

9.2 METHODS

9.2.1 PLANT SIMULATOR

Manca et al. (2013a) showed how a PS brings novelty to both training and assessment methods in the process industry. The proposed engineering solution consists of coupling a conventional OTS with an IVE (commonly referred to as the 3D environment) that simulates the field in all its immersive facets comprising spatial sounds. The departure of the PS by the so-called serious game solutions is granted by the presence of two real-time programs that work in the background: a dynamic process simulator and a dynamic accident simulator (for details, see Nazir and Manca, 2014).

Figure 9.2 shows the PS representation, whose structure can be conceptually summarized into three parts. Part 1 shows the consequences evaluated by the dynamic accident simulator. Part 2 reproduces the 3D stereoscopic plant environment

FIGURE 9.2 Schematic representation of the plant simulator. (Courtesy of Virthualis, Milan.)

(i.e., IVE) and reports the dynamic interaction between the accident and process simulators. Part 3 sketches the features of the dynamic process simulator. A performance assessment algorithm is also built into the PS, which runs in real time during the experiment. Parts 1 and 2 are meant for training field operators (FOPs), while Part 3 is used to simulate the DCS and train control room operators (CROPs). For the sake of clarity, the experiment focuses on Parts 1 and 2, and therefore is aimed at FOP training. The role of the CROP in the experiment is played by an expert (e.g., the trainer) to make it consistent and reproducible when applied to the tested trainees.

The process simulator evaluates the dynamics of the process units and the pipe network. However, it is not capable of describing what happens outside of the equipment (i.e., in the surrounding environment). The accident simulator receives information about the process variables from the process simulator and evaluates the possible releases, outflows, emissions, and leakages (in case of accident on the piping, unit rupture, or failure). It then evaluates in real time the accident dynamics (e.g., the features of a liquid jet and how it spreads or evaporates on the ground, the features of a pool fire after the ignition of the liquid pool, or the gas cloud dispersion after emission from the piping or equipment). The accident simulator also determines the variables that affect the dynamics of the process units (e.g., the thermal fluxes radiated to the equipment and the depressurization of a vessel in case of liquid or gas emission) and evaluates the consequences of the accident on the surrounding environment (both FOPs and equipment in terms of burns, exposure, dose, surface damage, and physical resistance). In short, the communication between the accident and process simulators is bidirectional, as they are influenced mutually.

The design phase included the conceptualization of the chemical process simulated in the dynamic process simulator (i.e., UniSim® from Honeywell) with the implementation of the spatial, stereoscopic, and interactive environment by a 3D engine. The accident event was conceptualized and simulated in the 3D IVE with AXIM™ (Manca et al., 2013a).

9.2.2 Participants

The experiment involved 24 undergraduate students (20 males; age 19–22 years, mean 20.8 years, standard deviation [SD] 1.03 years) of the third year of the bachelor degree in chemical engineering at Politecnico di Milano, Italy. The authors intentionally selected students belonging to the same semester in order to preserve the homogeneity of the experiment. The participants responded to an open call and no economic reward was provided. The participants were asked about their possible experience on similar experiments and their interest in 3D gaming to increase the experiment equalization and balance. The experiment was performed in accordance with the ethical standards laid down in the 1991 Declaration of Helsinki. All the participants signed a consensus form prior to their participation in the experiment.

9.2.3 Procedure

The industrial process implemented in the PS is a C3/C4 (i.e., propane/butane isomer) separation section of a refinery. C3/C4 separation is an essential section of any

FIGURE 9.3   (a) 3D representation of the C3/C4 separation section. (b) Liquid jet after the rupture of a flange. (c) Pool fire (see also the AVR feature in the inset reporting the thermal load on the FOP). (d) A participant (i.e., trainee) performing the experiment. (Courtesy of Virthualis, Milan.)

refinery, where the distillation of crude oil into different hydrocarbon fractions is obtained. Figure 9.3 shows a snapshot of the IVE. The C3/C4 separation section processes flammable and hazardous components. The accident scenario implemented in the PS replicates a real case study. Further details of the process are available in Brambilla and Manca (2011) and Nazir et al. (2012b). The simulated accident originates from an excavator working in proximity of the C3/C4 distillation tower. The excavator accidentally hits a pipe where liquid butane (C4) is flowing at a pressurized rate. Butane is a highly flammable and volatile hydrocarbon. The rupture of the pipe flange results in a C4 liquid jet that impinges on the ground and spreads into a pool (see Figure 9.3b and c).

For the sake of clarity, in real chemical plants, FOPs interact with physical devices distributed in the plant (e.g., valves, levers, and switches) and use senses such as sight, hearing, and touch (seldom smell and very seldom taste) to cross-check and broaden the perception formed by interpreting field data coming from sensors. Conversely, in real control rooms, CROPs are typically in front of an artificial representation of the process, where synoptic displays on the DCS report the most sensible process variables, which are often complementary to those experienced by FOPs. Table 9.1 reports the sequence of events of the experiment. For the sake of clarity, all the participants acted as FOPs, while an expert trainer, playing as the CROP, repeated exactly the same messages and sequence of actions for each participant (i.e., trainee). The exchange of information between the FOP and the CROP continued throughout the experiment to keep track of the process changes and required actions. Finally, the CROP asked the FOP to manually open a field-operated valve that controls the

TABLE 9.1

Sequence of Events Simulated in the Experiment

Event Sequence	Event Description
t_1	The FOP is at the C3/C4 separation section of the refinery (see Figure 9.3a).
t_2	The excavator hits a pipe and breaks a flange, which results in leakage (see Figure 9.3b).
t_3	The FOP reports the leakage to CROP.
t_4	The CROP suggests the FOP close a valve (valve I).
t_5	The leaked liquid creates a pool on the ground.
t_6	The pool is ignited, resulting in a pool fire.
t_7	The FOP communicates the fire ignition to the CROP.
t_8	The liquid emission is cut off, but the liquid level in the reboiler starts increasing and reaches the high alarm level.
t_9	The CROP asks the FOP to open a manually operated valve (valve II) to decrease the reboiler level.
t_{10}	The reboiler level decreases back to the correct value.

reboiler level, which on its turn has been upset by the abnormal situation resulting from the accident (see Table 9.1, t_9).

9.2.4 TRAINING

The 24 participants were divided into two groups of 12 elements each, equalized for both gender and game playing habits. Each participant was assigned an ID from 1 to 24. All the participants were amateurs, as they had no similar experience of real, in-the-field industrial process activities. After filling out and signing the consensus and basic information forms, they moved to separate rooms. The experiment was designed to avoid any interactions among the participants in order to preserve the results consistency. The two groups were trained with different methodologies. The first group (user IDs 1–12) was trained using the PS (IVE; see Figure 9.4a) and was called the immersive observers group (IO group). They were provided with 3D glasses for an immersive experience of the C3/C4 section of refinery, with limited details given by the trainer. The second group (user IDs 13–24) was trained by a conventional training method (i.e., classroom training with PowerPoint slides) and was called the conventional training group (CT group). The participants of the CT group were provided with 3D still images taken from the IVE and explained through a static slide presentation (see Figure 9.4b). A brief introduction to industrial accidents and safety-critical situations was given to all participants without mentioning the possible accident scenario that they would encounter in the experiment. The training time for both groups was 45 min. Both groups were provided with specific details about the C3/C4 section of the plant, abnormal situations and accidents in the process industry, general actions to be taken during an abnormal situation and accident

FIGURE 9.4 (a) Training of IO group in the IVE (i.e., PS). (b) Training of CT group with a slide-supported classroom lesson.

scenario, coordination between the control room and the field, and the significance of a quick response during an accident scenario.

Once the two separate and completely independent training sessions were given to both groups of trainees, the assessment of the trainees (and the online and real-time evaluation of their actions) was done in the 3D IVE of the PS. The trainees of both groups were asked to respond to an unexpected accident event.

9.2.5 Performance Assessment

The assessment of industrial operators based on a hierarchical and informed methodology allows evaluation of the effectiveness of training methods (Manca et al., 2013b). A robust, holistic, and real-time assessment based on the understanding of process, accident scenario (as in this case), and required cognitive skills allows improvement of the reliability and robustness of the training procedure. These features were considered thoroughly when designing and developing the automated performance assessment algorithm (for further details, see Nazir et al., 2013). Each action of the operator was recorded and analyzed, and thus performance was evaluated throughout the experiment, and not only at the end of the accident scenario (see Table 9.1). The assessment procedure was implemented in a computer program capable of not only evaluating the marks relative to the performance of operators, but also registering, storing, and analyzing the actions and decisions made by the operators during the experiment with the help of a dedicated and customized algorithm that runs in real time (Manca et al., 2014). The key performance indicators (KPIs) for the accident scenario were defined and compared by focusing on the operator's understanding of the process, accident scenario, and required skills. Figure 9.5 shows the KPIs devised for the performance assessment of the experiment, that is, the simulated accident scenario.

A brief detail of each KPI is discussed, within the context of the experiment summarized in Table 9.1:

Leakage identification: This is the identification of the leakage just after the excavator hits and breaks a flange (t_3 in Table 9.1).
Valve identification I: As per the suggestion of the CROP, the FOP (i.e., the trainee) is expected to close valve I (t_4 in Table 9.1).

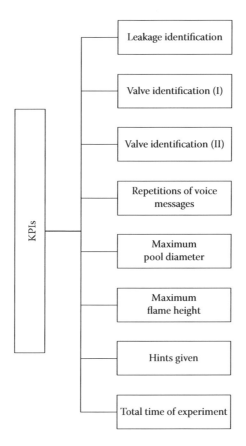

FIGURE 9.5   KPIs used for the performance assessment of the experiment.

Valve identification II: At t_9, the CROP asks the FOP to close valve II, which is the last action of the experiment.

Repetition of voice messages: These show the FOP's skill to comprehend the voice messages from the CROP with (or without) possible repetitions when required; that is, either the message is understood correctly the first time or some repetitions were necessary. During the training session, all the participants were made aware of the possibility of repetition of voice messages, once a clear request was made.

Maximum pool diameter: The maximum pool diameter produced along the experiment stands for the highest amount of liquid accumulated before the participant is able to respond and coordinate the leakage stoppage.

Maximum flame height: This KPI quantitatively determines the maximum height of the flame once the pool formed by the leakage is ignited. The higher the flame height, the higher the potential threats (including possible domino effects).

Hints given: During the experiment, the participants received hints in a systematic way. If the participant found himself or herself lost in the IVE, he or she was advised to go back to the accident scene by providing the direction where to go. The authors defined a boundary in the IVE so that if the field of view of the participant was not within the accident scenario, he or she was supposed to get lost. In this case, a hint was provided by explaining the correct direction. To preserve consistency, when needed, each hint was given after a specific and constant time interval (i.e., 60 s) for all the participants who exceeded that threshold.

Total time of experiment: The total time is the time taken by the participants to reach t_{10} as described in Table 9.1. In order to keep the study consistent, the triggering of the main events of the accident scenario was automated to make them reproducible for every trainee. For example, the time between the impact of the excavator on the pipe and the start of leakage was set at 2 s, and the time between stoppage of leakage and ignition of fire was set at 7 s.

9.3 EXPERIMENT RESULTS

This section analyzes the performance of trainees and participants involved in the assessment based on the eight KPIs of Figure 9.5.

The unpaired t-test for the maximum pool diameter is $t(22) = 2.45$, $p < 0.02$, and for maximum flame height it is $t(22) = 3.85$, $p < 0.001$. For all comparisons, differences were considered statistically significant at the 5% level (i.e., $p = 0.05$). The average value of maximum pool diameters produced by participants of the IO group is 1.56 m, while that of the CT group is 1.97 m; that is, the average maximum pool area of the IO group is 60% smaller than that of the CT group (the pool area plays a main role in evaporation both before and after its ignition). Similarly, the average maximum flame heights for the IO and CT groups are 4.2 and 5.9 m, respectively (i.e., a 40% difference). Together with the pool diameter, the flame height plays a main role in the heat radiation to the surrounding process units and FOPs who are involved in the emergency response to the accident event. See figure 9.6 for a graphical representation of the differences between the IO group and CT group on maximum pool diameter and maximum flame height.

With reference to the number of hints given, it is worth observing that those required by the CT group ($M = 2.42$, SD $= 1.38$) are larger than those of the IO group, but not significantly so ($M = 0.92$, SD $= 1$; $X^2 = 7.333$, df $= 5$, exact two-tailed $p = 0.189$). Specifically, in 41% of the cases, participants of the IO group did not require any hints. In summary, the participants of the IO group requested 11 hints, whereas those of the CT group requested 29 hints. This translates into 163% more hints needed by the CT group. Table 9.2 shows the distribution of the number of hints for the two groups.

Identification of valves, leakages, faults, and abnormalities is a skill required of an operator to run the process smoothly and cope with possible abnormal situations. In the case of the CT group, only 5 participants out of 12 were able to identify the leakage the first time without help, whereas 8 participants of the IO group successfully identified the leakage. In other words, two-thirds of IO group participants (i.e., 67%) were able to identify and report the leakage, while only 42% of the CT group were successful ($t(22) = 2.28$, $p < 0.02$). The valve identification by the participants

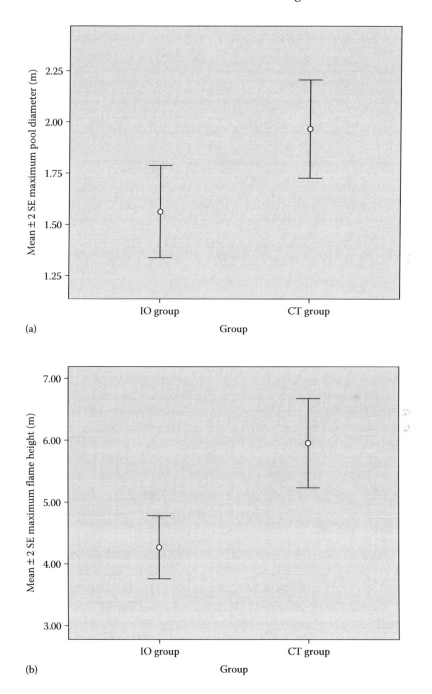

(a)

(b)

FIGURE 9.6   (a) Average values of maximum pool diameter for IO and CT groups (lower is better). (b) Average values of maximum flame height for IO and CT groups (lower is better). Error bars represent the 2 standard errors of the means.

TABLE 9.2

Distribution for Number of Hints Given for the Two Groups

	Number of Hints Given						
	0	1	2	3	4	5	Sum
IO group	5	4	2	1	0	0	11
CT group	1	2	3	4	1	1	29

of both groups reveals a higher number of correct identifications for the participants of the IO group. Six of the trainees (50%) of the CT group were unable to identify the valves (I and II), whereas for the IO group, the success rate was 9 (83%) for valve I and 8 (75%) for valve II. These KPIs (valve and leakage identifications) were evaluated before providing any help to the participants in order to keep the results consistent.

Regarding repetition of voice messages, in the case of the CT group, five trainees required the repetition of voice messages for their complete comprehension. Among them, four were unable to comprehend the incoming message with one repetition. Consequently, the message was repeated twice. On the other hand, only four participants of the IO group required the repetition, and more importantly, all of them were able to comprehend it the first time (see also Figure 9.7c). Again, the IO group outperformed the CT group by a 125% difference of message repetitions. Figure 9.7d shows the average time taken by both groups to finalize the experiment. On average, participants of the IO group spent 50 s less (i.e., 20% less time) than participants of the CT group. The results obtained from the unpaired t-test are $t(22) = 2.08, p < 0.02$. Table 9.3 summarizes the averaged performances of the IO and CT groups according to the quantitative assessment based on the eight KPIs of Figure 9.5. The last column of Table 9.3 takes as a reference point the KPI performance of the IO group.

9.4 DISCUSSION

This chapter examined the impact of distinct training methods on the performance of trainees in a simulated accident scenario. The training methods appeared to have an impact on the operator's performance. In fact, the IO group (i.e., immersive observers) performed better than the CT group (i.e., conventional training). That is, the trainees of the IO group reacted more quickly to the accident and made less errors. Therefore, the first hypothesis (i.e., trainees perform better within a simulated accident scenario after a training session in a 3D IVE) formulated in Section 9.1 proved true.

The results of this study are in line with the available literature and suggest an improvement in performance based on modification and update of conventional training methods (Burkolter et al., 2009; Kluge and Frank, 2014; Salas et al., 2012). The experimental work on the use of IVE in training has shown an improvement of achieved overall performance (Dalgarno and Lee, 2010; Damm et al., 2011;

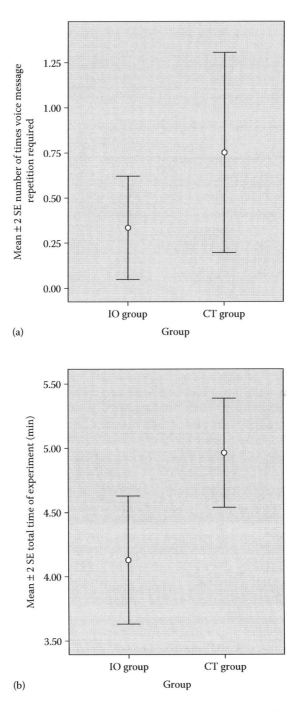

FIGURE 9.7 (a) Comparison of accuracy in recalling voice messages for each participant. (b) Average of the total time taken by each group to successfully complete the experiment (lower is better).

TABLE 9.3

Comparison of Averaged KPIs for IO and CT Groups

KPIs	IO Group Mean	CT Group Mean	Relative Difference with Regard to IO Group
Leakage identification (number of times), higher is better	0.667	0.417	−37%
Valve identification (I) (number of times), higher is better	0.833	0.500	−40%
Valve identification (II) (number of times), higher is better	0.750	0.500	−33%
Repetitions of voice messages required (number of times), lower is better	0.330	0.750	+127%
Maximum pool diameter (m), lower is better	1.562	1.966	+26%
Maximum flame height (m), lower is better	4.266	5.955	+40%
Hints given (number of times), lower is better	0.917	2.417	+163%
Total time of experiment (min), lower is better	4.127	4.959	+20%

Lele, 2013; Leu et al., 2013). Most of the studies on these topics (e.g., Burkolter et al., 2009; Kluge and Frank, 2014) were focused on CROPs, and the evaluation of performance was based on the final outcome of the process rather than logging the performance indicators at each step, necessary to accomplish a given task (as reported in this chapter). This study defined specific performance indicators at each notable event. In addition, a dedicated algorithm was used to keep the performance assessment consistent and coherent. This objective method (i.e., not based on self-report or questionnaire procedures) is of great importance in understanding the human reactions to accidents and planning adequate training methods. The findings of this work continue and expand the research line discussed in the existing literature (Burkolter and Kluge, 2012; Kluge and Frank, 2014; Salas et al., 2012). The results of Bouloiz et al. (2010) show a positive correlation on safety-related performance, with respect to the quality of training, and also support the outcomes of this work. All the adopted KPIs showed an improved performance of the IO group over the CT group. For the sake of consistency, it is worth highlighting that the KPIs were designed and selected before running the experiment. As can be gathered from Figure 9.5, the KPIs are intrinsically representative of both the process and accident dynamics.

In refinery processes, it is highly likely that an accident involves the formation of a pool of flammable liquid, which, if ignited, gives rise to a pool fire. It is of extreme importance for the FOP to promptly alert the CROP, without hesitation, to alleviate the possible consequences. In this study, the difference between average maximum pool diameters was 0.404 m for the IO and CT groups, as shown in Table 9.3. Even though this difference might appear negligible, the impact it can bring in real scenarios may be quite significant. It is reasonable to maintain that control of both the pool diameter and maximum flame height is a direct consequence of the cognitive skills (i.e., perception, allocation of attention, and effective communication) involved in processing the information coming from the field. During abnormal situations and accident scenarios, the margin of error for operators is very low, and it is expected that FOPs identify correctly and operate promptly the valves, comprehend the incoming messages without any misunderstandings, and take little time to finalize the required actions (Nazir and Manca, 2014). This study found that the IO group trained in the IVE outperformed the CT group trained with a conventional method. The IO group not only performed better in precision (e.g., identifications, requirement of hints, repetitions of voice messages, and identification of valves), but also was more responsive and prompt (i.e., total time of experiment). The inclusion of systematic hints in the experiment proved helpful in keeping the study consistent. Since the hints provided allowed the FOP to identify his or her physical position within the experiment, it is possible to assume that the training method of the IO group also enhanced the trainee's alertness (Plant and Stanton, 2013).

Training is crucial for many process industries, and the adoption of modern training methods (similar to the IVE method) can result in improved performance of operators. In addition, having systematic and real-time assessment procedures improves the evaluation reliability and consistency. In fact, it allows the measuring of data that the trainer would not be able to gather in other ways. The trainer is then also relieved from repetitive tasks and can focus on more important human-oriented activities. The training methods are mostly focused on normal operating conditions. Conversely, with IVEs, it is possible to simulate abnormal situations and industrial accidents that as a rule seldom occur and would be impossible to reproduce in a real environment due to safety reasons. Moreover, modifying already designed and operating chemical plants to further improve the process safety would be unfeasible in most cases due to well-defined and well-accepted control loops and boundary conditions (Burns et al., 2008; Wickens, 2000). An improved training methodology, however, can be a feasible step in increasing process safety and mitigating possible accidents. The encouraging results of this study can convince practitioners and decision makers to invest in advanced training simulators to assist industrial operators. This could also produce an improvement in terms of loss prevention and enhanced industrial safety.

9.4.1 LIMITATIONS OF THE STUDY

While the finding presented here shows a clear association between IVE training and performance in a given scenario, it is worth pointing out some limitations. First, the complete environment during the training (of the IO group) and assessment sections was IVE. In general, an ideal situation would consist of performing the assessment procedure at a real plant after being trained offline with either a PS or a conventional training method. Nonetheless, that would require huge resources to analyze the performance in a real environment with reference to the nature of the proposed accident event. In addition, this would involve an excessive risk for both the trainees and the plant, and thereby it would generate serious ethical issues. It is also important to note here that if the results obtained were due to the similarity between the kind of environment presented during the training methods and the actual accident scenario, it is reasonable to predict that performances in the real world should be better after being trained with a PS rather than with a conventional training method. Even though other literature experiments conducted with virtual simulators (e.g., driving simulators) were tested in real and live environments, the same could not happen with the experiment described in this work since it would involve a high risk and an extreme exposure of the plant ownership. Furthermore, Stone-Romero (2011) reports with considerable evidence that findings from experimental research conducted in "special purpose settings" (e.g., a simulated task environment) can be generalized and applied to real environment settings (p. 48).

Another limitation of this study is that the participants of the experiment were students and not real industrial operators; thus, they did not represent the actual population of target users of the PS. Nevertheless, they were all chemical engineering students having a common background and an equalized knowledge of chemical processes and relevant terminology. Somehow, they showed a close similarity to the potential users of the PS. Finally, the participants in the experiment were trained only once before facing the accident scenario, which is not the case for a real plant, where training is repeated periodically up to an acceptable threshold.

In the future, it would be of interest to analyze the amount of training sessions necessary to obtain a high level of accuracy within an accident scenario, by using either a PS or a conventional training method. Different results might also be expected as a function of the participants' skill in the field or in the control room, or a combination of both. This work can be considered the starting point for further studies to investigate the effectiveness of the proposed advanced training and assessment methodologies within different simulated scenarios, with real operators, and even in real plants (in case of normal operating conditions or mild abnormal situations). That is, the achieved results, showing the effectiveness of IVE training methods, pave the way to future research aimed at determining the extent of reproducibility and transferability of these preliminary results.

9.5   CONCLUSIONS

This chapter described the PS, a new tool to be used for training and assessment of industrial operators. Such a solution (which includes both a process and an accident simulator) can train the operators in a 3D IVE and test their performance, in real time, based on well-defined KPIs. The study compared the performance of participants who were trained using the PS with the performance of those who were trained with a conventional method. The assessment was based on the simulation of an accident event in a refinery section. The results clearly showed that training with a PS results in better performance (in terms of participants' reaction and awareness to an accident) than a more conventional training method. In conclusion, this study showed that by using specific training methods, it is possible to better understand and also improve the performance of operators, thus reasonably reducing those accidents that are primarily attributed to human factors.

REFERENCES

Aggarwal, R., Black, S. A., Hance, J. R., Darzi, A., and Cheshire, N. J. W. 2006. Virtual reality simulation training can improve inexperienced surgeons' endovascular skills. *European Journal of Vascular and Endovascular Surgery*, 31(6), 588–593.

Antonovsky, A., Pollock, C., and Straker, L. 2013. Identification of the human factors contributing to maintenance failures in a petroleum operation. *Human Factors: The Journal of the Human Factors and Ergonomics Society*, 56(2), 306–321.

Bouloiz, H., Garbolino, E., and Tkiouat, M. 2010. Contribution of a systemic modeling approach applied to support risk analysis of a storage unit of chemical products in Morocco. *Journal of Loss Prevention in the Process Industries*, 23(2), 312–322.

Brambilla, S., and Manca, D. 2011. Recommended features of an industrial accident simulator for the training of operators. *Journal of Loss Prevention in the Process Industries*, 24(4), 344–355.

Burkolter, D., and Kluge, A. 2012. Process control and risky decision-making: Moderation by general mental ability and need for cognition. *Ergonomics*, 55(11), 1285–1297.

Burkolter, D., Kluge, A., Sauer, J., and Ritzmann, S. 2009. The predictive qualities of operator characteristics for process control performance: The influence of personality and cognitive variables. *Ergonomics*, 52(3), 302–311.

Burns, C. M., Skraaning Jr., G., Jamieson, G. A., Lau, N., Kwok, J., Welch, R., and Andresen, G. 2008. Evaluation of ecological interface design for nuclear process control: Situation awareness effects. *Human Factors*, 50(4), 663–679.

Christianson, S.-Å., and Loftus, E. F. 1991. Remembering emotional events: The fate of detailed information. *Cognition and Emotion*, 5(2), 81–108.

Cockburn, A., and McKenzie, B. 2001. 3D or not 3D? Evaluating the effect of the third dimension in a document management system. Paper presented at Proceedings of the SIGCHI Conference on Human Factors in Computing Systems, Seattle, WA.

Colombo, S., Nazir, S., and Manca, D. 2012. Towards holistic decision support systems. Including human and organizational performances in the loop. *Computer Aided Chemical Engineering*, 31, 295–299.

Dalgarno, B., and Lee, M. J. W. 2010. What are the learning affordances of 3-D virtual environments? *British Journal of Educational Technology*, 41(1), 10–32.

Damm, L., Nachtergaële, C., Meskali, M., and Berthelon, C. 2011. The evaluation of traditional and early driver training with simulated accident scenarios. *Human Factors*, 53(4), 323–337.

Fitzsimons, G., Hutchinson, J. W., Williams, P., Alba, J., Chartrand, T., Huber, J., Kardes, F. R., et al. 2002. Non-conscious influences on consumer choice. *Marketing Letters*, 13(3), 269–279.

Gallace, A., and Spence, C. 2014. *In Touch with the Future: The Sense of Touch from Cognitive Neuroscience to Virtual Reality*. Oxford: Oxford University Press.

Gopher, D. 1996. Attention control: Explorations of the work of an executive controller. *Cognitive Brain Research*, 5(1–2), 23–38.

Kletz, T. 1999. Preparation for maintenance. In *What Went Wrong?* (4th ed., pp. 1–47). Houston, TX: Gulf Professional Publishing.

Kluge, A., and Frank, B. 2014. Counteracting skill decay: Four refresher interventions and their effect on skill and knowledge retention in a simulated process control task. *Ergonomics*, 57(2), 175–190.

Kluge, A., Nazir, S., and Manca, D. 2014. Advanced applications in process control and training needs of field and control room operators. *IIE Transactions on Occupational Ergonomics and Human Factors*, 2(3–4), 121–136.

Kluge, A., Ritzmann, S., Burkolter, D., and Sauer, J. 2011. The interaction of drill and practice and error training with individual differences. *Cognition, Technology and Work*, 13(2), 103–120.

Lele, A. 2013. Virtual reality and its military utility. *Journal of Ambient Intelligence and Humanized Computing*, 4(1), 17–26.

Leu, M. C., Elmaraghy, H. A., Nee, A. Y. C., Ong, S. K., Lanzetta, M., Putz, M., Zhu, W., and Bernard, A. 2013. CAD model based virtual assembly simulation, planning and training. *CIRP Annals—Manufacturing Technology*, 62(2), 799–822.

Manca, D., Brambilla, S., and Colombo, S. 2013a. Bridging between virtual reality and accident simulation for training of process-industry operators. *Advances in Engineering Software*, 55, 1–9.

Manca, D., Colombo, S., and Nazir, S. 2013b. A plant simulator to enhance the process safety of industrial operators. Presented at European HSE Conference and Exhibition, London.

Manca, D., Nazir, S., Colombo, S., and Kluge, A. 2014. Procedure for automated assessment of industrial operators. *Chemical Engineering Transactions*, 36, 391–396.

Naderpour, M., Nazir, S., and Lu, J. 2015. The role of situation awareness in accidents of large-scale technological systems. *Process Safety and Environmental Protection*, 97, 13–24.

Nazir, S., Colombo, S., and Manca, D. 2012a. The role of situation awareness for the operators of process industry. *Chemical Engineering Transactions*, 26, 303–308.

Nazir, S., Colombo, S., and Manca, D. 2013. Minimizing the risk in the process industry by using a plant simulator: A novel approach. *Chemical Engineering Transactions*, 32, 109–114.

Nazir, S., Kluge, A., and Manca, D. 2014a. Automation in process industry: Cure or curse? How can training improve operator's performance. *Computer Aided Chemical Engineering*, 33, 889–894.

Nazir, S., and Manca, D. 2014. How a plant simulator can improve industrial safety. *Process Safety Progress*, n/a–n/a.

Nazir, S., Sorensen, L. J., Øvergård, K. I., and Manca, D. 2014b. How distributed situation awareness influences process safety. *Chemical Engineering Transactions*, 36, 409–414.

Nazir, S., Totaro, R., Brambilla, S., Colombo, S., and Manca, D. 2012b. Virtual reality and augmented-virtual reality as tools to train industrial operators. *Computer Aided Chemical Engineering*, 30, 1398–1401.

Oulasvirta, A., Estlander, S., and Nurminen, A. 2009. Embodied interaction with a 3D versus 2D mobile map. *Personal and Ubiquitous Computing*, 13(4), 303–320.

Plant, K. L., and Stanton, N. A. 2012. Why did the pilots shut down the wrong engine? Explaining errors in context using schema theory and the perceptual cycle model. *Safety Science*, 50(2), 300–315.

Plant, K. L., and Stanton, N. A. 2013. The explanatory power of schema theory: Theoretical foundations and future applications in ergonomics. *Ergonomics*, 56(1), 1–15.

Rupasinghe, T. D., Kurz, M. E., Washburn, C., and Gramopadhye, A. K. 2011. Virtual reality training integrated curriculum: An aircraft maintenance technology (AMT) education perspective. *International Journal of Engineering Education*, 27(4 PART II), 778–788.

Saaty, T. L., and Shih, H. S. 2009. Structures in decision making: On the subjective geometry of hierarchies and networks. *European Journal of Operational Research*, 199(3), 867–872.

Salas, E., Cooke, N. J., and Gorman, J. C. 2010. The science of team performance: Progress and the need for more. *Human Factors*, 52(2), 344–346.

Salas, E., Tannenbaum, S. I., Kraiger, K., and Smith-Jentsch, K. A. 2012. The science of training and development in organizations: What matters in practice. *Psychological Science in the Public Interest*, 13(2), 74–101.

Stone-Romero, E. F. 2011. Research strategies in industrial and organizational psychology: Nonexperimental, quasi-experimental, and randomized experimental research in special purpose and nonspecial purpose settings. In Z. Sheldon (Ed.), *APA handbook of industrial and organizational psychology*, Vol 1: Building and developing the organization (pp. 37–72). Washington, DC: American Psychological Association.

Wickens, C., and McCarley, J. 2007. *Applied Attention Theory*. Boca Raton, FL: CRC Press.

Wickens, C. D. 2000. *Engineering Psychology and Human Performance* (3rd ed.). Upper Saddle River, NJ: Prentice-Hall.

Williams, B., Narasimham, G., Westerman, C., Rieser, J., and Bodenheimer, B. 2007. Functional similarities in spatial representations between real and virtual environments. *ACM Transactions on Applied Perception*, 4(2), 12.

Wood, M. H., Fabbri, L., and Struckl, M. 2008. Writing Seveso II safety reports: New EU guidance reflecting 5 years' experience with the directive. *Journal of Hazardous Materials*, 157(2–3), 230–236.

10 Knowledge Management for Counterbalancing the Process of Loss of Skills at Work
A Practical Study

*Raoni Rocha, Vitor Figueiredo,
and Ana Karla Baptista*

CONTENTS

10.1 INTRODUCTION

Currently, companies have countless procedures, operational rules, or task instructions that do not do justice to the diversities and difficulties experienced by workers while executing their activities. In most of situations, the social actor responsible for creating the procedures is far from the reality experienced in the field. One of the causes of that disconnection between what is written and what is real is the absence of spaces of discussion at work that allow the sharing of knowledge or the possibility to externalize strategies and actions that can be used when managing the difficulties in the field (Rocha et al. 2015).

The disconnection between the formal organization and the reality can have countless consequences, such as work accidents, an increase of the absenteeism index, productivity reduction, and difficulty in passing around knowledge in the organization (Daniellou et al. 2011). Beyond that, a significant distance between the reality in the field and the formal organization of the company can lead the company to lose skills with time, in the moving of generational changes that organizations experience.

The distance between rules and procedures imposed by the company regarding the real situations experienced in the field provides situations that make the learning process at work more fragile. On the one hand, experienced workers pass on their knowledge and use operatory strategies to go around the rules that does not contemplate the variety of the field; and on the other hand, inexperienced workers take ownership of the rules to try to fulfill the organization's demands, taking as reference only the situations presented in trainings or simulations.

The difficulties that inexperienced workers face and the strategies and knowledge of experienced workers when dealing with unpredicted situations in reality are not shared in the organization. From this condition, two questions emerge: How can the loss of skills caused by the process of generational changes in the organizations be managed? How is it possible to pass on the knowledge of practical field strategies and promote its dissemination to the operational actors in the organization?

The situation presented above fits the reality of the company studied in this research. It is a company responsible for distributing, producing, and commercializing electricity in the state of Minas Gerais, Brazil. The company has more than 8000 employees, and its participation in the Brazilian market can be summed up as follows: 12% in distributing electricity, 7% in generating electricity, 13% in transmitting electricity, 25% in the free consumer market, and 121,000 shareholders.

Given the size of this company, two units were chosen to develop research in the context exposed above. The goal of this research was to capture the experienced workers' skills by registering the real activity and creating a library with films and statements from the workers themselves and from the ergonomists involved in the situations experienced in the unit in question. The library was built with the workers and allowed the sharing of the gestures and attitudes of the experienced workers toward their respective units. A group of three ergonomist researchers participated actively in this process of building the visual material that was seen as the link between the situations experienced by the workers and the strategies chosen to reach the task's goal.

The theoretical framework used to ground this research is described below and can be divided into three topics: ergonomics schools, knowledge management, and acquisition of expertise.

10.2 THEORETICAL FRAMEWORK

10.2.1 ERGONOMICS AND ITS EYES ON THE JOB

Ergonomics, according to the approach based on human factors (Anglo-Saxon), is concerned primarily "with the physical aspects of man-machine interface, aiming to scale the workstation, facilitate the discrimination of information and displays the manipulation of controls" (Moraes and Mont'alvão 2000, p. 16). Expanding that concept, Montmollin (2005, p. 106) states, "The ergonomics centered on human component (Human Factors) do not have a need for an analysis of the work, which is replaced by the construction of a list of demands of the task, usually established by questions to heads from grids pre-established, based on the data on the human characteristics that supposedly come into action in tasks deemed."

The use of questionnaires or standardized checklists in ergonomics research, commonly ergonomics focused on human factors, simplifies the work and does not emphasize the singularities and particularities present in real activity. The questions previously developed do not portray the local reality of the research, and therefore are decontextualized and do not cover the variability of productive context and do not follow the dynamics of the progress of an action.

To propose a reductionist model of the human being, addressing only the visible elements in its interface with the machine, the Anglo-Saxon approach is limited to weave design recommendations that are incompatible with the adjustment possibilities and mobilization of the skills of workers, generating some inconsistencies between the object, the user, and the intended use (Assunção 2004). To analyze the work only under the prisms identified above, essential elements, such as the organization of work and the representativeness of real activity, are neglected, disregarding the expertise and collective settings of operators.

In spite of this Anglo-Saxon perspective being insistently repeated in several discussions, the focus on job and posture, currently, began to be only a complement in analyses of the work conducted in the approach focused on the activity. Therefore, these are aimed at a more in-depth look on the organization of work, on the collective settings established in the management of the work, and on the socio-economic impact of real activity. The contrast with the previous approach (Anglo-Saxon) is clear. Activity means here that the functions are no longer considered in an isolated manner, but the behavior and reasoning as they are in real situations of work, current or being made.

Although some authors consider the two approaches to be antagonistic, they may be considered hierarchical ergonomics focused on human factors, conceiving the technological devices as adapted to humans. It is a question of the ergonomics of first aid, a set of elements that is part of the job cut out of its context (Montmollin 2005).

The ergonomics focused on activity, covering work in the course of its action, in order to differentiate the prescribed from the real and highlight the strategies deployed by operators in the complex management of health and production. In this approach, the operator is analyzed not only from the perspective of the man–machine interface, but also as someone who "develops it activity in real time on the basis of this framework: the work activity is a strategy of adaptation to the actual working situation, the object of the prescription. The distance between the prescribed and the real is the concrete manifestation of contradiction always present in the act of work, between *what is required and what the thing asks*" (Guérin et al. 2006, p. 15).

The analysis of the work in this perspective, focused on activity, cannot be satisfied with the status of methods or preestablished model routes; it proposes the construction of specific models with specific theoretical frameworks. The activity is seen as a set of regulations and contextualized representations, in which they take part in the variability inherent in the environment and the worker himself or herself (Assunção and Lima 2003). The activity is also a set of contextualized adjustments, contemplating the variability of the context in which the work is performed and the own employee's variability.

In this sense, the distance to be identified and analyzed between the task and activity is a source of knowledge. The studies involved with this theme highlight the nature of the unpredictability of the activity, because it requires, at each moment, the creative intelligence of the operators and the mobilization of knowledge and skills (Weill-Fassina et al. 1993).

The methodological approach of the ergonomics of the activity is directed to the singularity of the act of work of each operator in specific contexts. In this process of analysis of the work and experience of the ergonomist in the field of research, it is important to try to find the meaning of the behavior of workers and the reason for their decision-making process. It is necessary to allow an offset of perspective: "put yourself in the place of another, rather than as a moral principle, but based on objective observations of subjective sense and explanation of reasons reasonable and interconnected" (Lima 2001, p. 139).

10.2.2 KNOWLEDGE MANAGEMENT

Currently, the change of generation of workers and the incentive policy for retirement in companies bring consequences for the management of the work, among them, the loss of knowledge and expertise, an increase in training costs, low potential for organizational learning, and security compromise.

A way to mitigate the loss of the operator's ability to succeed in organizations is to apply techniques of knowledge management. Thus, the knowledge acquired by workers and their skills are translated into forms of rules and manuals for application and introduced according to a series of rational principles of learning (Nonaka and Takeuchi 1995).

Despite an organization considering this transformation of the skills of operators by putting them in manuals for application as knowledge management, these techniques only put on paper the explicit knowledge of these workers; that is, standardized rules that are described in greater detail.

In short, what some organizations call knowledge management is, in fact, management of the explicit. The tacit knowledge is requested according to the situation; that is, the operator reacts differently to each task performed, and therefore depends on the context of the activity and the work to be performed (Collins 2007). Does the knowledge result from the practice and experience in different situations? Explicit and tacit knowledge are not completely separate entities, but are mutually complementary (Slack et al. 1998).

Ribeiro (2007) makes advances in this discussion and affirms that all explicit knowledge, to be used properly, requires tacit knowledge on the part of those who used it previously. The regression of the rules shows us the inadequacy of explicit knowledge, because "explicit" is not a characteristic of the type of knowledge, but of the encounter of a cultural being with a piece of encoded knowledge.

The essence of tacit knowledge is in its ability to fully participate in a way of life: "collective tacit knowledge" (Collins 2007). This means, for example, to be able to extrapolate or circumvent the existing rules, as well as act smoothly or improvise within a new culture, whether it is technical or not.

Figueiredo and Figueiredo (2012) show in their study that, even in "seemingly simple," repetitive activities, it is possible to identify the adjustment strategies and the mobilization of tacit knowledge between the rookies and the experienced operators of a textile industry to circumvent the limitations imposed by operational rules that depart from actual practice.

As Amaral (2002) stated, the management of knowledge is strongly related to the people within the organization who have skills and knowledge; and, from the direct interaction between them, have the ability to transmit such knowledge, thus contributing to an increase in creativity, innovation, and differentiation within the organization. Today's companies who know the importance of mobilization and knowledge sharing are looking for ways to transfer it so that this organizational knowledge is not lost.

The transmission of tacit knowledge among all members of an organization is becoming a strategic factor of companies. Therefore, inserting and encompassing the management of expertise in management systems of organizations increases the capacity of institutions to anticipate variations of "normality," increasing the stability and reliability of the organizational system as a whole (Rocha 2014). In this context, the problems and opportunities become occasions for the creation of knowledge and decision making. The results of the creation of knowledge are innovations or extensions of organizational capabilities (Choo 2002).

In addition, because practical knowledge is embedded in the person who has it or in the collective work, the transfer of it is a complex process. In many situations, when verbalizing their own work, the operator does not know how to explain what is in the course of their activity or does not have the scale of the countless operations performed.

The fact that the person is not able to explain by using rules does not mean that the work cannot be reproduced by another person (Collins 2001). It means that, in reality, the practical knowledge cannot be transferred, but learned. In other words, the transfer (or learning) of practical knowledge can only be given by means of the practice of activity, that is, by means of the "hand on earth."

A complementary way to improve the learning process in organizations, rather than approaches that treat only the management of the explicit, is the creation of spaces of discussion at work. The discussion of the activity in a real-life situation and the sharing of knowledge are practical and effective ways to highlight the strategies deployed in the management of situational variability, the involvement of the collective in the decision-making process, and the actions taken to overcome the difficulties faced in the field: "the space for discussion of real activity could not, therefore, bring reflections on the seriousness of the risk and on the actions most relevant in the management of an anomaly in the field ... These spaces were also sites of mutual learning, transferring of powers and influencing the health of the collective work" (Rocha et al. 2014, p. 68).

It is important to implement areas of discussion in organizations, because these spaces are tools that assist in the transfer of skills and knowledge among the workers.

They allow various moments of formal exchange of information between operators, of mutual learning between them, and the transfer of skills between the more experienced and the younger people (Rocha et al. 2015).

10.2.3 Acquisition of Expertise

The experienced operators mobilize their savoir faire thanks to different regulatory mechanisms, such as selective attention to information, automatism between rational action, appropriate trials, perception, and shortcuts. "The rookie looks for a little information everywhere, the experienced verifies the key points" (Daniellou et al. 2011, p. 44). The experienced person has a selective vision; his work is based on values and meanings lived and experienced throughout a working life, making it a singular vision and particular work activity.

In the trial of information, Ribeiro (2012, p. 29) shows differences between the beginners and experienced.

> … regarding the action, it was seen that those with low similarity usually cling to a prescriptive approach that leads to anti-economic situations, inefficient, and even dangerous. On the other hand, the fast acting with autonomy and trust is a registered trademark of employees of high similarity. Because of their ability to trial, they are able to assign meaning to the rules or to operational standards and apply them in the light of the specific characteristics of the situation. This makes the rules and standards 'explicit' to them, while they are only marks in a paper for some rookies.

Dreyfus and Dreyfus (1986) describe a model of acquisition of expertise that illustrates the perceptual differences between a novice and the experienced. The first stage of this model is the rookie; he breaks down the environment into decontextualized elements and the rules applied do not consider the environmental events, that is, are free of context. In the second stage, the advanced beginner understands certain complex rules and interprets situational facts, recognizing some differences in similar situations. In the third stage, the individual has the capacity to deal with a greater number of situational facts and can understand the context. Finally, in the last two stages, proficient and expert, individuals are able to make decisions apart from the rules and context; however, the expert acts more fluidly and intuitively.

The process of sharing knowledge between rookies and the experienced cannot merely be summed up in decontextualized trainings or the dissemination of institutional rules that are not consistent with the actual activity. "The central issue of learning is to become a practitioner and not learn about the practice" (Brown and Duguid 1991, p. 48). This quote refers to the concept of learning based on practice, which rejects models that prioritize knowledge isolated from practice. Learning is not a practice that can be particularly viewed and studied as a test object. On the contrary, learning must be seen as inevitably implicated in everyday interactions of the relationship between people and the world (Lave and Wenger 1991).

The organization must understand that learning is a process of social construction; that is, it is the result of the interaction between people manifested in daily behaviors (Wenger 2000). It is not the sum of the individuals that allow the learning, but the social practices in which they engage.

In this context, it can be said that the fundamental knowledge of an organization is not in its manuals and reports or centered on the individual, but in groups or communities that together, by interacting, make up the organization. The learning emerges from the social interactions, the meaning assigned to the data and information, and only has meaning when interpreted.

Thus, the process of learning and acquisition of expertise adopted in organizations should contemplate the different knowledges of social actors, beginners and experienced, in addition to contextualizing them with real situations experienced in the field. It is important that the rules relayed to rookies contain representative elements of real situations and not just subjective representations or simulations. The important aspect of this learning process is not only teaching the tricks, but also knowing the right time to use them.

10.3 CHARACTERIZATION OF THE CONTEXT AND THE DEVELOPED WORK

The work carried out by the teams in this study is related to the maintenance of medium-voltage power mains. Two units of power distribution were studied and are referred to in this chapter as units U1 and U2.

In general, the work on these units is as follows: Initially, the industry inspection of the electrical network notes its deficiencies and registers the system. Then the service description sector evaluates the service subscribed and directs it to the team of line-powered (or line-alive) units, as well as the team of the "de-energized" line. The technician identifies the system service, programs it, and passes it to the supervisors of the local units. The team performs the services, and the technician makes a recording of these services in the system. Another track of services in the units is originated by the department of emergency services, which prescribes services that are not prescheduled (such as cutting or redevelopment of energy) to electricians.

In total, 20 electricians work in units U1 and U2. In unit U1, there is a team with four electricians that performs its activities at 0700 to 1600. In unit U2, there are two work teams, one of them with 10 electricians who work from 0600 to 1500, and another with six electricians that perform their activities from 0800 to 1700. The electricians have 1 hour for lunch to be held in accordance with the work demand.

The organization of the teams in the field of work depends on the demand and may be composed of trios or quartets. Before leaving the units, a form of risk is filled in by components of these teams. When you arrive at the location of the intervention, a practice is systematically performed by members of all teams: the local programming of work. It is an analysis of the risk of the activity performed at the intervention site (next to the electrical pole).

During this practice, workers discuss the risks of services that will be implemented immediately and complete the form of prefilled risks in the units. After the activity, these electricians meet, again next to the electrical post where they were working, to discuss what was positive and what difficulties they encountered. This is called conversation post-task. Then the group returns to the unit or follows the other interventions.

10.4   METHOD BASED ON ACTUAL WORK

The research described in this chapter was developed by a team of three researcher-ergonomists who held two initial meetings with management in order to delimit the theme and geographical area of the study. Due to the demand of the company's managers, it was decided that the research would be carried out in two units of power distribution (U1 and U2).

The team of researchers then began a study of the structural organization of units U1 and U2, to understand how these sectors are divided, organized, and coordinated. Then a meeting was held with the teams of both units for the presentation and explanation of the project.

The next step of research was to organize visits to the two units, for monitoring and recording the operational activities with the aim to develop a library of real situations, which could be used in the training of novices in these units. The field visits were then carried out for 5 months, alternating between units U1 and U2: 1 day per week the team of researchers would be with unit U1, and the following week with unit U2.

Thus, the field follow-ups were carried out in 15 fields in 15 days in each of the two units. In total, there were 15 field visits, 8 in U1 and 7 in U2. The aim of the research was to register the real activity of electricians, but the team of researchers did not know exactly what were the most important moments to highlight during the filming, given the fact that they did not have experience with the electrical work and did not know the structural organization of the company. In this way, this question was openly placed with the workers in order to identify which moments were the most important of the activities and which deserved to be displayed in the training with the beginners. After a meeting with each of the two teams, three most important moments were identified for filming: the local programming of work, where the team of electricians meets at the side of the pole, immediately prior to the intervention, to talk about what needs to be done; the intervention itself; and the conversation post-task, where the team meets at the foot of the post to talk about the main points of the speech.

However, more than the film itself, the objective of this research was to generate learning situations between the workers, especially among the most experienced and the novice, in order to preserve the technical skills in the company. For this reason, the methods of confrontation (Mollo and Falzon 2004) play a key role because they allow the subject to express a part of his tacit knowledge. Thus, in addition to the footage of the activities themselves, the idea was to get the electricians to talk about what they were seeing in the filming. A fourth step of footage was then drawn up: the confrontation. This step was performed during 10 meetings over the course of 4 months in both units, and scheduled according to the availability of each of them. The videos of the three moments of previous footage were edited in Power Director 11 and placed in discussion with the group shot. The workers watched the videos, commented, and discussed the service performed in order to exchange experiences about the activity that appears.

At the beginning of the discussion, a volunteer was asked to comment on the service performed while viewing the video. After completing the video and terminating the reviews of the electrician volunteer, the other electricians were directed to do

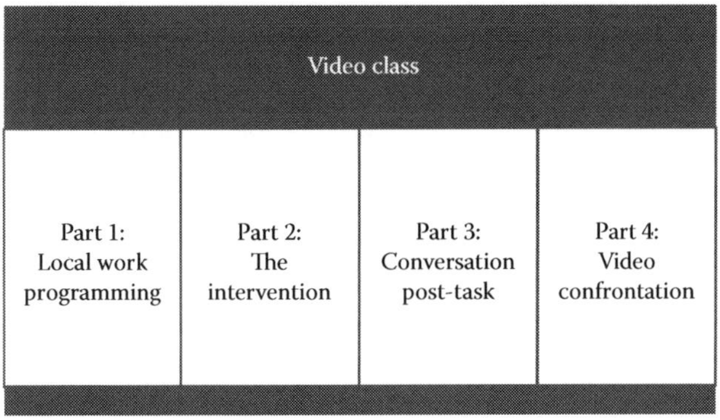

Video class			
Part 1: Local work programming	Part 2: The intervention	Part 3: Conversation post-task	Part 4: Video confrontation

FIGURE 10.1   Four parts of each video class.

the reviews that they judged necessary on the service. This time a discussion video of the field was also filmed, producing a complete step of confrontation. The material was again edited in Power Director 11 to produce a final video class that would have, in addition to the title of the task, the four moments explained above filmed, as shown in Figure 10.1.

The video classes were then recorded on DVDs and delivered in two pilot units. In total, 13 video classes were prepared, each containing the possible variations of activities found by electricians in the field (Figure 10.2). These video classes comprised the Library of Real Situations.

10.5   BENEFITS OF THE VIDEO LIBRARY OF REAL SITUATIONS

The methodology deployed produced video classes of everyday situations experienced by electricians, which fed a database called Library of Real Situations, allocated in the pilot units themselves. The results of this work are presented in two parts: first, we illustrate the composition of the Library of Real Situations, and then we show the benefits brought to units U1 and U2 by the methods developed.

10.5.1   COMPOSITION OF THE LIBRARY OF REAL SITUATIONS

As discussed above, the construction of the library was divided into two major steps: the acquisition of the savoir faire of electricians through videos of real situations in the field, and the discussion of these records with the workers in productive units. A practical case illustrates each of these steps below.

The team of electricians of units U1 or U2 goes out in the early morning, around 0730, to the site of the planned intervention. Arriving there, the electricians perform a visual inspection of the exact location of the operation to check the risks and opportunities for the implementation of the service (Figure 10.3). Then they plan the activity, so that the electrician that performs the service will comment on the steps that should be followed to achieve it, while the others discuss the implementation plan of intervention for it to be carried out efficiently and as safely as possible (such as the type of cover used and the best position in the truck). Then, the workers

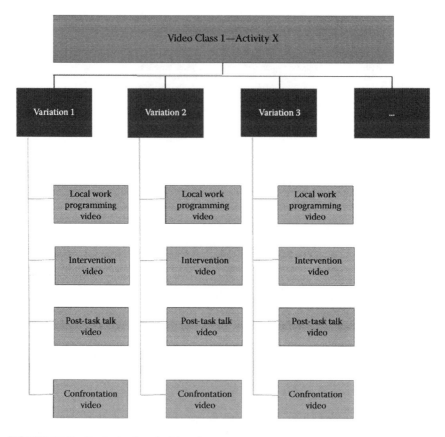

FIGURE 10.2   Contents of each video class.

FIGURE 10.3   Workers performing visual inspection of network (a), preparing the materials (b), running the activity (c), and conversing post-task (d).

FIGURE 10.4   Workers watching the edited videos to generate discussion.

prepare the materials and tools of work and climb the pole to run the scheduled service. After this, the electricians gather for the conversation post-task.

In all services accompanied by the researcher-ergonomists, the steps mentioned above were captured through videos and subsequently edited. Then they were displayed for the workers in the productive units in which they were made, so that they could be confronted with their own activity, and thus could discuss and generate knowledge for the group through that (Figure 10.4).

After that, the videos were reprinted so that this step could be added to the video class in question. At the end of the discussion of all the videos, they were recorded on DVDs, titled in accordance with the services monitored and recorded, and delivered to the production units.

In total, 13 services were accompanied by the researchers. In unit U1, the video classes built were of the following tasks: pruning of tree, sectioning of conductor, replacement of lozenge-shaped spacer, phase shift, transfer and distribution of driver, and withdrawal of temporary knife switch G1. In unit U2, the video classes built were of the following tasks: pruning of tree, installation of key SF6, installation of key temporary knife, jumper loose in structure EC3 with key knife, replacement of key fuse and lightning, withdrawal of key temporary knives G1 and G2, and withdrawal of key fuse with temporary change in the structure of M1 to M4 with installation of key knife.

For each of these activities, there were variations of the field not predicted. Consequently, strategies were deployed by electricians to manage these variabilities. Such strategies were only possible to perceive after the video confrontations, where the electricians could not explain part of the rules of the company and which were strategies of the group. Tables 10.1 and 10.2 illustrate the video classes built, with the name of each one (corresponding to the name of the task to be performed), standard procedures, and variables found in the strategies used by electricians to manage them.

10.5.2   BENEFITS FOR THE WORKERS AND THE COMPANY

The construction of the method and the confrontation with the images themselves brought a series of benefits for the local group, which was far beyond a collection

TABLE 10.1

Video Classes in Unit U1

Name	Standard Procedure	Variability and Used Strategies
1. Pruning of tree	• Check the location and positioning of the truck. • Evaluate the safety of the site (observe if there is any hornet's nest or electrical fence next to the tree). • Isolate the area. • Assess whether it is necessary to use the truck for one or two buckets and if you should use the hydraulic saw or handsaw. • Schedule the activity step-by-step. • Use the covers to isolate the network. • Perform the pruning using the hydraulic or manual saw, in accordance with planning. • Do the post-task talk.	• The electricians went out to perform the service without having all the information necessary for the activity or the site: the group was not aware, for example, of the size of the tree to be pruned. Then they went out with a truck and, when arriving at the site, had to use a handsaw instead of an electrical one to cut twig by twig. This is more time-consuming, but it is more secure, because they hold the branches without letting them touch the wires of low voltage (and thus to close short circuit).
2. Sectioning of conductor	• Check the exact location of the section of the conductor. • Prepare for the ascent by isolating the site from outside to inside. • Isolate the phase-out and the stage of the means. • Install a baton of in-phase sectioning inside. • Use the pliers voltmeter to see if there is load. • If there is no load, proclaim aloud that will cut. • After the cutoff, turn the cable ends to allow a perfect isolation. • Carry out the same procedure to the other phases. • Remove the covers. • Evaluate the service.	• The electricians realized that the street was closed, which would be inconvenient to the local traffic. As there was another scheduled service with a third company, they waited for the arrival of this team to see if there was any service in the post in front of the site of intervention. This way, the street would be closed only once and not at two different moments (by the electricians and the third team). • In the inspection, electricians observed that the spider is in good condition and did not need replacement. It was then decided that the section of the driver would be carried out in a previous post than the one programmed, where the spider was in precarious conditions. • During the execution of the service, a strong wind began. As the wind was not predicted, the electrician decided, for security, to install a spacer (material that prevents electrical wires from touching each other).

3. Replacement of lozenge-shaped spacer

- Carry out an initial thorough inspection: the cables that intervene usually have a problem in their upper part, a site that is difficult to see.
- Cover each of the stages with flexible cover for the conductor. Use a semiparted sheet so that the phases are covered, with the exception of the stage where the intervention will be carried out.
- Modify the sheet to protect phase by phase.
- Withdraw the linen and covers.
- Evaluate the service.

- The steel wire attached to the lozenge-shaped spacer was heavy, so it interfered with the completion of the arm swing of the spacer, causing a feeling of fatigue and tiredness. Then, he dropped the cable and put it on the pole, to lighten his load and thus be able to rotate the frame arm to perform the operation.

4. Phase shift

- Evaluate the conditions of the electrical network.
- Operate the key fuse with a baton and a load buster.
- Carry out the isolation of phase 3.
- Section phase 1.
- Replace the cable.
- Carry out the compression.
- Remove the isolation.
- Close the key fuse and proclaim it aloud.
- Evaluate the service.

- When they reached the site, the electricians observed that the electrical pole has lightning rods. According to them, in this case the activity is more complex because they must disconnect the lightning rods to perform the service.
- Another unusual situation is that the key, in this post, is different. It is a repeater key, and in these cases, it is necessary to have a greater concern in the wake of its opening, because technically, it is more difficult to open.
- The truck was parked on a street with a steep slope. This forced the electrician who was at the top of the bucket to perform the maneuvers with great care, because the bucket swayed too.

(Continued)

TABLE 10.1 (Continued)
Video Classes in Unit U1

Name	Standard Procedure	Variability and Used Strategies
5. Transfer and distribution of conductor	• Evaluate the conditions of the post. • Schedule the steps of the activity. • Isolate the network. • Section the conductor. • Clear the area for the third undertaking to complete the transfer of low voltage to another post. • Reconnect the conductor on the network.	• When arriving at the site, the electricians observed that there is a big box of bees in the post, making it infeasible to implement the activity. They evaluated, then, the possibility of carrying out the activity in other poles, realizing that they could accomplish this in a post different from that scheduled. Thus, the group contacted the company to reprogram the service in a different post. • The electrical pole in which the intervention would be held, although there has been no outbreak of bees, has a large amount of wireless telephone service. This makes it very difficult to move the bucket and increases the attention of the electrician, to avoid accidents on this harness in excess.
6. Withdrawal of temporary knife switch G1	• Ground the truck. • Install the bypass. • Withdraw key knife G1. • Reconnect the connector. • Carry out the procedure in 3 stages. • Isolate the previous phases with covers, to the extent that they are passing on to other phases. • Evaluate the service.	• There was no variable in the activity of the withdrawal of the temporary knife key of unit U1 on the day of the intervention. For this reason, the variable described in video class 6 of unit U2 was used.

TABLE 10.2

Video Classes from Unit U2

Video Class Name	Standard Procedure	Variability and Use Strategies
1. Pruning of tree	• Check the location and positioning of the truck. • Evaluate the safety of the site (observe if there is a hornet's nest or electrical fence next to the tree). • Isolate the area. • Assess whether it is necessary to use the truck for one or two buckets and whether it the hydraulic saw or handsaw should be used. • Schedule the activity step-by-step. • Use the covers to isolate the network. • Perform pruning using the hydraulic or manual saw, in accordance with planning. • Carry out the conversation post-task.	• There was no variable in this activity of unit U2 on the day of the intervention. For this reason, in this video class the variable described in the pruning of tree unit U1 was used.
2. Installation of key SF6	• Check the points by observing that there are no bad connections between them. • Raise with the key SF6. • Open key fuse. • Enter the circuit partitioning input cables of the key. • Isolate each of the phases with two whole sheets and two rigid covers per phase. • Remove the key phase. • Install 6 monophasic lightning rods, 3 on one side and 3 on the other. • Even with the isolated network, install the SF6 key and measure the jumpers. • Remove the covers and install the jumpers with alive claws. • Conduct the test with the Rypot. • Install the jumpers using a baton. • Put the push button in the "remote" position and evaluate the service.	• When the electricians were preparing the SF6 key to climb the pole, a grip was held on the key itself, before suspending it, instead of a grip on another product, as is prescribed by the company. This decreased the effort on the bushing key and, consequently, the force needed by the worker to press the drop-down key.

(Continued)

TABLE 10.2 (Continued)
Video Classes from Unit U2

Video Class Name	Standard Procedure	Variability and Use Strategies
3. Installation of key temporary knife	• Inspect the circuit. • Inspect the structure. • Isolate the wings outside and the means. • Brush the stage inside. • Install the key, performing a tightening of 30 N at each nut. • Install the spacer at this stage. • Proclaim aloud that the switch is installed and closed. • Evolve to the sectioning. • After authorization, make the cut and do the same procedure on all other stages. • Remove isolation. • Evaluate the service.	• This task is, in summary, performed by two electricians. However, the positioning of the bucket was compromised due to an excess of wires in the post. For this reason, the activity had to be performed by a qualified electrician.
4. Jumper loose in the structure of EC3 with key knife	• Inspect the safety conditions of the spans and structure. • Disconnect the jumpers using a baton. • Announce in a loud voice the disconnection of the jumpers. • If the distance is safe, the electricians don't have to isolate the network. Otherwise, the network must be isolated. • Replace the loose jumper. • Connect the jumpers on the network using a baton. • Announce in a loud voice the connection of the jumpers. • Remember that the entire time the cables are withdrawn or connected they must be witnessed by the staff of the soil. • Evaluate the service.	• The service was scheduled to repair a loose jumper. The electricians performed other services in addition to those programmed. After performing the inspection, they found the need to swap two key knives and maintain a third.

5. Replacement of key fuse and lightning

- Check the conditions of the structure.
- Open the key fuses.
- Disconnect the ray using a baton.
- Enter the structure by removing the top jumpers of the key fuse and isolating each phase.
- Use two rigid covers to protect the caliper of lightening rods.
- Remove the keys and defective lightening rods.
- Install new keys and lightening rods; reconnect the jumpers of the fuses, by removing the insulation from this side.
- Remove the insulation on the side of the lightening rods.
- Remove the low-voltage ground.
- Reconnect, keeping distance, the lightening rods using the baton.
- Close the fuse.
- Evaluate the service.

- After performing the inspection on the post, electricians realized the need to replace the insulator pin, which was burned, as well as the jumper. It was also observed that the conductor had signs of rupture, and therefore it was necessary to perform the shutdown of the medium-voltage grid in a previous pole, to minimize the risks.

6. Withdrawal of key temporary knives G1 and G2

- Ground the truck.
- Install the bypass.
- Withdrawal key knife G1 or G2 and reconnect the connector.
- Carry out the procedure in 3 stages.
- Isolate the previous phases with covers, to the extent that they pass on to the other phases.
- Evaluate the service.

- As electricians said, key G1 has a handle, which makes it more dangerous and difficult to remove, because if it is released by an electrician, it can meet another conductor and cause a short circuit.
- At the time of intervention, there were two contingencies. First, the electrician had difficulty positioning a sleeve on the cable, increasing the time of execution of the service. Electricians said that some types of cable hinder the placement of gloves. The other unexpected thing was the noise from the street, which made communication between the electricians who performed the operation and the support staff difficult. In this case, the strategy of electricians was to turn off the truck to facilitate communication.

(Continued)

TABLE 10.2 (Continued)
Video Classes from Unit U2

Video Class Name	Standard Procedure	Variability and Use Strategies
7. Withdrawal of key fuse with temporary change in the structure of M1 to M4 with installation of key knife	• Check the condition of the structure. • Isolate each of the stages. • Fit the suspension assembly. • Raise the 3 phases to release the spider. • Replace the spider of M1 with the one from M4. • Drop the phases for the new structure. • Remove the assembly using a semiparted linen cloth to untie and two whole sheets to cover the structure. • Install the disk insulators. • Raise the first phase. • Announce and install the bypass. • Announce and cut the stage. • Flip the cable ends in the direction of the line. • Isolate using the entire sheet. • Install the key knife. • Connect the ends of the cable and make the compression. • Remove the covers. • Carry out the same procedure on the middle and external stages. • Install the identification plate. • Evaluate the service.	• An unexpected thing identified was that the line had signs of rupture, called x. It was necessary, then, to repair this problem along with the scheduled service. • Another unexpected event occurred during the intervention: to make the change in the structure of M1 to M4, the sleeve remained toward the handle, which cannot happen. Then, this part of the task needed to be redone.

of videos to be used by experienced workers and rookies in productive units. The group of electricians of the two units were able to rethink the way of performing activities, begin discussing the matter of know-how and strategies used by them, help the beginners understand the dynamics of the fieldwork and their variability, and think about what was needed to pass on the knowledge, strategies, and essential elements that characterize the activity for rookies. Such benefits were identified by the workers themselves during the discussions and confrontations of the images of the work.

It was observed, in the course of the project, that the records on the everyday tasks made the teams of units U1 and U2 rethink the manner of implementation of the activities, especially regarding the safety of the process.

> The routine makes us forget the danger that is our activity, and often we do not think about how we are doing it. With everyone seeing what we do in the field, everyone ends up thinking more on how to do it.
>
> **—Electrician of U1**

The moments of discussions also generated knowledge for local groups, once the discussions about the know-how and strategies used in the activities were motivated.

> This time to watch the videos makes us evaluate if the service was well executed, if it was performed according to the standard, and if you have something that you should change. And it is great that people can think about this.
>
> **—Electrician of U1**

The library was also designed for new electricians coming to the sector that do not have experience in the field. These videos, with the help of technicians and supervisors, will help beginners understand the dynamics of the fieldwork and its variability, as reported by experienced workers of the productive units.

> Even if it is the same task, there is not a service that is equal to the other. The conditions, the streets, and the poles change, and these videos will give an idea to the person who is arriving to the company.
>
> **—Electrician of U2**

> For a rookie, the videos will be good, because they show the reality of the field. In the company, formal training is much easier because we don't see the problems of day-to-day street.
>
> **—Electrician of U2**

> Everyone makes things very easily, for example, tightening a nut, put a bolt, but who is starting has difficulty of how to get a nut with the sleeve 17, such as putting a clamp to secure the sheet right. And the library will greatly assist the rookies on it.
>
> **—Electrician of U2**

Finally, the developed method has instigated the workers to think about the need to pass on their knowledge, strategies, and the essential elements that characterize the activity for rookies who will come to the units, as a way to control the change of generations in the company.

10.6 DISCUSSION

The discussion of this work demonstrates at least two major points of reflection. First, we discuss how the method developed can be thought of as a practical tool for knowledge management to be reproduced in other organizations with similar characteristics to those described in this chapter. Then, we discuss the role of researcher-ergonomists for which this research would generate fruitful results.

10.6.1 Tool for the Management of Knowledge

In addition to the context of various organizations today, the company in this study faces a difficulty related to the change of generations and the loss of technical skills of operational workers, which happens at the same time that these workers retire. Also, the preparation and the replacement of experienced workers are not carried out properly, which can lead the company, in a few years, to lose the qualification of its technical body.

The training of rookie workers in the company studied was always carried out on the basis of theoretical situations and ideals of intervention. These trainings are not sufficient for the inclusion of electricians in real situations. The actual context of action involves various situations of variabilities and contingencies, to which, generally, a rookie had no contact during the course.

In virtually all activities accompanied by researchers, there were variations of activity previously scheduled due to unforeseen circumstances found in the places of intervention. Materials damaged, poles with excessive wires, poorly executed services by third parties, and unfavorable urban environments (narrow streets and with a slope, excessive noise, presence of boxes of bees, etc.) are some of the examples of the variables routinely encountered by electricians. In the face of such situations, they should manage the variabilities to continue production with efficiency and safety. Thus, they put into practice a series of strategies before and during the execution of the service: modification of materials used, maintenance of parts and damaged equipment, change of location of the intervention in the electrical network, and disconnection of cables not provided for are some of the examples of strategies used to increase the quality and safety of the operation.

For the reasons mentioned above, the methodological device built by workers could not capitalize on some real operational situations bringing benefits in the fields of safety and health and the development of workers' skills. To discuss a real-world situation, the workers analyze the risk, look at past experiences, and reflect on the possibilities of management. The confrontation of the videos was the moment in which the workers reflected on their work and discussed the environmental conditions, on the logic that was used to make the work plan, on the vagaries and difficulties faced during the implementation of the service, on the strategies used, and on the safety of the operation.

The training in the company is now based on practical situations experienced by operators. Different from the majority, these trainings do not address ideal situations produced in the laboratory, but discuss real-life situations: on the basis of the parts that are missing in the pole at the time of intervention, the solar reflex that prevents the display of a material in the conversation next to the pole, or the tool that does not find space to be used due to the excess of wires in the pole.

This way, the implanted device in this research can be an inspiration to the organization, to the extent that it allows constant learning among workers during the discussions of the views of the films. Considering the levels of Dreyfus and Dreyfus (1986), the purpose of the method was not to make a beginner (novice) an experienced worker through the discussion, but to make him or her more advanced and competent through the sharing of concrete situations.

10.6.2 ROLE OF THE ERGONOMIST DEVELOPING THE METHOD

One of the goals of research in ergonomics and human factors must be a reflection on the issues that arise from the practice and the researcher. This question arises through repeated demands of researchers faced with difficulties in the field and should be better explored in current research on the subject (Rocha 2014; Daniellou 1992). Thus, this item aims to reflect on what was important, from the point of view of the conduct of research by the researchers, and the results of this research.

The methodology was only successful because the team of three researcher-ergonomists participated actively in the process of the construction of the method, from the initial negotiations with the managers of the company and the units under study, through the constant presence in the operational field and confrontations in the units, to the construction of the library (filming, editing, and recording of videos). For this reason, and because it is an innovative method, barely studied or developed in work related to ergonomics, the method was being gradually discovered by researcher-ergonomists, in so far as the research has progressed and taken shape. How and what they shot, what content to edit, and how to confront the images with the workers were issues slowly addressed by the team on the basis of the errors and hits that appeared. Given the importance of this issue, some conditions will be exposed here for the implementation of this method and the role that the ergonomist must take in projects of this magnitude.

The process of filming went through various adjustments, and some general conditions were determined by the team of researchers for the implementation of the filming. If initially the staff tried to determine a maximum time of footage for each video class, during the execution of the work, it was noticed that this was very difficult; because the field activities are very private between the researchers and some of them, a longer period of filming was required.

Another important reference of the footage was what the workers (electricians and supervisors) considered important to shoot. If, initially, the researchers had some reference to the movies, they were modified in accordance with what the workers thought was important to film.

In addition, the three moments of filming (before, during, and after the intervention) had some peculiarities that are worth mentioning here. During the shooting of preparation of the work site, it was necessary to remain close to the electrician who speaks, to get better audio quality and vary the camera between the group and the intervention site. During the filming of the intervention, it was important to find the best angle at which professional gestures were visible, as well as vary the zoom between the professional gestures and the open camera, picking up the overall situation. Already during the conversation post-task, it was necessary to remain near the electrician who was speaking for the best audio quality, and vary the camera between the group and the intervention site already performed.

Finally, during the shooting of confrontation in the productive unit, it was found that it was important to film in front of the group in order to capture the expressions of the workers, but also in the back to see what they pointed at during the video and emphasize certain situations. In addition, it was necessary to encourage them or confront them, as triggering the initial discussion. At the beginning of the confrontation, the participants had a tendency to remain silent.

10.7 CONCLUSION

The results generated by this research allow us to discern the device co-built with workers as an inspiration for the organization, to the extent that it causes various learning situations between workers during the discussions. If several electricians are retiring and taking with them technical knowledge and tacit consent of the work, the library, and its method of preparation, an inverse process can be generated where the skills of the workforce are reflected and discussed among themselves, and thus will be developed within the organization studied.

Finally, the method developed was not born spontaneously and must be well structured before it is started. Practical experience has shown that at least three conditions are necessary for the implementation of the method of this research, conditions that, if not met, certainly jeopardize the benefits demonstrated in this chapter. These conditions are the engagement of the direction of the company in the project; the creation of a climate of confidence between the workers, and between workers and the team of researchers; and efficient communication between the production units and the sectors responsible for training in the company for the organization of the material to be used in training.

REFERENCES

Amaral, D. C. 2002. Arquitetura para gerenciamento de conhecimentos explícitos sobre o processo de desenvolvimento de produto. Doctoral thesis, Universidade de São Paulo, Brazil.

Assunção, A. A., and F. P. A. Lima. 2003. A contribuição da ergonomia para a identificação, redução e eliminação da nocividade do trabalho. In Mendes, R. (ed.), *Patologia do Trabalho*. 2nd ed. atualizada e ampliada. São Paulo: Atheneu, vol. 2, parte III, 1767–1789.

Assunção, A. A. A. 2004. Cadeirologia e o mito da postura correta. *Revista Brasileira de Saúde Ocupacional*, 29(110), 41–55.

Brown, J., and P. Duguid. 1991. Organizational learning and communities-of-practice: Toward a unified view of working, learning and innovating. *Organization Science*, 2(1), 40–57.

Choo, C. W. 2002. Sensemaking, knowledge creation, and decision making: Organizational knowing as emergent strategy. In Choo, C. W., and Bontis, N. (eds), *The Strategic Management of Intellectual Capital and Organizational Knowledge.* New York: Oxford University Press.

Collins, H. M. 2001. What is tacit knowledge? In Schatzki, T. R., Cetina, K. K., and Savigny, E. V. (eds), *The Practice Turn in Contemporary Theory.* London: Routledge, 107–119.

Collins, H. M. 2007. Bicycling on the moon: Collective tacit knowledge and somatic-limit tacit knowledge. *Organization Studies,* 28 (2).

Daniellou, F. 1992. Le statut de la pratique et des connaissances dans l'intervention ergonomique de conception. Thèse d'Habilitation à Diriger des recherches, Bordeaux, Éditions du LESC.

Daniellou, F., M. Simard, and I. Boissières. 2011. *Human and Organizational Factors of Safety: A State of the Art.* Number 2011-01 of the Cahiers de la Sécurité Industrielle. Toulouse, France: Foundation for an Industrial Safety Culture.

Dreyfus, H., and S. Dreyfus. 1986. *Mind over Machine: The Power of Human Intuition and Expertise in the Era of Computer.* New York: Free Press.

Figueiredo, V. G. C., and M. K. Figueiredo. 2012. Do controle físico à exigência cognitiva? A transição do saber-fazer entre o fordismo e o toyotismo e sua aplicabilidade na indústria têxtil. In *Congresso Brasileiro de Engenharia de Produção,* Ponta Grossa-PR, Brazil.

Guérin, F., A. Laville, F. Daniellou, et al. 2006. *Understanding and Transforming Work—The Practice of Ergonomics.* Lyon, France: ANACT.

Lave, J., and E. Wenger. 1991. *Situated Learning: Legitimate Peripheral Participation.* New York: Cambridge University Press.

Lima, F. P. A. 2001. A formação em ergonomia: reflexões sobre algumas experiências de ensino da metodologia de análise ergonômica do trabalho. In Kiefer, C., Fagá, I., and Sampaio, M. R. (orgs), *Trabalho–educação–saúde: um mosaico em múltiplos tons.* São Paulo: Fundacentro, 133–148.

Mollo, V., and P. Falzon. 2004. Auto- and allo-confrontation as tools for reflective activities. *Applied Ergonomics,* 35(6), 531–540.

Montmollin, M. 2005. Ergonomias. In Castilho, J., and Villena, J. J. (eds), *Ergonomia: Conceitos e Métodos.* 1st ed. Lisboa: Dinalivro, 103–111.

Moraes, A., and C. Mont'Alvão. 2000. *Ergonomia: conceitos e aplicações.* 2nd ed. Rio de Janeiro: Editora 2AB.

Nonaka, I., and H. Takeuchi. 1995. *The Knowledge-Creating Company: How Japanese Companies Create the Dynamics of Innovation.* New York: Oxford University Press.

Ribeiro, R. 2007. *Knowledge Transfer.* Doctoral thesis, School of Social Sciences, Cardiff University, UK.

Ribeiro, R. 2012. Tacit knowledge management. *Phenomenology and the Cognitive Sciences,* 12(2), 337–366.

Rocha, R. 2014. Du silence organisationnel au débat structuré sur le travail: les effets sur la sécurité et sur l'organisation. Doctoral thesis, University of Bordeaux, France.

Rocha, R., F. Daniellou, and V. Mollo. 2014. O retorno de experiência e o ligar dos espaços de discussão sobe o trabalho: uma construção possível e eficaz. *Trabalho e Educação,* 23(1), 61–74.

Rocha, R., V. Mollo, and F. Daniellou. 2015. Work debate spaces: A tool for developing a participatory safety management. *Applied Ergonomics,* 46, 107–114.

Slack, N., S. Chamber, C. Hardland, et al. 1998. *Operations Management.* 4th ed. London: Pitman.

Weill-Fassina, A., P. Rabardel, and D. Dubois. 1993. *Représentations pour l'action.* Toulouse, France: Éditions Octarès.

Wenger, E. 2000. Communities of practice and social learning systems. *Organization,* 7(2), 225–246.

11 Human Factors Analysis and Behavior Modeling for the Simulation of Evacuation Scenarios

*Verena Wagner, Konrad Wolfgang Kallus,
Norah J. Neuhuber, Michael Schwarz, Helmut
Schrom-Feiertag, Martin Stubenschrott, Martin
Pszeida, Stefan Ladstätter, and Lucas Paletta*

CONTENTS

11.1 INTRODUCTION

Mathematical models for the simulation of crowd behavior have been investigated as quantitative tools for demonstrating evacuation performance in the case of an emergency (Gwynne et al. 2009; Helbing et al. 2002; Sagun et al. 2011; Schadschneider et al. 2009). Different factors can influence an emergency situation—the main factors are the building, the environment, and human factors, and all of these should be taken into account in an evacuation simulation. Despite existing regulations, incidents have shown that taken measures might not be sufficient in the case of fire, and as discovered by Kobes et al. (2010), human factors play a major role in the outcome of evacuations and research shows growing interest in human factors and psychology (Hofinger et al. 2014). Therefore, the design and organization of evacuation systems can be seen as crucial factors to gain a positive outcome of such critical events. Within conflict situations, people's behavior deviates from the expected behavior, for example, neglecting to leave the danger zone as soon as possible. Lewin (1935) defines *conflict* as "a situation in which oppositely directed, simultaneously acting forces of approximately equal strength work upon the individual" (p. 122). Conflict situations within evacuation situations are, for example, other individuals who are still in the danger situation, personal belongings and valuables, and so forth.

Novel agent-based approaches that reflect decision-making processes and behavioral factors of individuals are based on assumptions due to the lack of empirical data on human behavior in evacuation situations (Veeraswamy et al. 2011). The presented work attempts to significantly improve evacuation simulation by parameterizing the human agents' behavior by means of a model that is based on human factors analyses on data captured in critical situations during evacuation exercises. A theoretical framework to describe behavioral aspects of human beings in critical situations represents the theory of behavioral inhibition (BIS) and behavioral activation (BAS) systems (Gray and McNaughton 2000). The theory is based on research of neurobiological processes of anxiety, whereby conflicting aspects of a situation are able to inhibit ongoing behavior of an individual (e.g., stopping and hesitant evacuation behavior). On the other hand, the BAS is responsible for actively facing difficult situations (e.g., the initial step to start with the evacuation process). Following Fowles (1980, 1988), the BIS can be associated with electrodermal activity. Additionally, the BAS can be associated with cardiovascular activity.

In addition, results of different research groups show that mobile eye tracking can be applied in order to study eye movements and attention for the purpose of revealing behavior-relevant cues (Assad 2003; Hayhoe and Ballard 2005; Onat et al. 2014), in particular, during critical situations in the evacuation scenario (Paletta et al. 2014).

Furthermore, the behavior of evacuation assistants (e.g., poise, unsafe behavior, and unsafe actions of the evacuation assistant) may have a decisive role on evacuees' behavior. Allison (1992) suggests that people tend to orientate toward persons who are perceived as powerful, successful, and representing high status. Accordingly, the more reliable the person appears, the more information about the appropriate behavior for the specific situation that is subjectively provided by him or her. In addition, Cohen and McKay (1984), as well as Cohen and Wills (1985), indicated positive effects of social support during stressful situations.

An experimental study (Wagner et al. 2015) investigated the research question of how the occurrence of an evacuation assistant influences the behavior, as well as the emotional state, of evacuees while acting in different conflict situations. The findings on human factors were used to develop an agent-based simulation model allowing an egress analysis or evacuation analysis that is able to reflect the frequently observed human social behaviors, through simulating the cognitive processes of individual agents and interactions among neighboring agents in the simulation environment.

This chapter is organized as follows: First, the method and the results of the experimental study are described and discussed (Section 11.2). Section 11.3 explains the perception and behavior modeling based on human factors analysis and the utilization of an agent-based simulation to simulate individual and social behavior. A set of experimental tests (simulation study) were conducted to validate the final agent-based simulation by comparing the simulation results with empirical observations of the field study. Section 11.4 summarizes the chapter and presents potential fields of application.

11.2 FIELD STUDY

In this experimental study (for more details, see also Wagner et al. 2015), 23 untrained volunteers (11 women, 12 men) participated. The participants were between 19 and 56 years old (average age of 28.09 years; standard deviation [SD] = 10.27). A multilevel approach was chosen to combine assessments of the subjective emotional state (adjective checklist [Janke et al. 1986] and a short questionnaire), as well as objective psychophysiological responses of the study participants (cardiovascular and electrodermal activity and eye tracking). Therefore, more than half of the participants were equipped with the portable VARIOPORT system of Becker Meditec to record cardiovascular (electrocardiogram [ECG]) and electrodermal activity. Also, four participants were equipped with wearable eye tracking glasses (ETGs) 1.0 from SensoMotoric Instruments™ (SMI). The exact distribution can be seen in Wagner et al. (2015, p. 1797).

To stay as close to reality as possible, the experimental study was investigated in a realistic environment. The participants were naïve to the research question in that they did not know they would have to evacuate before the fire alarm was presented. The intensity of the conflict situation was varied (two different conflict situations: low and high). As described by Wagner et al. (2015), the first trial was intended to simulate a conflict situation that should activate the BIS, whereas the second trial was intended to simulate a situation in which an active response was required and therefore should activate the BAS. Furthermore, the occurrence of the evacuation assistant was manipulated within the different trials (e.g., with or without evacuation assistant, and without evacuation assistant or with evacuation assistant with unsafe or poised occurrence).

11.2.1 Study Results and Discussion

11.2.1.1 Behavioral Inhibition System (BIS)

One parameter of electrodermal activity, the frequency of nonspecific spontaneous fluctuations (NS.SCR), indicates an activation of the BIS if participants faced a conflict situation of high intensity without an evacuation assistant, $F(1,9) = 8.85, p = 0.016$ (interaction conflict × evacuation assistant) (Wagner et al. 2015). Participants who

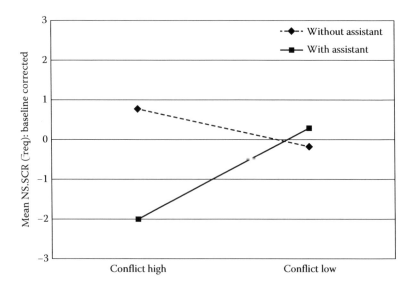

FIGURE 11.1 Interaction conflict × evacuation assistant: mean values of NS.SCR. (From Wagner, V., et al., *Procedia Manufacturing* 3:1799, 2015.)

faced a high-conflict situation without an evacuation assistant showed a significantly higher frequency of nonspecific spontaneous fluctuations than participants who faced the same situation with an evacuation assistant. This effect was not observable under the low-conflict situation (see Figure 11.1).

The results are in line with the research results of Fowles (1980, 1988), who postulates an association between BIS and electrodermal activity.

The subjective ratings support the results of the psychophysiological parameters: participants with an evacuation assistant reported significantly better mood, tended to be more balanced, and perceived less "anger and frustration" and "goal blocking" than participants that evacuated without an evacuation assistant. Overall, the results suggest that the occurrence of an evacuation assistant acted as a stabilizing factor and safety signal in this uncertain and ambiguous situation. The detailed analyses are described by Wagner et al. (2015, p. 1799).

11.2.1.2 Behavioral Activation System (BAS)

The second part of the experimental study was designed to show a possible activation of the BAS during evacuation situations. Therefore, the participants were faced with a conflict situation in which a (active) departing from the group was required due to tasks, although an evacuation assistant was present. The results of the cardiovascular activity, as well as the subjective data, support a stronger activation of the BAS in this situation (for more details, see Wagner et al. 2015).

11.2.1.3 Occurrence of the Evacuation Assistant

As presented by Wagner et al. (2015), the behavior of the evacuation assistant seems to be an important factor for evacuation situations. The results of the electrodermal

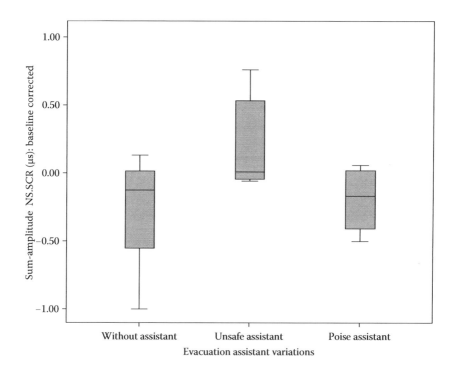

FIGURE 11.2   Effect variation of the occurrence of the evacuation assistant: sum-amplitude of NS.SCR.

activity (parameter sum-amplitude of NS.SCR) indicate a stronger activation of the BIS if participants evacuated with an evacuation assistant with unsafe occurrence than if participants evacuated without an evacuation assistant (see Figure 11.2).

Also, the study of Kinateder et al. (2014) already demonstrated that social influence may have positive and negative effects. Especially, the passive behavior of others can hinder a safe evacuation.

11.2.1.4   Visual Perception

To investigate the influence of evacuation assistance on the visual attention, a measurement methodology for relating fixation distributions to object perception and behavioral relevance, as proposed by Paletta et al. (2014), was implemented. The approach calculates the distribution of visual fixations for the participants, gathered by data from the eye tracking glasses, relative to evacuation assistance relevant objects such as exit signs, path markings on the floor, and evacuation assistants. Analysis has shown that participants focus on other persons, like evacuation assistants or other participants, while exit signs are mostly ignored (see also Wagner et al. 2015) (Figure 11.3).

The fixation distribution of participants relative to specific objects of interest in the environment (person, exit sign, ground floor) was visualized using Gaussian-filtered scatter plots (see Figure 11.4). These plots statistically represent the visual attention of participants by means of the distribution of the collected fixation data

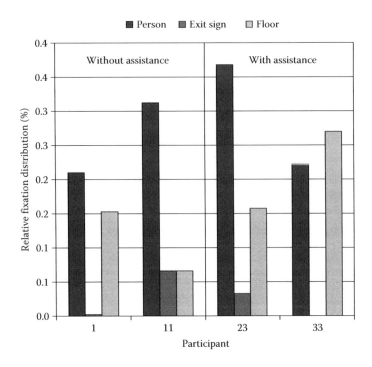

FIGURE 11.3 Relative fixation distribution for the evacuation assistance objects (person, exit sign, and floor). (From Wagner, V., et al., *Procedia Manufacturing* 3:1801, 2015.)

relative to either exit signs or persons. For example, Figure 11.4a depicts the filtered distribution of fixations with respect to the occurrence of an exit sign in the field of view (FOV). In total, the frequency on exit signs is rather low in comparison with the frequency on persons, as depicted in Figure 11.4b.

11.2.2 SUMMARY: FIELD STUDY

Overall, the results give important indications to improve evacuation situations by avoiding critical situations in which persons tend to become unconfident and therefore become incapable of action (Wagner et al. 2015). The results strongly advise against the assignment of evacuation assistants who show unsafe occurrence and unsafe actions during the evacuation. To ensure a safe evacuation process, the application of well-trained evacuation assistants is emphasized. However, further studies should address the possible negative effects of an evacuation assistant with unsafe occurrence.

Analysis of eye tracking data shows that fixations are focused on social cues, such as on the appearance of persons, especially on evacuation assistants that have been introduced as authorized persons before the study. This suggests that a generalization of the outfits of evacuation assistants is highly important to make them recognized as authorized persons, and from this focus, the attention of the evacuees is on appropriate social cues.

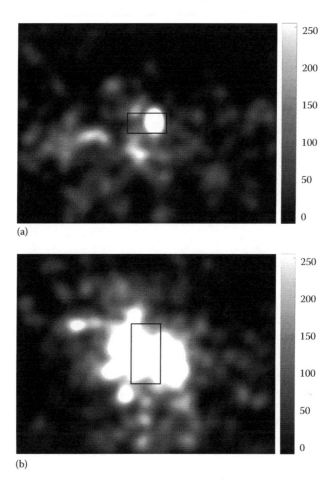

FIGURE 11.4 Gaussian-filtered scatter plots of distributions of points of regard (PORs) relative to artificially centered areas of interests—(a) exit sign and (b) person—in the FOV. The visualization shows brighter regions as areas where participants tended to focus their gaze, with gray values to the right, respectively.

11.3 AGENT-BASED SIMULATION

The findings on human factors analysis served as a basis for the modeling of an evacuation simulation by parameterizing the human agents' behavior with information on human factors.

In the simulation, an individual is represented as an autonomous agent who interacts with the virtual environment and with other agents. The autonomous agent simulates individual behavior, follows the sense–plan–act paradigm (Gat 1998), and is composed of three subsystems: a perception model describing the perception of the surroundings (sense), a behavior model for the selection of a situation-aware behavior (plan), and a locomotion model as the realization of the behavior resulting in the agent motion (act).

LISTING 11.1   BASIC ALGORITHMIC STEPS
OF THE SIMULATION LOOP FOLLOWING
THE SENSE–PLAN–ACT PARADIGM

Main Loop

while an assembly point is # or the simulation time has run out
not reached:

update sensor;	# create list of surrounding objects and agents
perceive surrounding,	# calculate attention probabilities for the objects
make the decision;	# select appropriate behavior and action
update next destination;	# compute new position to move
execute the movement;	# check collision and update agent's location

end while

The basic algorithmic steps of the cognitive agent and its interactions among neighboring agents and obstacles in the virtual environment are shown in Listing 11.1.

The perception and behavior models were developed and parameterized by the correlation between the multisensory perceptual and psychophysiological data collected during the field study and are described in detail in the following sections.

11.3.1 PERCEPTION MODEL

The implementation of well-grounded cognitive aspects for the agent-based simulation model was computed using several parameters from the field study. A model of human cognition, based on human perception, visual attention, and orientation aspects, was used to map these parameters to the autonomous agent in the simulation. Figure 11.5 shows the overall scheme and specific components of the developed human perception model that interfaces to the autonomous simulation agent. Perception of the agent is in principle constructed from the actual view of the

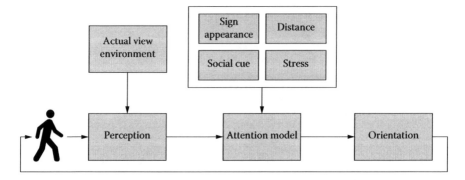

FIGURE 11.5 Interface between the human perception model and the agent-based simulation.

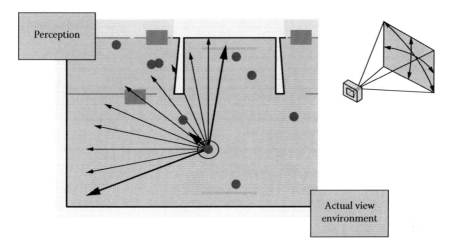

FIGURE 11.6 Sample representation of a perception model of the agents (FOV). The representation is displayed as a two-dimensional abstraction of the three-dimensional frustum.

autonomous agent on its simulation environment (see Figure 11.5). Depending on objects in the surroundings (e.g., exit signage and other persons), their distance to the agent, the occurrence of social cues (single person or group of persons), and eventually, indication of stress of the agent, the attention model is parameterized and a final distribution of saliency on the FOV toward the simulation environment is generated. The direction of maximum saliency is then determined and forms the orientation recommendation that is delivered to the agent for further processing.

The orientation as a key output of the perception model was continuously computed for a filtered FOV of the agent. Figure 11.6 shows a sample of this filtered FOV representation, as a two-dimensional abstraction of the three-dimensional frustum. The perception model is based on the visibility and the distance to an object, as well as social cues. Visual constraints for the agent were set using the ETG constraints, distance constraints from learned parameters of the field study. For example, an exit sign may be visible in the FOV of the agent, but if it is more than 12 m away, it is possibly too small to perceive and therefore filtered from the FOV. The goal of the human perception model was to assign a saliency value, which means a value of importance regarding the agents' perception to each object in the FOV of the agent.

For the simulation, the attention-based activation within the FOV of an agent was represented by a 20-bin histogram of angular orientations, as shown in Figure 11.7. The entire FOV was constrained to 160° horizontally; therefore, each bin of the FOV histogram represented 8° of the FOV. For each object in the FOV and within a distance limit, a saliency distribution was activated, being centered at the bin that connects to the object center, and contributes with its saliency distribution to an overall final distribution of saliency with respect to the bin. Figure 11.7 depicts a sample development of one-dimensional saliency based on a Gaussian FOV distribution and several object-specific distributions. The process results in an aggregated probability distribution for gazing at specific objects, such as signs and person, in the FOV. For each bin, the final probability is stored and the vector is transferred to the agent.

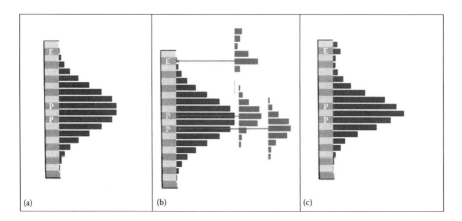

FIGURE 11.7 Construction of a sample one-dimensional saliency distribution. Based on (a) center-based Gaussian saliency distribution with respect to the FOV elements, one can add, based on the occurrence of objects (exit signs) and persons, (b) several object-specific distributions. (c) Integration and normalization of the resulting distribution leads to an aggregated distribution of the current saliency.

11.3.2 BEHAVIOR MODEL

Based on the findings from the field study, a set of decision rules that describe human decision making in emergencies was extracted and modeled in a decision tree providing a systematic reasoning structure. The resulting decision tree, as shown in Figure 11.8, extends the decision tree of Pan (2006) with behavior rules for exit signs. A decision process can be described as choosing the particular behavior based on sensory input and internal psychological and sociological factors. The nonleaf node represents a condition, and a leaf node represents a behavior decision. The rules are encoded in the following order: First, if the agent sees an assembly point, it walks toward the assembly point. If the agent is accompanied by a leader, it follows the leader. If the agent sees an exit, it goes toward the familiar exit or the closest exit if several exits are available and queues at the exit if it is jammed. If the agent cannot see any exit, it searches for exit signs and follows the instructions. If no signage can be found, the agent starts to explore the environment by searching from room to room for an exit, leader, or exit sign.

The results of the field studies have shown that trained leaders are preferred over signage because it is assumed that leaders have complete knowledge about the environment and know how to evacuate from the danger zone. Especially under stress, this behavior is reinforced because in such circumstances persons tend more to need and accept leadership than in normal conditions.

Each behavior decision will trigger the execution of a specific behavior routine. *Go to goal point* steers an agent toward a goal point. *Follow leader* lets an agent move forward by following another agent. *Queue at exit* lets an agent move to an exit and queue up when the exit is jammed. *Follow instructions* lets an agent follow

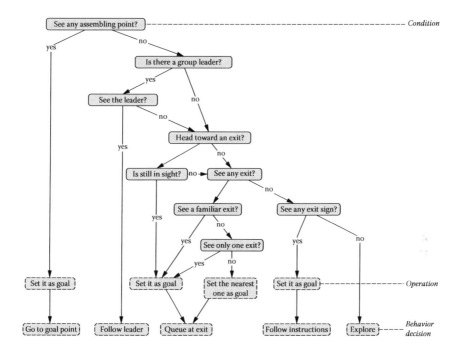

FIGURE 11.8 Decision tree as systematic behavior reasoning structure.

the instructions recognized from the signs. *Explore* lets an agent walk in the virtual environment looking for clues like exits, leaders, or signs.

11.3.3 LOCOMOTION MODEL

The locomotion model executes the behavior routine, which corresponds to a behavior decision selected by the behavior model, like *go to goal point*, *follow leader*, *queue at exit*, and so forth. A behavior routine is composed of a sequence of steering and motion behaviors, to navigate the cognitive agent among neighboring agents and obstacles through the virtual environment. Pedestrian motion is handled on two different levels. First, a path to the next point of interest is found by building a regular grid of the infrastructure and searching for the quickest path using the Theta* algorithm (Nash et al. 2007), which yields obstacle-free waypoints to get to the goal. Second, the movement to each waypoint is modeled by a simple social force model after Helbing and Molnar (1995), where opposing forces from other pedestrians and walls are combined with an attractive force toward the next waypoint to steer the agent.

11.3.4 SIMULATION STUDY

The integration of the perception, behavior, and locomotion models in a cognitive agent made it possible to handle the physiology, stress, perception, emotion, and reproduced human behavior that arise during emergencies. Such advanced simulation models can exploit their full potential only if they are linked to empirical investigations of emergency situations that would provide the appropriate parameters to describe behavior. Therefore, the conducted field studies were modeled in the simulation environment and the simulations results were compared with the empirical observations. The simulation should be able to reproduce the impact of the appearance of an evacuation assistant and also reproduce the behavior in conflict situations, such as departing from the group of evacuees or reaching an exit blocked by dense smoke.

11.3.4.1 Simulation Scenario 1

In Scenario 1 of the field study, the participants had to find a safe place independently without guidance from an evacuation assistant, using the signage only. Figure 11.9 shows the layout of the test area and the expected path of the test planning. Exit U was blocked by dense smoke, and one participant had to depart from the group of evacuees to evacuate another person from the bathroom. In the simulation, a group of agents is placed at random near the starting point in area B (see Figure 11.9). Each agent has to find its way to a safe place on its own without any collective behavior. A single agent was placed in the bathroom (WC), and another agent was instructed to find the bathroom and evacuate the agent from the bathroom to a safe place. In that scenario, Exit U was blocked with dense smoke and barrier tape.

The field study of the scenario without an evacuation assistant was conducted two times with different groups of four people at starting point B and one person at WC. Figure 11.10 shows the escape routes of participants in the field study. The tracking was done using a iBeacon indoor positioning system providing an position accuracy of 1 m. Both groups successfully found a safe place (Exit W) in a direct way. Also, the person in the bathroom was quickly found and evacuated to the safe place W. An interesting observation of group 1 was that a single participant ignored both smoke

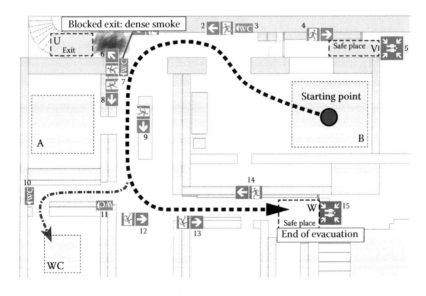

FIGURE 11.9   Description of simulation Scenario 1.

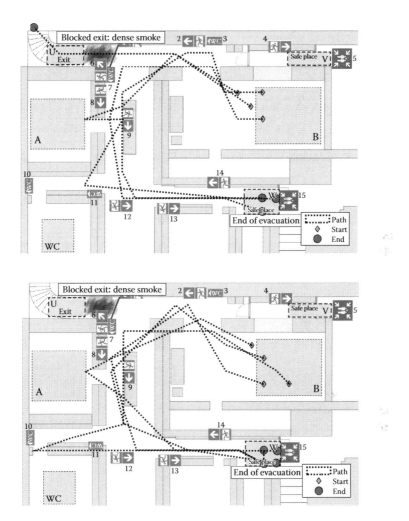

FIGURE 11.10   Observed escape routes of Scenario 1 from two participant groups of the field studies.

and the barrier tape and tried to escape the building via Exit U. In the simulation, we did not consider this option, but closed the exit completely for this scenario.

The simulation run of this scenario was repeated 10 times to see the impact of the stochastic variables in the simulation, which include slightly different start positions, different perception thresholds for the exit signs, and different desired speeds of the pedestrians. Additionally, the number of active agents was increased from four to eight to get even more varying results. Two exemplary results are shown in Figure 11.11. While all participants in the field study evacuated to safe place W, in the simulation it could be observed that between one and four agents evacuated alternatively to safe place V. The reason is the agent behavior in the exploration mode, not only looking forward, but also looking around them every few seconds. The left picture in Figure 11.11 shows that some agents start heading toward Exit U, but once they

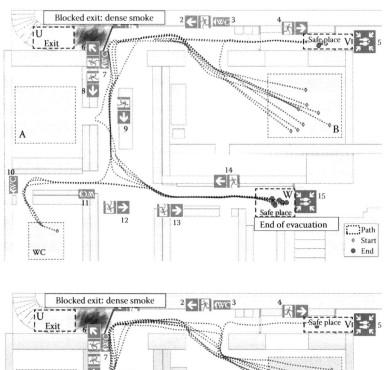

FIGURE 11.11   Simulation results of Scenario 1.

see that that exit is blocked, they take a look around and see safe place V. Most agents (even more visible in the simulation run in the right picture of Figure 11.11) continue and follow the instructions of Signs 7, 12, and 13 (see numbering of sign position in Figure 11.11) toward safe place W. On average, only one person per simulation run took a quick peek into room A to find information or a safe place, which is consistent with the observations of the field study. Agent 8, assigned with the task of finding the bathroom and evacuating the person (Agent 9), had no problems finding safe place W.

Listing 11.2 shows the detailed strategy of these two agents. All relevant signs toward the bathroom (WC) were seen with a high probability, between 98% for

**LISTING 11.2   SEARCH STRATEGY OF AGENT
ASSIGNED WITH THE TASK TO EVACUATE
ANOTHER AGENT FROM THE BATHROOM**

0:07.700 Agent 8 saw sign 3 'WC' (x:5.69,y:-1.92,z:0.20,angle:-71.38) with attraction:0.000

0:10.900 Agent 8 saw sign 3 'WC' (x:3.11,y:-0.88,z:0.20,angle:-74.24) with attraction:0.982

0:10.900 Agent 8 saw sign 2 'Exit' (x:2.80,y:-1.17,z:0.20,angle:-67.32) with attraction:0.000

0:10.900 Agent 8 found clue on sign 3 (dist:4.19,attr:0.98,needed:0.38) to get to WC: Go left

0:14.100 Agent 8 saw sign 7 'WC' (x:4.44,y:2.41,z:0.20,angle:-28.47) with attraction:0.000

0:14.100 Agent 8 saw sign 1 'Exit' (x:3.46,y:0.26,z:0.20,angle:-85.71) with attraction:1.000

0:14.100 Agent 8 saw sign 8 'Exit' (x:4.31,y:2.89,z:0.20,angle:-33.86) with attraction:0.000

0:17.200 Agent 8 saw sign 7 'WC' (x:1.96,y:0.71,z:0.20,angle:-19.78) with attraction:0.857

0:17.200 Agent 8 saw sign 8 'Exit' (x:2.03,y:1.20,z:0.20,angle:-30.68) with attraction:0.000

0:17.200 Agent 8 found clue on sign 7 (dist:3.06,attr:0.86,needed:0.00) to get to WC: Go right

0:20.200 Agent 8 saw sign 9 'Exit' (x:2.45,y:0.26,z:0.20,angle:6.06) with attraction:0.719

0:20.200 Agent 8 saw sign 12 'Exit' (x:8.90,y:0.13,z:0.50,angle:-89.15) with attraction:0.000

0:23.200 Agent 8 saw sign 12 'Exit' (x:6.00,y:-0.66,z:0.50,angle:-83.73) with attraction:0.000

0:26.200 Agent 8 saw sign 11 'WC' (x:1.66,y:-1.78,z:0.20,angle:-59.18) with attraction:0.680

0:26.200 Agent 8 saw sign 13 'Exit' (x:3.33,y:3.63,z:0.20,angle:-42.55) with attraction:0.287

0:26.200 Agent 8 saw sign 12 'Exit' (x:2.96,y:-0.64,z:0.50,angle:-77.85) with attraction:1.000

0:26.200 Agent 8 found clue on sign 11 (dist:3.25,attr:0.68,needed:0.28) to get to WC: WC

0:34.100 Agent 9 found agent 8 to follow: following 'guide' (agent 8,distance 2.97)

0:46.300 Agent 8 saw sign 14 'Exit' (x:6.18,y:-12.47,z:0.20,angle:-26.36) with attraction:0.000

0:46.300 Agent 8 found TARGET 'Exit W' (dist:18.37,attr:0.70,needed:0.08)

Sign 3 and 68% for Sign 11. When exiting the bathroom with the evacuated agent, safe place Exit W was also found with a high probability of 70%, which was over the threshold of 8% for this particular agent and exit.

11.3.4.2 Simulation Scenario 2

Scenario 2 demonstrates an evacuation assistant with an unsafe occurrence. Figure 11.12 shows the layout of the test area with the expected escape route of the test planning. Exit U was again blocked by dense smoke, and the previously known safe place W was obviously made unsafe. In this simulation scenario, four agents start in area A and have to evacuate to a safe place whereby one agent acting, as an evacuation assistant (trained leader), leads the others to the presumably safe place W. When arriving to W, they notice an acrid smell, which leaves the evacuation assistant confused. Upon this, the three other agents lose confidence in the evacuation assistant and start searching for an alternative safe place. The evacuation assistant itself changes the role from a leader to a follower and starts following one of the other agents (see Figure 11.12).

The field study of this scenario was again carried out with two groups of four people each. The escape routes can be seen in Figure 11.13, showing that both groups did not have difficulty in finding alternative safe place V. One participant in this study ignored the dense smoke and left the building through Exit U.

Therefore, in this simulation scenario, we allowed a possible exit through the smoke via the stairs. Two representative simulation runs are depicted in Figure 11.14. In the image on the left, all four persons left the infrastructure through the smoke to Exit U. In the simulation run on the right, only one person left the simulation through that exit and three (including the evacuation assistant) gathered in the safe place V.

Listing 11.3 shows the important landmarks that were seen by the three cognitive agents in this simulation run (Agent 4 is not shown, as he just followed Agent 3).

The following phenomena can be observed when looking at the decision making of the agents:

1. Both Agents 1 and 3 see and follow Sign 9, which leads them back to where they have come from. The agents still follow that advice, as there might be another safe place down that way. After searching the bathroom, they continue to the top of the room, where they finally see Exit V. Although the agents have seen Sign 1, which points to the nearer Exit U, they continue to Exit V because that can been seen directly by the agents, and seeing the target directly always takes higher precedence over following signs that just deliver clues to the target.

2. At minute 1:31.1, Agent 1 sees both Signs 1 and 8 with different directions (the order "Go left" is interpreted relative to the sign orientation). The current approach is to take the information on the nearest sign, as that usually shows the information to the nearest exit.

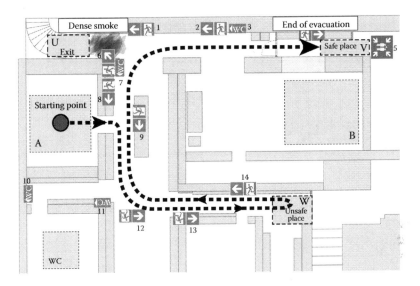

FIGURE 11.12   Description of simulation Scenario 2.

11.3.5   DISCUSSION SIMULATION RESULTS

In Scenarios 1 and 2, cognitive agents have to find the safe place on their own, selecting a behavior based on the processed sensory input by the perception model or simply following an evacuation assistant. The results are promising and have shown that the cognitive agent is capable of reproducing empirically observed behavior, and thus allows simulation of the behavior of safe and unsafe evacuation assistants. The stochastic nature of the simulation model also leads to different simulation results for multiple simulations, which also matches the behavior seen in the field study.

The main challenge is to understand decision making in situations where agents do not find any signs and experience uncertainty of being lost. This depends on context, situation, and environment, but also on the ability of people to quickly learn and get a basic understanding of space—cognitive mapping—even in unfamiliar places (Golledge 1999). In the simulation, every room is equally interesting for the agent and is searched with the same likelihood in the exploration mode. This is in contrast to reality, where people are attracted by certain areas, like wider or brighter hallways (Vilar et al. 2009), or it is more likely that people try to find the entry where they came from to exit the building (Graham and Roberts 2000; Sandberg 1997), but also avoid certain areas; for example, people have learned that it is virtually ruled out that an escape route leads through a dark box room. While it would be possible to manually assign a positive or negative attractiveness to such rooms beforehand, it must be ascertained that the values are appropriate for all kinds of different contexts. For instance, in Scenario 1, the task to find the bathroom requires a different strategy of exploring the area than searching for one of the exits, which are usually located along hallways. Additionally, the attractiveness is not constant during the entire evacuation and may change dynamically due to the ability of people to learn the understanding of space quickly.

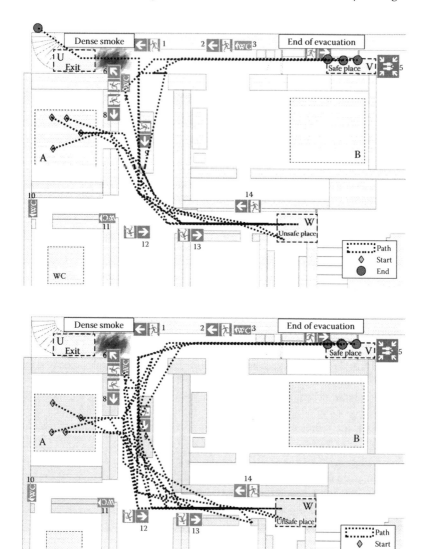

FIGURE 11.13 Observed escape routes of Scenario 2 from two participant groups of the field studies.

11.4 OVERALL SUMMARY AND DISCUSSION

In a field study, human behavior in evacuation situations was investigated by means of innovative human factors analyses. The results of the study delivered quantitative measures describing the behavior and the visual perception in a realistic environment. The results of the field study strongly advise the assignment of well-trained

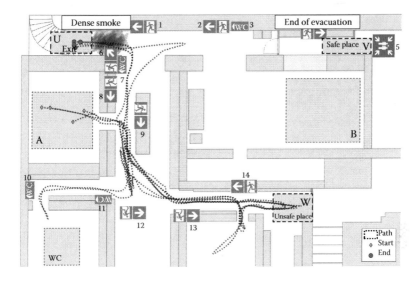

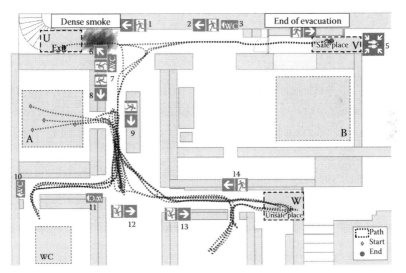

FIGURE 11.14 Simulation results of Scenario 2.

evacuation assistants to reduce the risk of critical situations and prevent persons from becoming unconfident and therefore incapable of action.

Furthermore, human factors analysis served as a basis for the modeling of an evacuation simulation by parameterizing the cognitive agents' behavior with information on human factors.

The developed autonomous agent, composed of a perception model, behavior model, and locomotion model, was capable of reproducing empirically observed human behavior and enabled simulation of scenarios with a high degree of realism to evaluate alternative evacuation scenarios and strategies efficiently, especially the

LISTING 11.3   AGENT DECISIONS FOR
SIMULATION OF SCENARIO 2

1:15.7 Agent 2 found clue on sign 8 (dist:6.57,attr:0.66,needed:0.03) to get to
Exit: Go left

1:20.5 Agent 1 found clue on sign 9 (dist:3.48,attr:0.60,needed:0.34) to get to
Exit: Go right

1:21.8 Agent 3 found clue on sign 9 (dist:2.69,attr:0.69,needed:0.54) to get to
Exit: Go right

1:21.8 Agent 3 found clue on sign 1 (dist:8.66,attr:0.77,needed:0.20) to get to
Exit: Go left

1:31.1 Agent 1 found clue on sign 8 (dist:7.86,attr:0.84,needed:0.76) to get to
Exit: Go left

1:31.1 Agent 1 found clue on sign 1 (dist:10.32,attr:0.75,needed:0.04) to get to
Exit: Go left

1:33.1 Agent 3 found clue on sign 8 (dist:6.88,attr:0.96,needed:0.46) to get to
Exit: Go left

2:01.8 Agent 2 found clue on sign 1 (dist:4.62,attr:0.72,needed:0.57) to get to
Exit: Go left

2:06.6 Agent 2 found TARGET 'Exit U' (dist:3.82,attr:1.00,needed:0.53)

2:22.2 Agent 3 found clue on sign 2 (dist:5.60,attr:0.96,needed:0.63) to get to
Exit: Go left

2:22.7 Agent 1 found TARGET 'Exit V' (dist:12.33,attr:0.99,needed:0.19)

2:22.7 Agent 1 found clue on sign 4 (dist:9.72,attr:0.72,needed:0.05) to get to
Exit: Go right

2:28.6 Agent 3 found TARGET 'Exit V' (dist:13.98,attr:1.00,needed:0.57)

assessment of the impact of evacuation assistants. As described by Pelechano and Badler (2006), trained people know how to evacuate a dangerous location because they are familiar with the environment and other people will follow them. This results in a decrease of evacuation time as the number of trained agents in the environment increases, and an optimal percentage of trained people, with only 10%, were identified.

The developed perception model of the agent computes the visual perception of the environment, which is very beneficial to evaluate the placement of guidance signs. This will enable us to test sign placements for various scenarios, for example, different locations where the crisis could start, by means of evacuation simulation.

As a result, the simulation supports improved placement of guidance components, impact assessment of evacuation assistants, capacity analysis of escape routes, the identification of critical areas, and the duration of evacuation.

ACKNOWLEDGMENT

This work has been partially funded by the Austrian Research Promotion Agency by grant n°836270 (EVES) and grant n°832045 (FACTS).

REFERENCES

Allison, P.D. 1992. The cultural evolution of beneficent norms. *Social Forces* 71(2):279–301.

Assad, J.A. 2003. Neural coding of behavioral relevance in parietal cortex. *Current Opinion in Neurobiology* 13:194–97.

Cohen, S., and G. McKay. 1984. Social support, stress and the buffering hypothesis: A theoretical analysis. In *Handbook of Psychology and Health*, ed. A. Baum, S.E. Taylor, and J.E. Singer, 253–67. New York: Hillsdale.

Cohen, S., and T.A. Wills. 1985. Stress, social support, and the buffering hypothesis. *Psychological Bulletin* 98(2):310–57.

Fowles, D.C. 1980. The three arousal model: Implications of Gray's two-factor learning theory for heart rate, electrodermal activity, and psychopathy. *Psychophysiology* 17(2):87–104.

Fowles, D.C. 1988. Psychophysiology and psychopathology: A motivational approach. *Psychophysiology* 25(4):373–91.

Gat, E. 1998. On Three-layer architectures. In *Artificial Intelligence and Mobile Robots*. eds D. Kortenkamp, P. Bonasso, and R. Murphy. Menlo Park: AAAI Press.

Golledge, R. 1999. *Wayfinding Behavior: Cognitive Mapping and Other Spatial Processes*. Baltimore, MD: Johns Hopkins University Press.

Graham, T.L., and D.J. Roberts. 2000. Qualitative overview of some important factors affecting the egress of people in hotel fires. *International Journal of Hospitality Management* 19(1):79–87.

Gray, J.A., and N. McNaughton. 2000. *The Neuropsychology of Anxiety: An Enquiry into the Functions of the Septo-Hippocampal Systems* (2nd ed.). New York: Oxford University Press.

Gwynne, S., Kuligowski, E., Kratchman, J., and J. Milke. 2009. Questioning the linear relationship between doorway width and achievable flow rate. *Fire Safety Journal* 44(1):80–87.

Hayhoe, M., and D. Ballard. 2005. Eye movements in natural behavior. *Trends in Cognitive Sciences* 9(4):188–94.

Helbing, D., Farkas, I.J., Molnar, P. and T. Vicsek. 2002. Simulation of pedestrian crowds in normal and evacuation situations. *Pedestrian and Evacuation Dynamics* 21(2):21–58.

Helbing, D., and P. Molnar. 1995. Social force model for pedestrian dynamics. *Physical Review E* 51:4282–86.

Hofinger, G., Zinke, R., and L. Künzer. 2014. Human factors in evacuation simulation, planning, and guidance. *Transportation Research Procedia* 2:603–11.

Janke, W., Debus, G., Kallus, K.W., Hüppe, M., and L. Schmidt-Atzert. 1986. *Befindlichkeitsskalierung nach Kategorien und Eigenschaftswörtern (BSKE (EWL)-ak-24)*. Würzburg: Julius-Maximilian-Universität.

Kinateder, M., Müller, M., Jost, M., Mühlberger, A., and P. Pauli. 2014. Social influence in a virtual tunnel fire—influence of conflicting information on evacuation behavior. *Applied Ergonomics* 45(6):1649–59.

Kobes, M., Helsloot, I., de Vries, B., and J.G. Post. 2010. Building safety and human behaviour in fire: A literature review. *Fire Safety Journal* 45(1):1–11.

Lewin, K. 1935. *A Dynamic Theory of Personality*. New York: McGraw-Hill.

Nash, A., Daniel, K., Koenig, S., and A. Felner. 2007. Theta*: Any-angle path planning on grids. In *Proceedings of the AAAI Conference on Artificial Intelligence*, 1177–83.

Onat, S., Açık, A., Schumann, F., and P. König. 2014. The contributions of image content and behavioral relevancy to overt attention. *PLoS One* 9(4):1–19.

Paletta, L., Wagner, V., Kallus, K.W., Schrom-Feiertag, H., Schwarz, M., Pszeida, M., Ladstaetter, S., and T. Matyus. 2014. Human factors modeling from wearable sensed data for evacuation based simulation scenarios. In *Advances in Applied Digital Human Modeling*, ed. V. Duffy, 70–78. USA: AHFE Conference.

Pan, X. 2006. Computational modeling of human and social behaviors for emergency egress analysis. PhD dissertation, Stanford University.

Pelechano, N., and N.I. Badler. 2006. Modeling crowd and trained leader behavior during building evacuation. *IEEE Computer Graphics and Applications* 26(6):80–86.

Sagun, A., Bouchlaghem, D., and C.J. Anumba. 2011. Computer simulations vs. building guidance to enhance evacuation performance of buildings during emergency events. *Simulation Modelling Practice and Theory* 19(3):1007–19.

Sandberg, A. 1997. Unannounced evacuation of large retail-stores. An evaluation of human behaviour and the computer model Simulex. Master thesis, Lund University, Sweden.

Schadschneider, A., Klingsch, W., Klüpfel, H., Kretz, T., Rogsch, C., and A. Seyfried. 2009. Evacuation dynamics: Empirical results, modeling and applications. In *Encyclopedia of Complexity and Systems Science*, ed. R.A. Meyers, 3142–76. New York: Springer.

Veeraswamy, A., Galea, E.R., and P.J. Lawrence. 2011. Wayfinding behavior within buildings—an international survey. *Fire Safety Science* 10:735–48.

Vilar, E.P., Rebelo, F., and L.M.B. Teixeira. 2009. The influence of guidance system in indoor wayfinding using virtual reality. In *Proceedings of the First International Conference on Integration of Design, Engineering and Management for Innovation IDEMI09*, 1–9, Porto, Portugal.

Wagner, V., Kallus, K.W., Neuhuber, N.J., Schwarz, M., Schrom-Feiertag, H., Ladstaetter, S., and L. Paletta. 2015. Implications for behavioral inhibition and activation in evacuation scenarios: Applied human factors analysis. *Procedia Manufacturing* 3:1796–803.

12 Development of an Interactive Educational Game to Learn Human Error

In a Case of Developing a Serious Game to Acquire Understanding of Slips

Midori Inaba, Ikuo Shirai, Ken Kusukami, and Shigeru Haga

CONTENTS

12.1 INTRODUCTION

Human errors are believed to contribute to most accidents or incidents in diverse activities, such as aviation (Wiegmann and Shappell 2001), driving (de Winter and Dodou 2010), rail (Edkins and Pollock 1997; Reinach and Viale 2006), and health care (Arnstein 1997; Carthey et al. 2001). Moreover, many of these human errors comprise well-trained acts that may be appropriate in other situations. Previous studies have reported that such skill-based errors account for 61% of all human errors (Davies et al. 2000; Reason 1990). One type of skill-based error, a slip, is an error in which individuals make a response incorrectly, while the other type of skill-based error, a lapse, is a lack of required reactions (Reason 1990). Because a slip is fundamentally a correctly trained act, further training is reported to have little effect in preventing the error. Instead, the first step against making slips is to comprehend what they are and their mechanisms (Airbus 2005).

However, there seem to be few programs in which individuals can effectively learn the mechanisms of a slip. People often acquire knowledge about human error in conventional ways, such as by attending lectures or reading textbooks. Compared to a lecture-type approach, Kolb (1984) suggested that an "experiential learning approach" is superior in terms of helping learners comprehend the mechanism of events. An experiential learning program provides educational content to learners by presenting an opportunity to have an experience that is relevant to the content and that engages them in the learning process. Nevertheless, there are some problems with application of the experiential learning approach to an educational program about slips. First, learners must encounter danger or trouble when they make a slip in the program, as is true when a slip is made in a real-life situation. Moreover, the program must create an environment for learners to make a slip naturally. Learners may become vigilant and avoid making slips, as oftentimes they may feel uncomfortable making an error. On the other hand, the setting of the program to force them to make an error, such as high task difficulty, might lead learners to attribute why they make slips just to the task difficulty.

To cope with these problems, we introduced the format of a serious game into the development of experiential learning programs regarding slips. Serious games are a genre of games used for purposes other than solely enjoyment (Arnab et al. 2012; Pourabdollashian et al. 2012; Susi et al. 2007). Recently, serious games aimed at safety education have been developed (Haga et al. 2013; Tesei et al. 2012). Serious games present several advantages, one of which is that they enable learners to learn about risks without exposing themselves to real danger (David and Watson 2010; Gee 2003; Aldrich 2009; Gibson et al. 2006). In addition, learners can become immersed in a game that allows them to react to risks themselves. These learner-centered experiences are expected to more effectively raise learners' awareness about the consequences of their actions (Garris et al. 2002). With the goal of designing a game that anyone can use to learn about slips through direct experience, the present study examined what factors in a game cause learners to make slips.

12.2   BASIC FEATURES OF THE GAME AND
EXPERIMENTAL FACTORS EXAMINED

Our game is named *Detect Aliens!* In the game, learners play the role of an officer at a security checkpoint, and must detect a few alien invaders who are trying to conquer the earth. The learners' task is to match a presented image against a list of specific visual characteristics of the aliens, decide whether the image is an alien or a human, and press the appropriate button. The target error in this game was erroneously pressing "human" when an alien appeared. Previous studies refer to slips as a failure in attentional management, not a failure in judgment (Reason 1990). Additionally, Rasmussen (1983) suggested that slips are caused when the routine for actions toward a given stimulus is activated toward a different stimulus. Such reduced attentional control over routine actions is often considered to occur when responding to infrequent stimuli among frequent stimuli (Norman 1981). Therefore, the relative frequency of aliens in the game was reduced compared to those for humans. In detail, our game consists of several sets of trials, and each set includes a trial in which an image of an alien is presented. In the other trials in the set, images of humans are presented.

In addition to reducing the frequency of aliens, this study manipulated three factors relative to the game settings that can contribute to the occurrence of a slip.

12.2.1   TASK DIFFICULTY

According to the Yerkes–Dodson law, learners are predicted to focus attention on a given task, resulting in fewer slips when the task difficulty is moderately high (Teigen 1994). An excessively easy task may reduce the learner's arousal and attention; however, easy tasks are also accompanied with a lower error rate in general. We carried out an exploratory investigation of the appropriate level of task difficulty for inducing slips in the game. Task difficulty was controlled using two components of the game settings. The first was the categorization rules to decide whether a character was an alien or a human. Each character image was composed of four facial features: eyes, nose, mouth, and hair. For each feature, several types of image were prepared (Figure 12.1a), and certain criteria regarding the facial features were used to decide whether the character was an alien or a human. Making these criteria more complicated was assumed to increase the task difficulty. The other component influencing task difficulty was the visual discriminability of aliens from humans, that is, the similarity of each of the facial features that composed the character images. Less discriminability among images increases task difficulty.

In addition, the probability of the slips was assumed to become higher when learners feel the task is easy and become inattentive to the differences between the

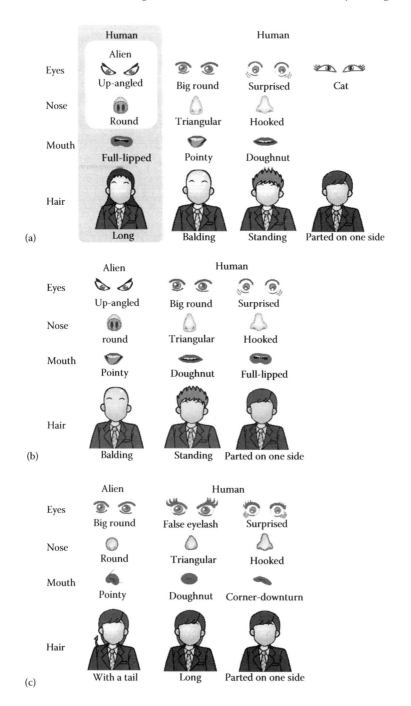

FIGURE 12.1 Facial features of an alien or a human. (a) Experiment 1, (b) Experiment 2, (c) Experiment 3.

humans and aliens. Thus, we additionally examined the impacts of consistency of task difficulty within the game, that is, whether the task difficulties are the same or different among sets of trials. Conditions where the visual discriminability of the sets of trials differed were compared.

12.2.2   LENGTH OF TIME LIMITS FOR EACH TRIAL

Baars (1992) suggested that a slip sometimes reflects the speed–accuracy trade-off, and therefore that increasing speed of responses contributes to slips. Time pressure is predicted to cause participants to emphasize speed over accuracy in their responses and miss the aliens. We set 3 s as the control condition for time pressure in each trial.

12.2.3   CYCLE OF PRESENTING IMAGES OF ALIENS

Slips in response to infrequent events are considered to be due to the difficulty in remaining attentive for when these infrequent events happen. Even if the frequency of the events is the same, however, the difficulty in keeping vigilant to them is likely to differ according to the cycle by which the infrequent events occur. For example, participants might be reminded to stay alert to aliens when aliens appear infrequently, yet regularly. However, participants' attention to the aliens is predicted to decrease if aliens appear frequently at first and then become more infrequent toward the end of the game. In this study, we manipulated the interval of aliens' appearances and examined the effects of the cycle of images.

12.3   EXAMINATION OF GAME SETTINGS TO INCREASE NUMBER OF PARTICIPANTS' SLIPS

12.3.1   PILOT STUDY

The main purpose of the pilot study was to select the image sets to be used in Experiment 1 and that would have a higher probability of inducing slips in participants. We compared the probability of slips in several patterns of image sets.

12.3.1.1   Methods

As Figure 12.1a shows, we set the following rules regarding categorization of aliens and humans. When a character had (1) up-angled eyes or (2) a pig nose, it should be categorized as an alien, but there were also two exception rules; if characters had (3) a full-lipped mouth or (4) long hair, they should be categorized as human, even if they also had up-angled eyes or a pig nose. We created 9 types of alien faces and 20 types of human faces that suited these criteria by combining facial features. These facial images were assigned to nine image sets. Thus, each set consisted of 10 images: 9 humans and 1 alien.

The nine image sets were shown to participants once or twice each over a total of 300 trials. The alien image in each of the nine image sets was presented once in every 10 trials (Figure 12.2). In each trial, a stimulus and a bar showing the time limit to respond were displayed on the screen. The time limit bar shortened as time

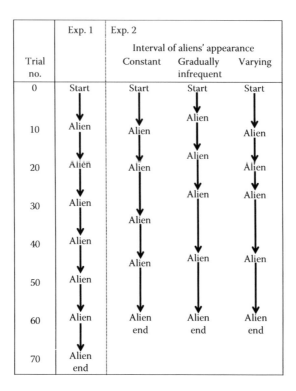

FIGURE 12.2 Interval of aliens' appearance. The image of a human was presented in trials between images of aliens (depicted by arrows).

passed, and the next trial would begin automatically if the time limit was run past with no response from the participant. In the pilot study, the time limit was 3 s.

A total of 58 undergraduate students (age range 18–24 years old) participated in the pilot study. Participants were instructed to choose, using buttons, whether the image was an alien or a human.

12.3.1.2 Results

Results showed that 80% of participants incorrectly categorized an alien as a human at least once. The percentages of participants who made incorrect responses to an alien were over 20% in most image sets. However, this percentage was below 5% for the two image sets where only using rules 1 and 2 applied (i.e., where the exceptions did not apply). These results suggest that application of the exception rules increased participants' probability of responding to aliens incorrectly. Based on this result, we decided to use the rest of the seven image sets that used all four rules and had a higher probability of incorrect responses to aliens in the next experiment.

12.3.2 EXPERIMENT 1

Experiment 1 was carried out to confirm if the game setting decided in the pilot study would cause participants to make a slip error in the game. In this experiment, we measured three variables. One was the percentage of participants with incorrect responses toward aliens. The second was the average percentage of correct and incorrect responses toward aliens and humans. We estimated the percentage of responses toward each of the two types of images for each participant, and calculated the average of these percentages. If a slip was caused by a failure of attentional control over routine acts, the participants were predicted to make few incorrect responses to humans that appeared frequently. The third variable was the response time. If the observed errors were slips, the response time for incorrect responses to aliens was predicted to be shorter than the time for correct responses, as making correct responses toward infrequent aliens was assumed to require much more cognitive processing to inhibit the more frequent act of judging an image as human.

We also examined the impact of the time pressure. We predicted that greater time pressure would promote participants' tendency to emphasize speed in their response and contribute to more failures of attentional control. This may be reflected in an increase in the number of participants making incorrect responses to aliens and a shortened response time in these responses.

12.3.2.1   Methods

Seventy images of aliens and humans selected in the pilot study were presented. Similar to the pilot study, seven images of an alien were provided, once in every 10 trials (Figure 12.2). The time limit for each trial was 3 or 2.5 s. Seventy-eight adults participated (52 men and 26 women, age range 20–59 years old) in either a 3 s or 2.5 s condition. Methods other than those mentioned above were the same as in the pilot study.

12.3.2.2   Results

Figure 12.3a shows the percentage of participants who committed an error in detecting aliens at least once. For the 3 s time limit condition, 67% of participants categorized an alien as a human at least once. However, against our prediction, the percentage was lower in the 2.5 s condition (28%). The average error rates toward aliens and humans are shown in Figure 12.4a. This figure shows the error rates for both aliens and humans were over 60%. In addition, we counted time-out errors and found that the percentages of these errors were 10% for the 3 s condition and 46% for the 2.5 s condition.

Figure 12.5a depicts the average response time for correct and incorrect responses to images of humans and aliens. Participants responded more slowly in correct responses than in incorrect responses toward aliens for the 3 s condition. For the 2.5 s condition, the response times were shorter in general, and there was no significant difference between correct and incorrect responses toward aliens.

12.3.2.3   Discussion

The accuracy of discrimination was generally low. In particular, the error rate for humans was higher than that for aliens, even though the responses to frequent images of a human should have been much easier than those to infrequent images of an alien. Many participants remarked that the exception rules were too logically complicated to discriminate aliens and humans in a short duration. The participants

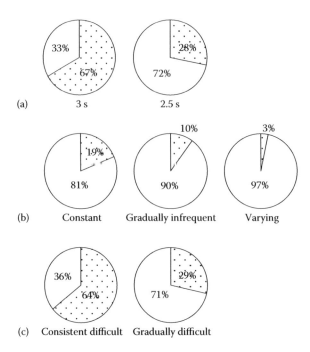

FIGURE 12.3 Percentage of participants who committed an error in detecting aliens at least once. (a) Experiment 1, (b) Experiment 2, (c) Experiment 3.

in the present experiment were adults, whereas those in the pilot study were under-graduate students, who might have been more accustomed to thinking about the game's logical puzzles. Because the target users of our game are adults working in risky operations, we decided to modify the categorization rules to be simpler in future experiments, without using exception rules.

Two types of impacts were observed with regard to the time limit. First, the shorter time limit generally lowered the accuracy of discrimination for 28% of the participants in the 2.5 s condition. They made many incorrect responses and time-out errors. The shortened time limit appeared to heighten the confusion of the par-ticipants when they struggled with the difficult task. The other type of impact of the shorter time limit is reflected in the result that 72% of the participants in the 2.5 s condition never missed an alien. However, several participants, 70%, miscat-egorized a human as an alien. Considering the discrepancy from the percentage that responded "aliens" to images of humans in the 3 s condition (20%), the participants might not actually be judging these images to be aliens, but might rather be adopting a different strategy than we expected given the shorter time limit. Considering the framework of signal detection theory (Salind 2006; Stanislaw and Todorov 1999), participants might have loosened their criterion to press "alien." The story of the game was to detect aliens as an officer for the purpose of safety and protecting humans' lives. Therefore, in line with the game's story, the action to be most avoided as the officer was to fail to correctly detect aliens. Due to the greater time pressure, it is possible that participants considered the priority of the task according to the story and took a strategy to avoid missing the aliens. Because this strategy decreases the probability of making incorrect responses to aliens, it was necessary for the design of our game to prevent participants from adopting this strategy. Therefore, we concluded the 3 s condition to be superior to the 2.5 s condition for inducing slips in participants.

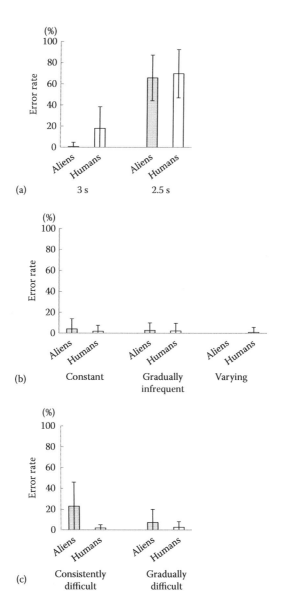

FIGURE 12.4 Average error rates toward aliens and humans. (a) Experiment 1, (b) Experiment 2, (c) Experiment 3.

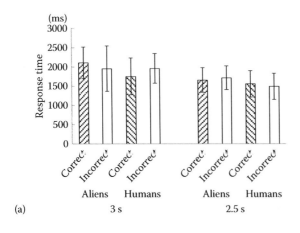

(a)

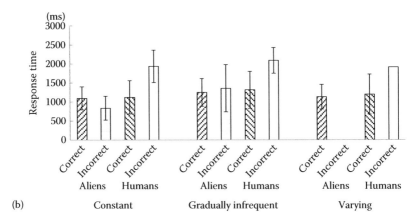

(b)

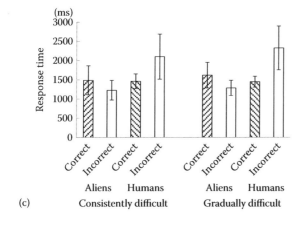

(c)

FIGURE 12.5 Average response time for correct and incorrect responses toward aliens and humans. (a) Experiment 1, (b) Experiment 2, (c) Experiment 3.

It should also be noted that we thought that the number of trials should have been reduced to better provide a serious game that learners can easily play. A few participants took more than 15 min to play the game with 70 trials and became exhausted. It might disturb participants' learning of the mechanism of slips if participants attribute their errors made during the game to simple fatigue. However, a certain number of trials are necessary to make participants feel the routine of responding to images of a human. We reduced this number to 60 trials in total, and also reduced the probability of alien images appearing.

12.3.3 EXPERIMENT 2

In Experiment 2, we addressed two factors that may increase the chance of participants making a slip. One factor was the categorization rules, which were modified to be simpler. The second was the cycle of presenting alien images, which was controlled by the interval that aliens would appear. We hypothesized that the greater the intervals between aliens, the less alert participants would be, resulting in more slips. The measurements to examine the effects of these factors were the same in Experiment 1.

12.3.3.1 Methods

We defined an alien as having (1) up-angled eyes, (2) a pig nose, (3) pointy lips, or (4) a balding head (Figure 12.1b). Five types of alien faces were created by composing facial features that included one of the parts mentioned above. Images of humans were made using features other than the four alien-specific features. We prepared five image sets. The number of each image set differed depending on the conditions of the interval of alien appearances, but every set included one image of an alien, and other images were humans. The five sets were presented once for each participant, for a total of 60 trials.

There were three conditions of the intervals of aliens' appearances (Figure 12.2). One was a constant condition, in which the number of images in each image set was nearly 10. In the other two conditions, the gradually infrequent condition and the varying condition, two of five image sets included less than 10 images and two included more than 10 images, with the remainder including 10. The order of presentation differed for the two conditions. In the gradually infrequent condition, the size of the image set increased consistently from the beginning to the end of the game. In the varying condition, the largest image set was posited in the middle of the overall trial.

A total of 88 participants (81 men and 7 women, age range 20–59 years old) were divided into three groups. Each group was assigned one of the three conditions of interval setting. The time limit in each trial was 3 s. Methods other than those mentioned above were the same as in Experiment 1.

12.3.3.2 Results

As shown in Figure 12.3b, only 19% of participants categorized the image of an alien as a human in at least one trial in the constant condition. The percentage was even lower in the other two interval conditions. Figure 12.4b depicts the average error rates to aliens and humans. Accuracy generally increased compared to Experiment 1. In addition, the average percentage of time-out errors for all participants was 0.009%.

The pattern of response time for each response type was similar among the conditions (Figure 12.5b). In particular, participants in the constant condition pressed the button faster for incorrect responses than correct responses to aliens.

12.3.3.3 Discussion

The modified categorization rules reduced participants' erroneously categorizing humans as aliens. Compared to Experiment 1, the percentage of participants who made incorrect responses to aliens was greatly decreased. However, we recognized that rules in the constant condition have the potential to cause players to make a slip. The average response time for incorrect responses was shorter than the response time for correct responses when an alien appeared, while the response time for correct responses to humans was shorter than for incorrect responses to these humans. This finding indicated that participants needed to increase their cognitive processing load to respond correctly to the alien differently from the human. Such cognitive processing is considered to involve attentional control over infrequent events. Without it, the participants might have responded "humans" to aliens. This kind of incorrect response appears to fit Reason's idea that slips may be caused by a failure of attentional control (Reason 1990).

The results about the impact of interval between aliens' appearances did not support our hypothesis that participants in the gradually infrequent group would make the most slips among three interval conditions. One reason could be that the increased alertness caused by more frequent aliens in the earlier sets of the game lasted until the end of the game. Alternately, it is also possible that the participants in the constant condition became less vigilant to aliens because they might have found it easy to respond to aliens appearing in a fixed cycle. However, the cycle of aliens' appearances was not exactly constant. There were small differences in the number of trials among the four intervals and five images of an alien. Such differences seemed to induce participant slips. Thus, we concluded that the constant condition for the interval between alien appearances was the most preferable condition for causing participants to make a slip.

Thus, the final remaining problem was the low percentage of participants who were able to experience a slip. In the next experiment, we examined the effects of task difficulty depending on the discriminability of stimuli, and the consistency of the difficulty throughout the game.

12.3.4 EXPERIMENT 3

In the final experiment, we investigated the impact of the consistency of a set amount of task difficulty. We continued to use simple categorization rules without exception rules, and adjusted the discriminability of images between aliens and humans. Impacts of the consistency of task difficulty were examined by comparing two conditions. One was a consistently difficult condition, in which the discriminability of aliens and humans was stable among sets of trials. The other was the gradually difficult condition, where aliens and humans were more easily discriminated in earlier sets of the trials, and images became more similar as the trials progressed. We predicted that the latter condition would induce slips more than the former. Learners might feel it is easy to discriminate between humans and aliens in the earlier sets, and this lingering, reduced vigilance might result in slips particularly in the later sets, where aliens that are less distinguishable from humans appear.

12.3.4.1 Methods

The aliens' facial features were designed by modifying one of two types of human facial features (Figure 12.1c). Aliens' faces were composed of four parts, including any one part unique to aliens. We defined hard-to-discriminate human faces as those that included any facial part that resembled an alien part, and defined easy-to-discriminate human faces as those that lacked such a part. In the consistently difficult condition, the number of hard-to-discriminate humans was greater than the number of easy-to-discriminate humans in every image set. In the gradually difficult condition, the earlier image sets mostly contained easy-to-discriminate humans and the later image sets contained mostly hard-to-discriminate humans. A total of 60 adult participants (55 men and 5 women, age range 18–58 years old) were randomly allocated to either of the two conditions. Other settings, such as the number of trials in each set and the interval between alien appearances, were the same as in the constant interval condition of Experiment 2.

12.3.4.2 Results

The percentage of participants who incorrectly responded to aliens was 64% in the consistently difficult condition and 29% in the gradually difficult condition (Figure 12.3c). As shown in Figure 12.4c, the participants seldom categorized humans incorrectly. The average response time for each response type is shown in Figure 12.5c. Similar to the results for Experiment 2, the response time for aliens was shorter for incorrect responses than for correct responses in both conditions.

12.3.4.3 Discussion

We interpreted participants' incorrect responses to aliens observed in this experiment to be slips, based on the accurate responses for humans and the shorter response time toward aliens for incorrect responses than correct responses. The decrease in discriminability between aliens and humans increased the probability that participants would make slips in the game. Task difficulty is generally considered to increase participant errors, but it also can motivate them to focus their attention on the task, ultimately resulting in fewer errors. The participants appeared to be less sensitive to control of the task difficulty when visual discriminability was manipulated than when the categorization rules were made more complex.

A clear impact of consistency of task difficulty was apparent in the percentage of participants making slips, with the percentage of slips higher for the consistently difficult condition than for the gradually difficult condition. This finding did not support our prediction that participants would have reduced concentration in the later trials of the game when they felt the discrimination task was easy in the earlier trials. Examining the results in detail, we found that many participants made consistent slips under two conditions. First, many participants responded incorrectly to the first of the five aliens (Figure 12.6). This result may show that the repetitive responses

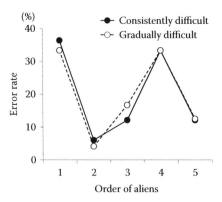

FIGURE 12.6 Comparison of percentages of incorrect responses to each alien between conditions (Experiment 3).

as "humans" in the first 10 trials made participants make an error the first time they were required to make a different response.

The percentage of missing aliens was also higher for the fourth alien. Misses to this alien seem to occur due to lowered vigilance toward infrequent stimuli. The accuracy in trials toward aliens increased after participants responded incorrectly to an alien. Even though the correct answer was not shown to them after every trial, they might have realized that they had missed an alien immediately after they made their response, and become more alert toward the next alien. However, it may be difficult for the participants to sustain their increased alertness after repeated success, such as the correct detection of the second and third aliens.

Nevertheless, these interpretations do not explain the impacts of the consistency of task difficulty. We speculate that the higher ability of the consistently difficult condition to induce slips is attributed to the average higher task difficulty of this condition, in which hard-to-discriminate humans were more numerous than in the gradually difficult condition.

12.4 TEST OF THE EFFECTS OF EXPERIENCING A MENTAL SLIP ON TRAINING FOR SLIP ERRORS

We introduced a serious game into the learning program about slips because we assumed that learners' firsthand experience of making a slip would promote their learning about this type of error. Here, we examined the degree of participants' knowledge about slips, depending on whether they made slips in the game. In more detail, the participants in Experiment 3 were provided with a short explanation about the slips after they performed the game and saw the results of their performance. At the end of the experiment, participants took an unannounced short test about the content of the error explanation. We analyzed the relationship between participants' performance on the test and their experience in making slips during the game.

Although confirming the effects of our game on inducing slips can be performed with individuals in any field, the effects related to learning should be tested with the expected users of our program, and the explanation should include commentary specific to those users' field. The main users of programs about human error are expected to be individuals engaged in safety-sensitive jobs. In this study, we gained the participation of railway employees, and tailored our explanation of slips to be more specific to railway workers.

12.4.1 Methods

Participants were 60 railway employees, whose performances in the game were reported earlier in Section 12.3.4 on Experiment 3. Following the game, the number of participants' missed aliens was shown on the monitor, and the short explanation about slips was subsequently presented. This explanation described slips and their mechanism in reference to previous studies (Reason 1990; Norman 1981). A part of the explanation was, for example, "a slip, the error you made in the game, is considered to be caused by a failure to control a routine action toward frequent 'humans' when infrequent 'aliens' appear." We also showed a few examples of railway incidents or accidents caused by slips, such as an incorrect action during repetitive button presses on a tablet for setting a track closing, as all participants in this experiment were the employees of the same railway company. After 1 h, participants were asked to answer a few multiple-choice (four options) questions about the content of the explanation.

12.4.2 Results and Discussion

We compared the average percentage of correct answers to three questions about what participants learned in the short explanation following the game (Figure 12.7). The percentage was higher for participants who had the experience of making a slip error than for those without that experience. This result may indicate that experiencing a slip in the game promotes learning about the explanation of its mechanism. The increased motivation not to make the same error again can be considered to promote participants' learning about slips. It is also possible that a firsthand experience of a

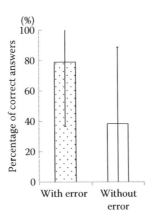

FIGURE 12.7 Average percentage of correct answers to three questions concerning the short explanation following the game.

slip increased the relevance of the explanation for participants and facilitated their learning.

12.5 CONCLUSION AND LIMITATIONS OF THIS STUDY

This study investigated the effective setting of a serious game designed to provide learners with the chance to experience a mental slip. Participants often made slips when aliens and humans that resembled each other were required to be categorized by logically simple rules. Although time pressure is generally thought to increase slips (Baars 1992), the present results showed the possibility that learners take strategies to reduce the slips when placed under greater time pressure. Moreover, the consistency of game settings, such as the interval between aliens' appearances, increased the probability of participants making slips. Finally, using the game design along with these findings, we confirmed that having a firsthand experience of slips in the game promoted subsequent retention of the explained content about the slip. Thus, our attempt to provide learners with the experience of making a mental slip was believed to develop the program that enables them to learn about slips effectively.

When using this program for safety education in a given domain, it may be preferred to customize part of the explanation to promote learners associating the slip with their own domain. Fauquet-Alekhine (2011) suggested that learners' motivation to use the serious game mainly depends on how much they understand their personal benefit from learning that content. To help participants link what they learned to their work in the real world, examples of accidents or incidents caused by railway employees were added in the short explanation in Experiment 3. While one of the advantages of our game was that anyone can use it to learn about slips, regardless of factors such as age or job, the customization of the explanation following the game is expected to promote the motivation to learn the error for individuals engaged in risky operations in any field.

We describe three limitations of this study. First, to clearly show that the serious game promotes learning, it might be important to compare performance on the knowledge test between participants who learned about the slip using a serious game and those who learned it via another medium, for example, a textbook (Wouters et al. 2009). Such comparison was not carried out in with the railway employees in the present study, as they participated in the experiment during the limited time of their shift and there was no chance to test whether the game compensated their learning if they had understood the context less in the textbook than in the game. One thing we might say is that participants with an experience of making a slip in the game are more likely to learn its mechanism than those who learned that error from a textbook. This prediction is based on the finding that participants with the experience of making an error recognized its mechanism more correctly than those who did not make slips and then read the short explanation. Nevertheless, to confirm the positive effect of the game on learning about slips even when participants do not experience this error in the game, it will be necessary to obtain such participants' knowledge test data.

Second, we did not show how much participants acquired the knowledge after they used the game than before. Questions about slips that could be asked in advance of the game can alert participants to the content of the game and therefore prevent them from experiencing slips in the game. In this study, we compared the accuracy of answers about the short explanation on the assumption that there was not a difference in knowledge about slips among participants before the experiment.

Finally, this study did not cover how much our serious game prevents slips in actual situations or accidents due to these slips, even though such issues appear commonly in many safety educational programs. We seldom obtain evidence of improved safety after an intervention including education because the success of prevention of human errors is shown via the absence of errors that occur rarely and discontinuously (Reason et al. 1998). It may even be true that the number of accidents is insensitive to safety improvements by education. For example, Rasmussen and Vicente (1989) regarded that human errors are not stochastic events that can be reduced or removed through training programs, but rather that they are the results of complex interactions of various factors. These interactions are expected to be attenuated by a multilateral approach that involves, for example, a transfer of routine work from individuals to automated systems, work planning to prevent the individual from feeling that his or her operations are unchanged, and support of individuals to take measures to avoid failures in attentional control. The importance of such a multilateral approach may not be accepted unless individuals understand the mechanism of slips, in which concentration or effort to maintain one's attention is no longer a countermeasure against attentional failure. Learning human errors such as slips does not reduce the occurrence of these errors, but we believe that comprehension of these human errors underlies appropriate actions against errors and the resultant accidents.

REFERENCES

Airbus. 2005. Human Performance Error Management. Flight operations briefing notes. Blagnac, France: Airbus. http://www.airbus.com/fileadmin/media_gallery/files/safety_library_items/AirbusSafetyLib_-FLT_OPS-HUM_PER-SEQ07.pdf.

Aldrich, C. 2009. *Learning Online with Games, Simulations, and Virtual Worlds: Strategies for Online Instruction.* Boston: Wiley.

Arnab, S., Berta, R., Earp, J., de Freitas, S., Popescu, M., Romero, M., Stanescu, I., and Usart, M. 2012. Framing the adoption of serious games in formal education. *Electronic Journal of e-Learning* 10: 159–171.

Arnstein, F. 1997. Catalogue of human error. *British Journal of Anaesthesia* 79: 645–656.

Baars, B. J. 1992. *Experimental Slips and Human Error.* New York: Plenum Press.

Carthey, J., de Leval, M. R., and Reason, J. T. 2001. The human factor in cardiac surgery: Errors and near misses in a high technology medical domain. *Annals of Thoracic Surgery* 72: 300–305.

David, M. M., and Watson, A. 2010. Participating in what? Using situated cognition theory to illuminate differences in classroom practices. In *Directions for Situated Cognition in Mathematics,* ed. A. Watson and P. N. Winbourne, 31–57. New York: Springer.

Davies, D. R., Matthews, G., Stammers, R. B., and Westerman, S. J. 2000. *Human Performance: Cognition, Stress and Individual Differences.* East Sussex, UK: Psychology Press.

de Winter, J. C. F., and Dodou, D. 2010. The driver behaviour questionnaire as a predictor of accidents: A meta-analysis. *Journal of Safety Research* 41: 463–470.

Edkins, G. D., and Pollock, C. M. 1997. The influence of sustained attention on railway accidents. *Accident Analysis and Prevention* 29: 533–539.

Fauquet-Alekhine, P. 2011. Human or avatar: Psychological dimensions on full scope, hybrid, and virtual reality simulators. In *Proceedings of the Serious Games and Simulation Workshop*, 22–36. Observatoire de Paris, France.

Garris, R., Ahlers, R., and Driskell, J. E. 2002. Games, motivation and learning: A research and practice model. *Simulation Gaming* 33: 441–466.

Gee, J. P. 2003. *What Video Games Have to Teach Us about Learning and Literacy*. New York: Palgrave MacMillan.

Gibson, D., Aldrich, C., and Prensky, M. 2006. *Games and Simulations in Online Learning: Research and Development Frameworks*. Hershey, PA: IGI Global.

Haga, S., Onodera, O., Yamakawa, A., Oishi, A., Takeda, Y., Kusukami, K., and Kikkawa, T. 2013. Training of resilience skills for safer railways. Developing a new training program on the basis of lessons from tsunami disaster. In *Proceedings of the 5th Symposium of the Resilience Engineering Association*, 173–178. Soesterberg, Netherlands.

Kolb, D. A. 1984. *Experiential Learning: Experience as the Source of Learning and Development*. Englewood Cliffs, NJ: Prentice-Hall.

Norman, D. A. 1981. Categorization of action slips. *Psychological Review* 88: 1–15.

Pourabdollashian, B., Taisch, M., and Kerga, E. 2012. Serious games in manufacturing education: Evaluation of learners' engagement. *Procedia Computer Science* 15: 256–265.

Rasmussen, J. 1983. Skills, rules, and knowledge; Signals, signs, and symbols, and other distinctions in human performance models. *IEEE Transactions on Systems, Man and Cybernetics SMC* 13: 257–266.

Rasmussen, J., and Vicente, K. J. 1989. Coping with human errors through system design: Implications for ecological interface design. *International Journal of Man-Machine Studies* 31: 517–534.

Reason, J. 1990. *Human Error*. Cambridge: Cambridge University Press.

Reason, J., Parker, D., and Lawton, R. 1998. Organizational controls and safety: The varieties of rule-related behavior. *Journal of Occupational and Organizational Psychology* 71: 289–304.

Reinach, S., and Viale, A. 2006. Application of a human error framework to conduct train accident/incident investigations. *Accident Analysis and Prevention* 38: 396–406.

Salkind, N., ed. 2006. *Encyclopedia of Measurement and Statistics*. Thousand Oaks, CA: Sage.

Stanislaw, H., and Todorov, N. 1999. Calculation of signal detection theory measures. *Behavior Research Methods, Instruments, and Computers* 31: 137–149.

Susi, T., Johannesson, M., and Backlund, P. 2007. Serious games—an overview. Technical Report HS-IKI-TR-07-001. Skövde, Sweden: School of Humanities and Informatics, University of Skövde.

Teigen, K. H. 1994. Yerkes-Dodson: A law for all seasons. *Theory and Psychology* 4: 525–547.

Tesei, A., Barbieri, A., and Kessel, R. 2012. Survey on serious games applied to security, safety, and crisis management. *Procedia Computer Science* 15: 320–321.

Wiegmann, D. A., and Shappell, S. A. 2001. Human error analysis of commercial aviation accidents: Application of the human factors analysis and classification system (HFACS). *Aviation, Space, and Environmental Medicine* 72: 1006–1016.

Wouters, P., van der Spek, E., and van Oostendorp, H. 2009. Current practices in serious game research: A review from a learning outcomes perspective. In *Games-Based Learning Advancements for Multi-Sensory Human Computer Interfaces: Techniques and Effective Practices*, ed. T. Connolly, M. Stansfield, and L. Boyle, 232–250. Hershey, PA: IGI Global.

Section III

Models and Other Topics

13 Transitional Journey Maps

Reflections on Creating Workflow Visualizations

Reinier J. Jansen, René van Egmond, and Huib de Ridder

CONTENTS

13.1 INTRODUCTION

Our daily lives are filled with interruptions and transitions from one task to another, resulting in a fragmented workflow. These can be students who knock on our doors when we are writing a paper, or traffic updates that require us to reschedule our route to work. Consider nurses who sequentially divide their attention between patients (e.g., Potter et al., 2004). Or consider a team of police officers, who just transported a suspect to the police station after a demanding pursuit. They are about to process the corresponding paperwork at their office when they receive an urgent call, after which they start driving to the reported incident location. The historical profiles of task transitions have been associated with recuperation in task performance (Matthews and Desmond, 2002) and mental workload (Morgan and Hancock, 2011). Furthermore, there is a substantial body of research that investigates the impact of interruptions on our work and well-being (e.g., Monk et al., 2008; Bailey and Iqbal, 2008). However, as Baethge (2013) argues, these studies typically focus on isolated interruptions, thereby neglecting the accumulation of many interruptions throughout a day. As a result, she continues, an understanding of isolated interruptions cannot be generalized to a working day. In addition, Randall et al. (2000) argue that theoretical constructs based on findings in one domain may not be generalizable to another domain. These notions of limited ecological validity and generalizability have resulted in a move outside of the familiar laboratory environment, judged by the increasing amount of field studies in living labs (e.g., Keyson et al., 2013; Vastenburg et al., 2009; Niitamo et al., 2006). Changes in research methodology cause changes in the way we present and, consequentially, interpret our data. Data visualization facilitates exploration by transforming large amounts of textual or numeric data into graphical formats (Kondaveeti et al., 2012; Segelström, 2009; Card et al., 1999). Yet, to our knowledge, there are no guidelines regarding data visualization of workflows.

We were approached by two organizations with the request to study human information processing activities at work. The Dutch National Police was in the process of updating information technologies in their vehicles. They were interested in knowing how much information police officers can process in various work situations. This knowledge was to be translated into a set of requirements to aid in the selection of appropriate information technologies. Next, the European Space Operations Centre (ESOC, Darmstadt, Germany) wanted an improvement of the alarm sound design in their satellite control rooms. An evaluation of how operators deal with these signals in their workflow was used to inform the subsequent alarm design process. Although these contexts appear very different at first sight, the two case studies presented in this chapter show that both workflows are characterized by frequent task transitions and interruptions.

Our background in informational ergonomics was one of the reasons why we were approached. Informational ergonomics is about understanding how people use information, but also about understanding how to communicate information through design (e.g., visualizations). Thus, in both studies, workflow analyses were performed as input for subsequent research and design activities. Consequentially, the act of creating workflow visualizations became part of the design process.

Throughout our investigations, we encountered several theoretical and practical questions on how to interpret the data as function of categorization and visualization. The objective of this chapter is not to provide a final answer to all these questions. Rather, it is our hope that our way of dealing with these questions will foster critical reflection among those who wish to perform future studies on workflow-based information processing contexts.

13.1.1 LEVELS OF ABSTRACTION

The problem of highly fragmented workflow lies in the fact that (1) one cannot finish an activity before a transition to another activity is required, and (2) it takes time to change one's mind-set back to the original activity (Monk et al., 2008). Zheng et al. (2010) define workflow fragmentation as the rate at which operators switch between tasks. Alternatively, González and Mark (2005) quantify workflow fragmentation as the average time continuously spent on an activity, before a transition takes place. In both cases, increased levels of workflow fragmentation are found at decreasing durations of activity segments. An important question from an information design perspective is when to best provide an operator with an information item. Since some activities typically last longer than others (e.g., reading vs. writing a paper), it makes sense to calculate workflow fragmentation separately for each activity category. This notion favors the time expenditure–based perspective on workflow fragmentation.

The next question, then, is at which abstraction level activities should be defined in order to measure their durations. We will explain the consequences associated with this question through an example of driving a car. Michon (1985) describes driving behavior on three levels: strategic, tactical, and control. The strategic level concerns general plans, such as route choice and scheduled destination time. The tactical level concerns planned activity patterns, such as overtaking and merging. Finally, the control level concerns automatic activity patterns, such as lane keeping and breaking. Figure 13.1 depicts transitions between activity categories over time. The strategic, tactical, and control levels are related to each other, in that driving a complete route at the strategic level encompasses a sequence of maneuvers at the tactical level (e.g., s1 consists of t1-t3-t4-t1), each of which in turn consists of sequential activity at the control level. Note that there is no one-to-one relationship between the levels; actions at the control level can be part of several maneuvers at the tactical level (e.g., steering actions can be found in overtaking, but also in merging). The

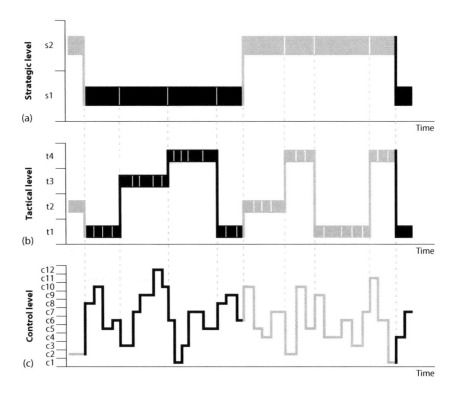

FIGURE 13.1 Three levels of driver behavior as introduced by Michon (1985): strategic (a), tactical (b), and control (c). In each panel, the horizontal axis represents time. Hypothetical data illustrate how segments within each level can be categorized on the vertical axis, and how clusters of segments on a lower level are the basis for transitions on higher levels.

execution of a plan at the strategic level can take up to several hours, unless circumstances necessitate a change of plans (e.g., traffic updates). The duration of maneuvers on the tactical level take several seconds, whereas activities on the control level are described in terms of milliseconds. As a result, capturing all transitions at the control level requires a higher sample rate than those at the strategic level. Moreover, the example in Figure 13.1 shows how describing the same workflow at a higher level of abstraction results in longer activity segments, and less apparent workflow fragmentation. This notion raises a related question: Which level of abstraction results in a meaningful categorization of activities?

The relation between abstraction level, sample rate, and fragmentation not only is relevant for traffic research, but also should in fact be considered in any domain studied by the human factors, ergonomics, and human–computer–interaction communities. In these communities, two common frameworks to describe work at different levels of abstraction are activity theory (e.g., Nardi, 1995) and Rasmussen's (1983) abstraction hierarchy. Michon's (1985) strategic, tactical, and control driving behavior levels are comparable with, respectively, the working sphere (e.g., addressing purpose), action (e.g., goal) and operation (e.g., automatic condition) levels of activity theory as described by González and Mark (2005). Alternatively, they are related to the abstract function (e.g., addressing why), generalized function (e.g., what), and physical process (e.g., how) levels of the abstraction hierarchy. In the two case studies presented next, we have categorized workflow according to the overarching goal of the corresponding activities, which corresponds with the action or generalized function level. Data were collected through what eventually became transitional journey maps, a new method to visualize workflow. By describing intermediate visualization stages, we will show that finding a meaningful level of abstraction can be the outcome of an interpretation process, rather than the starting point.

13.2 CASE STUDY 1: DUTCH NATIONAL POLICE VEHICLES*

The Dutch police force is currently looking for ways to improve the information system of their police vehicles, including pushing information (e.g., neighborhood updates and on-board training) to the vehicle. A central question is how much information officers can process in various work situations. Streefkerk et al. (2006) argued that a mobile police information system should be context-aware (i.e., involving time, location, environmental, and social factors) to prevent cognitive overload. This implies that the dynamics of police work should be taken into account for the development of such a system. For example, indications on average time spent on an activity and the corresponding mental workload may assist in determining the length and appropriateness of an information event (i.e., a moment during which information is presented). As it turns out, a detailed description of work dynamics is lacking in police literature. Therefore, the goal of this study is to better understand cognitive demands imposed on police officers by capturing the dynamics of operational policing.

13.2.1 FRAGMENTATION IN POLICE WORK

Tromp et al. (2010) describe the work of Dutch police officers in terms of three activity categories: static, dynamic preventive, and dynamic reactive. In the static activity category, police officers are not assigned to a specific call, and they are working either at the office or in a parked vehicle. The dynamic preventive category concerns surveillance activities in a moving vehicle. Finally, police officers are said to operate in the dynamic reactive category when they are assigned to an urgent call while in their vehicle. Lundin and Nuldén (2007) identified five ways in which Swedish officers used their patrol car: on their way to an incident, on their way from an incident, at the site of an incident, for general surveillance when driving around or parked at a specific location, and parked at the station handling detained people or paperwork. A comparable categorization was found in a study on British police officers interacting with mobile technology (Sørensen and Pica, 2005). Here, the researchers distinguished five primary activity types: waiting in the car before an incident, driving to an incident, taking action at the incident, driving from the incident, and waiting in the car after an incident. Furthermore, they emphasized that this

---

* The case study presented here is a revised version of Jansen et al. (2014).

so-called generic cycle of operational policing can be interrupted and rearranged due to intermediate events (e.g., incoming calls with a higher priority). Borglund and Nuldén (2012) share this statement, identifying work rhythm as a problem area in the Swedish police force: "Much of police work is characterized by interruptions. Planned and ongoing activity can be discontinued at any time. Current routines and access to computer-based systems create a somewhat fragmented work situation for the officers" (p. 97). Similar accounts have been reported for the U.S. (Straus et al., 2010) and Dutch (Bouwman et al., 2008) police forces. Thus, the notion of fragmented work seems acknowledged in literature on operational policing.

Given the continuous switching between activities, it is important to not only focus on stationary mental workload during an activity, but also consider the effects of transitions between activities on mental workload. Yet, detailed investigations into police routines are typically represented through activity statistics using a full work shift as the time window (e.g., Anderson et al., 2005; Frank et al., 1997; Smith et al., 2001). These statistics do not provide information on whether an activity is executed without interruptions, or about patterns of fragmentation. Moreover, these investigations do not reflect police officers' subjective experiences related to these activities. While attempts to characterize police work fragmentation using scenarios (Borglund and Nuldén, 2012) or narratives (Sørensen and Pica, 2005) do include subjective experiences, they fail to quantify fragmentation. Therefore, the present study aims to unite a quantitative description of work dynamics with subjective experiences related to cognitive demands.

13.2.2 METHOD

A series of ride-alongs with Dutch police officers were arranged. Based on the method of contextual inquiry (Beyer and Holtzblatt, 1997), officers were interviewed and observed in their natural work environment, where they provided explanations as their work unfolded.

13.2.2.1 Participants

Ten officers (eight males, two females) volunteered to be accompanied in their patrol cars. Each officer had at least 2 years of experience with operational policing. Four ride-alongs were arranged, including three full 8-hour shifts and two shift changes in total. Hence, the vehicle was chosen as the central focus during ride-alongs, while personnel configurations changed from shift to shift. The ride-alongs included solo (two cases) and dual (four cases) patrol. With durations varying between 4.5 and 11 hours, in total 28 hours of data were collected. Colleagues of the officers often asked the researcher about his presence during stops at the police station. Their comments on work dynamics and organization are treated as part of the study results.

13.2.2.2 Apparatus

Data were collected with pen and paper, featuring timestamps, descriptions of the current activity, events in the officer's information environment that caused a transition to another activity (e.g., incoming calls and comments following an officer's observation), and utterances related to cognitive demands. All data were initially logged on a template with three rows of predefined activity categories. As requested by the client, these activity categories corresponded with the classification of Tromp et al. (2010) (i.e., static, dynamic preventive, and dynamic reactive).

13.2.2.3 Procedure

Before the ride-along began, the researcher explicitly stated that the study was not intended to judge the officers' performance. Agreements were made on safety and privacy. During the ride-alongs, the researcher tried to minimize hindrances by discretely observing what was going on. This nonparticipatory research approach was at times violated, for example, when an officer asked for details about a recent call. Existing studies recommend that the relationship with the officer should not be sacrificed for the sake of minimizing reactivity (Stol et al., 2004; Spano and Reisig, 2006). Interestingly, such a question can be regarded as a verbalization related to high cognitive demands. Officers were occasionally asked to explain what happened during transitions, but only if the work demands allowed for such concurrent reports. Otherwise, they were asked to give a retrospective report shortly after the event.

13.2.3 RESULTS

A new method to visualize workflow will be introduced. The method is used to report findings on cognitive overload, and differences between solo and dual patrol.

13.2.3.1 Activity Categorization

The left panel of Figure 13.2 displays the first page of the original field notes (in Dutch) of the first ride-along. The horizontal and vertical axes correspond with time and activity category, respectively. The text fields concern observations of and statements by a team of two police officers. The police officers were initially surveilling the neighborhood, until they were assigned to an incoming call. This was noted with "to incident" (Dutch: *naar melding*) in the dynamic reactive category. A few minutes later the call was cancelled by the dispatcher, as noted with "cancel" in the dynamic preventive category. Two arrows were drawn to connect the sequence of notes, thereby creating a sense of order and time. As a result, two transitions between the dynamic preventive and dynamic reactive activity categories were visualized. Next, an alarm sound (*whiew*) of the automatic license plate detector (Dutch: ANPR) was heard. One officer asked about the location of the detected car (Dutch: "Waar is-ie?"), to which the other officer replied that the car went in the opposite direction (Dutch: "Tegengestelde richting"). The officers' active search response was interpreted as a transition from the dynamic preventive to the dynamic reactive category. When it turned out that the detected car could not be intercepted, an arrow was drawn to indicate a transition back to the dynamic preventive category.

A section of the field notes of the third ride-along is shown in the right panel of Figure 13.2. Compared to the former field notes, there are differences in visualization style, the number of activity categories, and the arrangement of the activity categories. There is a continuous line that represents the activity category in which the police officers momentarily operated, and transitions between activity categories. This continuous line is augmented with text fields, whereas previously the text fields were augmented with arrows when there were transitions. Thus, the visualization

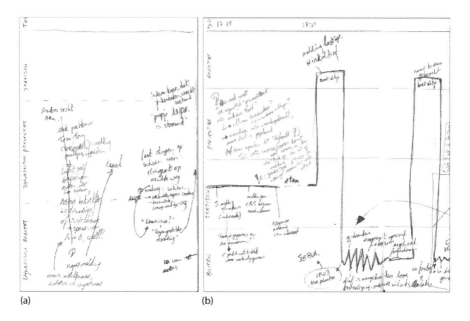

(a)   (b)

FIGURE 13.2 Original field notes of the first observation session (a) and the third observation session (b).

style evolved as result of the perceived necessity to connect events in an orderly way. Furthermore, dynamics within activity categories were captured through different line styles. For example, a dashed line was drawn when the police officers were having a short break (Dutch: *eten*), and the aggressive behavior of a drunken shoplifter was represented through a zigzag line.

An additional activity category was introduced after finding a recurring activity that was not represented by the definitions of the existing categories. Although the static category covers situations in which police officers have left their vehicle, it does not include engaging at an incident. The capacity to interact with information technology is unlikely comparable between, for example, office work and handcuffing a drunken shoplifter. Such reactive behavior appears to be covered within dynamic reactive. However, that category does not include situations outside of the vehicle. Therefore, the outside (Dutch: *buiten*) category was introduced to represent police officers who have left their vehicle to engage at an incident (e.g., to catch the drunken shoplifter).

The order in which the activity categories were presented was changed twice. Whereas static, dynamic preventive, and dynamic reactive were originally visualized from top to bottom (see left panel of Figure 13.2), this order was reversed during the third ride-along (see right panel of Figure 13.2). This reordering was based on comments by police officers, who associated high driving speed levels during emergency situations with high adrenaline levels and low information processing capacities. Presenting the dynamic reactive activity category provided a better visual indication of the mental workload experienced by police officers.

The second rearrangement concerned the placement of the outside activity category. There was a logical reason why this activity category was originally presented at the bottom: the upper two categories were always related to a moving car, and the lower two categories were the only ones related to activities outside of the car. However, we observed that transitions to the outside category typically originated from the dynamic reactive and dynamic preventive categories. The visual appearance of the sudden drop from dynamic reactive to outside in the right panel of Figure 13.2 suggests that the workflow was disrupted, while capturing the shoplifter was actually a logical step after driving to the incident location. In addition, many of these outside activities are likely associated with higher levels of mental workload than, for example, office work in the static category. Therefore, the outside activity category was eventually presented on top of the other categories.

Following a similar rationale as with the introduction of the outside category, the original framework of Tromp et al. (2010) was refined into six activity categories. Static was subdivided into "parked at the station" and "parked surveillance." Dynamic preventive was subdivided into "driving surveillance" and "driving to the station." Finally, dynamic reactive was subdivided into "driving to the incident" and "engaging at the incident" (formerly labeled "outside"). These six activity categories correspond with an adapted version of the framework by Lundin and Nuldén (2007), in that driving and nondriving surveillance activities were categorized separately.

13.2.3.2 Transitional Journey Maps

We refer to the graphical representation of interconnected objective data (e.g., observations) and subjective data (e.g., statements) as *transitional journey map*. Four transitional journey maps were constructed, one for each ride-along. An example can be found in the lower part of Figure 13.3. The vertical axis displays six activity categories, whereas time is found on the horizontal axis. The main actors are represented through three thick lines: the police vehicle (dark gray), the driver (medium gray), and in the case of dual patrol, the co-driver (light gray). A journey through activity categories is created as the actors cross the underlying framework. Additional lines are used in case other actors come into play (e.g., the case of the copper thief, here represented in black). Stationary vehicles are depicted with a dashed line. Similarly, dashed lines are used when officers are taking a break. Upon entering their car, officers' corresponding lines are joined with the vehicle's line. Segments of activities are demarcated by the time between adjacent transitions.

A transition is defined as a change from an activity category to another one. In Figure 13.3, transitions are labeled with hexagonal boxes, a character for the corresponding ride-along, and a number for the order of occurrence. For example, C23 refers to a segment of previous activity at the police station, and marks the transition from "parked at station" to "driving surveillance." Descriptions for ongoing activities are depicted in white boxes for quick reference. Observations and utterances about an ongoing situation are depicted in a regular font style, whereas a an italic font style is used for retrospective accounts. Because of its dominant role in police work, instances of radio communication can be found in a separate row. The thin alternating arrows in Figure 13.3 show how messages are going back and forth between the officers and the dispatcher (e.g., the call of the missing girl).

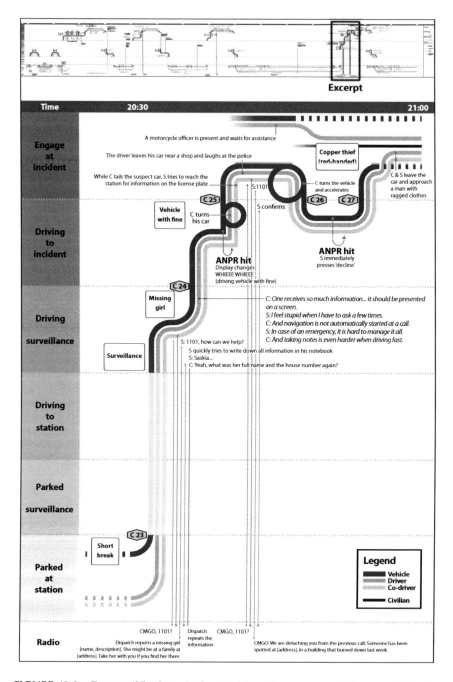

FIGURE 13.3 Excerpt (35 minutes) of a transitional journey map (450 minutes). The horizontal and vertical axes display time and six activity categories, respectively. Additional details are described in the text.

13.2.3.3 Applying Transitional Journey Maps to Operational Policing

Visual inspection of a full transitional journey map confirms the notion that police work is fragmented. The lower part of Figure 13.3 shows periods of many short activity segments, followed by relatively long stretches of paperwork at the police station. This is reflected in the boxplots of Figure 13.4, which show the durations of activity segments per activity category, including all ride-alongs. Outliers in the "engaging at incident" category were cases where victims or suspects were questioned, namely, theft (A7, C21) and domestic violence (B27, C13). All of these cases required more than half an hour of paperwork, with an outlier at 2 hours (C23). However, officers were often interrupted by incoming calls before finishing their office work, as reflected by the median duration of 17.9 minutes. Other outliers refer to picking up remote colleagues (A9), surveillance while bringing the researcher to the train station (Aend), and surveillance across a deserted national park (B8).

The categories "parked at station" and "driving surveillance" seem to take longer than "driving to station," "driving to incident," and "engaging at incident," which seem to have similar segment durations (see Figure 13.3). Given the skewed distributions, nonparametric tests (SPSS v20) were used to compare between activity categories. As only one instance of parked surveillance occurred, this category was excluded from further analysis. Segment duration is significantly affected by activity category ($H(4) = 23.71$, $p < 0.001$). Seven Mann–Whitney tests were used to follow up this finding. Therefore, a Bonferroni correction was applied, and all effects are reported at a 0.007 level of significance. The duration of activities in

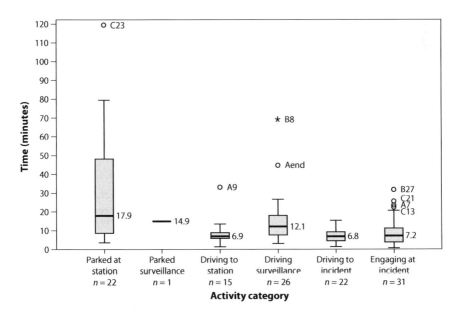

FIGURE 13.4 Boxplots of time spent in each activity category, summarized over all ride-alongs. Median values are shown next to each box. Whiskers depict the lowest and highest data within a 1.5 interquartile range of the lower and upper quartiles, respectively. Outliers labeled with an asterisk or a circle concern solo and dual patrol, respectively.

the "parked at station" category was generally significantly longer than "driving to station" ($U = 70$, $r = -0.48$), "driving to incident" ($U = 90$, $r = -0.54$), and "engaging at incident" ($U = 148$, $r = -0.48$), but not longer than "driving surveillance" ($U = 208$, $r = -0.23$). Furthermore, activities performed in the "driving surveillance" category took significantly longer than "driving to incident" ($U = 139$, $r = -0.44$) and "engaging at incident" ($U = 235$, $r = -0.36$), but not longer than "driving to station" ($U = 108$, $r = -0.37$). It can be concluded that the most time for an informing event can be found when officers are working at the police station, or during surveillance while driving. Based on this dataset, an informing event should take less than 6.8 minutes if at least half of these events are to be fully processed in any activity category before a next transition takes place. However, these statistics do not address whether officers have spare capacity to successfully process the information.

13.2.3.4   Reports of Cognitive Overload

Comments by police officers regularly contained descriptions of situations witnessed during other ride-alongs, which were indicators for cognitive overload. For example, compare the following anecdote with Figure 13.3:

> An incoming call instructs the officers to advance to a car that was broken into. L. takes a notebook from her pocket to record the address: "This way you don't have to ask again." A. responds: "On the group radio one often hears colleagues asking for a repetition of the suspect description. At the time they receive a call and they have to move as fast as possible, their mind set is already preoccupied." (Field notes from ride-along 1)

Because of the activity descriptions and their characteristic visual pattern, the layout of a transitional journey map facilitates remembering and retrieving events with related comments. Furthermore, the content of a comment dictates in which activity category it should be placed (e.g., a colleague at the station talking about an arrest belongs to "engaging at the incident"). Thus, an overview of information processing issues within an activity category can be obtained by scanning along the corresponding row in the transitional journey maps. This approach resulted in the identification of an information processing paradox.

On the one hand, police officers not only monitor the radio for messages addressed to themselves, but also want to stay informed of the whereabouts and tasks of their colleagues. One reason is safety: "If there is a call with violence, it's good to know if colleagues are nearby … then you know if and how long you should wait before stepping in." Vice versa, officers may offer assistance. Second, there are functional implications: "Those officers are busy over there, so I'll compensate by patrolling more centrally in this area." Finally, it is part of a social system: when returning to the station after an emotionally demanding call, officers find support from colleagues that listened in. One officer commented that he was missing too much information, even though three channels were concurrently monitored (i.e., car radio and two earpieces).

On the other hand, police officers have trouble processing all information. As described above, incoming calls regularly contain too much information to remember. This is further inhibited by situational and state-related factors: "If a situation is dangerous, you feel the adrenaline, stress, fatigue and tension, and this affects your ability to concentrate. In those situations, it is hard to hear something amidst other voices." Messages are often hard to comprehend due to auditory masking by the police vehicle (e.g., when driving at high speed, often accompanied by a siren) and signal degradation in the communication system. In the meantime, the continuous monitoring and filtering of radio messages takes its toll. Up to 26 messages were counted in a time span of 5 minutes. Officers complained about high volumes, occasional feedback beeps, and fatigue: "My left ear is deaf for other sounds because of the earpiece. After a busy shift, I still hear the voices at home."

Comments on the necessity of monitoring radio communication were found in all activity categories, except for "parked surveillance." However, the representativeness of this exception is doubtful, since action in this category was observed only once. Comments on auditory masking were found in all activity categories that involved driving. Comments on overload were found in all activity categories, except for "driving to station." Overall, the observations and comments suggest that police officers want more information than they can handle with the current system.

13.2.3.5   Comparison of Solo and Dual Patrol

All outliers in Figure 13.4 were cases of dual patrol, except for B8. This suggests a considerable difference in time spending between solo and dual patrol and, as a result, more time for information events during dual patrol. Nonparametric tests were performed per category. Using a Bonferroni correction, the effects were compared with an alpha level of 0.008. None of the tests on time spending reached statistical significance. Nevertheless, police officers did mention differences between solo and dual patrol modes. The biggest impact is the opportunity to distribute tasks among officers in the case of dual patrol. Generally, the driver only concentrates on driving, whereas the co-driver is responsible for communication and surveillance tasks. Many officers commented that it is hard to operate the mobile data terminal while driving solo. Additionally, there are organizational differences between the patrol modes: "If you're patrolling solo, you only get a call when the others cannot handle it. In cases of violence, we always operate with couples." This suggests that differences may be found between the distributions of transitions.

Figure 13.5 depicts state diagrams for solo and dual patrol. An arrow line represents each cause for a transition between two activity categories. Thicker lines are used if the same cause was observed more than once. The total times spent observing solo and dual patrol were 12.6 and 15.1 hours, respectively. The relative time spent in each activity category is represented by the size of the corresponding circles. The two figures reveal that solo patrol involves relatively more driving surveillance activity than dual patrol (36% vs. 15%). Solo patrol involved more transitions from "driving surveillance" to "engaging at incident" (10 vs. 2), but less transitions from "driving to incident" to "engaging at incident" (5 vs. 12). Interestingly, in both patrol modes, 15 transitions were counted toward "engaging at incident." However, relatively more time on "engaging at incident" was spent in dual patrol (20% vs. 12%). This was caused by the longer times spent investigating incidents with violence (see outliers in Figure 13.2). Additionally, dual patrol involved more time

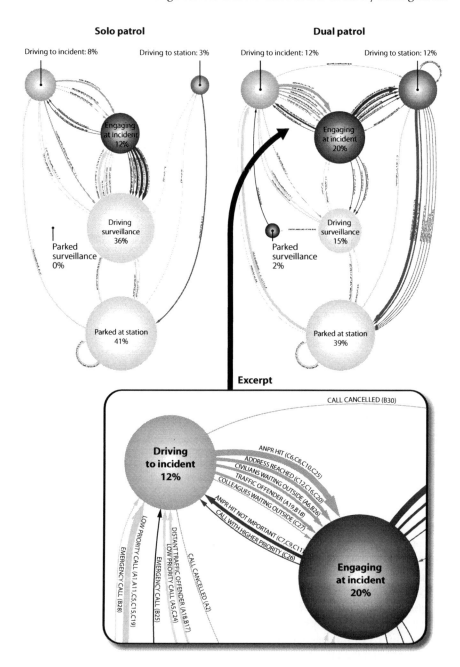

FIGURE 13.5   State diagrams of transitions during solo patrol (top left, 12.6 hours observed) and dual patrol (top right, 15.1 hours observed). The excerpt shows observed causes for transitions between activity categories. Codes in parentheses refer to the hexagonal boxes (i.e., transitions) in the transitional journey maps.

spent on "driving to station" (12% vs. 3%), which may be due to the large amount of paperwork after serious incidents, and a higher likelihood of transporting victims or suspects afterward. In sum, the state diagrams on solo and dual patrol reflect the organizational differences uttered by the police officers.

13.2.4 DISCUSSION

The information environment of a police officer is embedded in an unpredictable workflow, and includes numerous visual and auditory information channels. It has been argued that knowledge of work dynamics and related cognitive demands is beneficial for the development of an information and communication system. Regarding work dynamics, the current data suggest that in most cases, an information event that takes place while driving may not take longer than 6.8 minutes if it is to be fully processed before a next transition takes place. However, this finding does not address cognitive demands. The comments made by officers suggest that the demands of concurrent driving, surveillance, and monitoring force them into a permanent state of task-related effort (De Waard, 1996). Therefore, although no complete performance breakdowns were observed, the continuously experienced high workload was exhausting in the long run. "Driving to station" is the only activity category that may be used for additional information events, given the absence of comments on cognitive overload. However, the applicability would be limited during solo patrol. Overall, analyses of workflow fragmentation and officers' comments on cognitive overload suggest that the police vehicle's cockpit should be improved, before pushing additional information can be considered. Particularly, alleviation of the auditory channel is needed. The current results warrant further application of the transitional journey map method in other contexts, such as information processing by other emergency services, control rooms, and tracking group dynamics for crowd management.

13.3   CASE STUDY 2: ESOC SATELLITE CONTROL ROOMS*

The European Space Operations Centre (ESOC) requested us to evaluate and improve the auditory signals in their satellite control rooms. One of the issues was that warning beeps occur so often that they are ignored by the space controller (spacon). Also, irrelevant alarms disrupted the activities of colleagues working at a nearby workstation. There have been cases where the spacon would turn off those alarms, because their presence was causing too much stress. In both cases, the result is an increased risk of missing critical events.

These issues reflect characteristics associated with the safety hazard known as alarm fatigue. Alarm fatigue occurs when operators are continuously exposed to a large number of false alarms. When operators start to distrust these disruptive alarms (Cvach, 2012), they may disable, silence, or ignore them (Korniewicz et al., 2008). This has resulted in patient injury and death in the medical context (Sendelbach and Funk, 2013). The general strategy to counteract alarm fatigue is to reduce the number of unimportant alarms. For example, standards on alarm systems in process

---

* The case study presented here is a revised version of Jansen et al. (2015).

industries (ANSI/ISA 18.2, 2009; NEN-EN-IEC 62682, 2015) specify targets for average and peak alarm rates per operator console. These standards describe a rationalization stage, with the purpose of ensuring that alarms are only implemented if they are actionable (e.g., requiring a response).

However, there are potential side effects to alarm reduction. In communication between a satellite and ground control, a loss of signal is normally considered to be an abnormal condition, and should result in an alarm. Woods (1995) describes an example in which operators complained that the same alarm went off during each scheduled transponder switchover. When the system engineers removed the alarm, however, the operators complained that there was no indication that the event was occurring as expected. Rauterberg (1999) investigated the effect of auditory feedback on monitoring a simulated industrial plant with multiple machines. Each machine generated both auditory and visual feedback to inform the operator of its events. The auditory signals were redundant, as they referred to the same information as the visual signals. However, removal of the auditory signals resulted in decreased plant performance. Another study in the medical context shows that auditory signals may not always result in a response, but nurses use these signals as indicators for a patient's status (Bitan et al., 2004). In sum, the removal of redundant signals does not necessarily improve performance, if these alarms provide meaningful information (Stanton et al., 2000).

Moreover, these examples show two facets of auditory signal interpretation: (1) operators anticipate system state changes within a certain context, and (2) alarm signals are sometimes used as feedback signals to confirm anticipated state changes. Therefore, alarm fatigue prevention may benefit from identifying which alarms are anticipated. The role of anticipation in alarm responses will be explored in the ESOC satellite control rooms.

13.3.1 Interpretation of Auditory Signals

In a typical control room, the system evaluates whether a set of process parameters are within specified limits. When a parameter crosses its limit, the system will generate a visual or auditory "out of limits" signal. Operators require contextual background knowledge to interpret the meaning of these signals (Seagull and Sanderson, 2001; Stanton, 1994). On the one hand, this knowledge determines whether the operator anticipated the event to which a signal corresponded. On the other hand, this knowledge allows the operator to determine whether the situation related to a signal is actionable. As a result, signals can be interpreted in four different ways, as shown in Table 13.1.

The event state [actionable/nonactionable] separates actual alarms from signals that are generally experienced as nuisance. The situation state [anticipated/unanticipated] further refines these interpretations. A signal related to an anticipated event and an actionable situation will remind, rather than alert, that something is going to happen. Therefore, the related signal functions as feedforward.

Signals that relate to an anticipated event in a nonactionable situation can have two interpretations. In the first interpretation, they can be a nuisance. For example, an operator is informed that a core temperature crosses a threshold by hearing an auditory signal. However, this signal may also sound during maintenance on the

TABLE 13.1

Interpretations of an Auditory Signal, as a Function of the Actionability of a Situation, and Anticipation toward Events

Situation/Event	Anticipated	Unanticipated
Actionable	Feedforward	Alarm
Nonactionable	Nuisance or feedback	Nuisance

system's ventilator, even though the temperature is temporarily allowed to cross the threshold. When the operator is aware of this situation, he or she may choose to ignore the signal. This illustrates how known malfunctions and planned system changes can turn an alarm signal into a nuisance signal. In the second interpretation, the example of Woods (1995) on loss of signal in communication between a satellite and ground control illustrates how an alarm signal is used as a feedback signal. This means that the value of anticipated nonactionable signals needs to be examined on a case-by-case basis.

13.3.2 SYSTEM DESCRIPTION SATELLITE CONTROL ROOMS

ESOC accommodates several control rooms, each of which is related to one or more satellite missions. The distance to earth determines how long the contact with a satellite can be. This contact is referred to as a pass. All passes are scheduled. The dynamics of missions are different because of the period of contact and contact loss, as well as the distance between the satellite and earth. If the satellite is close to earth (mission type: earth observers), the pass duration, as well as the period between passes, is short. Contact with the satellite is almost instantaneous. Another option is a satellite with a fixed position in space (mission type: astronomy). This type of satellite is permanently in contact with the control rooms, but the antenna picking up the signal (antennas are located at three different places on earth) may change. If a satellite is at a long distance from earth (mission type: interplanetary), passes are long (e.g., 8 hours), but then the contact is also lost for a long time. Because of the long distance, it can take up to 20 minutes to send an instruction to the satellite, and an equal duration to receive a confirmation message from the satellite. Consequentially, there may be differences in anticipation toward events among the missions.

Each satellite is operated by a spacon, who monitors incoming data and events. When an error occurs, the spacon is informed by an auditory signal and an error message on the screen. For each mission, there is a specific protocol that the spacon has to follow. If necessary, an engineer is involved in resolving the problem.

13.3.3 METHOD

Three ESOC mission control rooms were visited (e.g., earth observer, astronomy, and interplanetary). The spacons on duty were interviewed and observed in their natural work environment, based on the method of contextual inquiry (Beyer and Holtzblatt, 1997). The primary focus was on auditory signals, as opposed to visual signals, given ESOC's request.

13.3.3.1 Participants

Four experienced male spacons were involved in this study. Each spacon was specialized in a single mission. In one mission, an experienced spacon was coaching an apprentice. In another mission, one spacon substituted for another spacon at the end of his shift.

13.3.3.2 Apparatus

Data were collected with pen and paper and two Roland R-05 portable field recorders. The notes consisted of timestamps, descriptions of the current activity, events in the spacon's information environment (e.g., auditory signals), utterances on work dynamics, and the presence of colleagues. The field recorders were primarily used to transcribe ongoing conversations between spacons, colleagues, and researchers, as well as voice loop communication.

13.3.3.3 Procedure

Spacons were informed about the presence of the researchers prior to the observations. After setting up the equipment, the researchers tried to minimize hindrance by discretely observing what was going on. Ongoing discussions between spacons and colleagues facilitated understanding of the situation. Spacons were occasionally asked to explain, for example, what they were working on, or what an auditory signal meant. These questions were only asked if the work situation allowed for such concurrent reports. Otherwise, spacons were asked to give a retrospective report shortly after the event. Engineers occasionally entered the satellite control rooms. Their comments on the ongoing mission status, as well as on auditory signals, are treated as part of the study results.

13.3.4 Results

In total, 6.6 hours of data were collected, during which 140 auditory signals were recorded (earth observer: 100 minutes, 38 signals; astronomy: 109 minutes, 34 signals; interplanetary: 185 minutes, 68 signals). Thirty-one auditory signals continued to ring until a spacon acknowledged them (e.g., ti-lu-li-ti-lu-li). Spacons labeled these sounds as alarms. Additionally, 109 signals were labeled as warnings (e.g., beep) or feedback signal (e.g., printer sounds). As these signals did not require an acknowledgment, it was not always possible to determine if they were actually heard by spacons. Therefore, only signals labeled as alarms are analyzed, unless stated otherwise.

The collected data of each mission are represented as a transitional journey map. A state transition diagram is constructed to summarize transitions occurring in all missions. The two representations are used to perform a Bayesian inference on alarm anticipation, and to distinguish feedback signals from nuisance signals.

13.3.4.1 Activity Categorization

The construction of a transitional journey map requires a grouping of activities into a fixed number of activity categories. However, there was no documentation of activity categories at the start of this study. The first observation took place in the Cluster mission control room, where one out of four earth observer satellites could be controlled at a time. Our first approach in categorizing the data was according to which satellite was momentarily controlled, analogous to how Potter et al. (2004) describe nurses' workflow as transitions from patient to patient over time. Figure 13.6 displays the field notes of the first observation. In this section, the spacons finished the pass of one satellite (e.g., sc1) and faced difficulties in establishing a connection with the next satellite (e.g., sc2). Auditory signals were represented as peaks in the continuous line. In some cases, the spacons expressed that there was no need to react to an alarm (e.g., "pass is finished"), whereas in other cases, alarms initiated problem-solving behavior (e.g., consulting protocols). Thus, there were a variety of responses to the same auditory signal. Unfortunately, the chosen categorization did not allow for visualizing this variety of responses in terms of transitions. Another problem was that no transitions were observed in the other missions, either because of the long pass duration (e.g., interplanetary) or because the mission did not feature passes (e.g., astronomy).

In our second approach, we compared the original system description with activities that were observed or described in all mission control rooms. Four activity categories were derived. First, starting or ending a pass refers to activities such as requesting a new connection or breaking the existing connection. Second, monitoring can be characterized as a vigilance task: spacons monitor incoming data, ready to respond to potential problems. Third, commanding is about controlling the satellite, for example, by sending a list with maneuvers. Although this state involves monitoring activities as

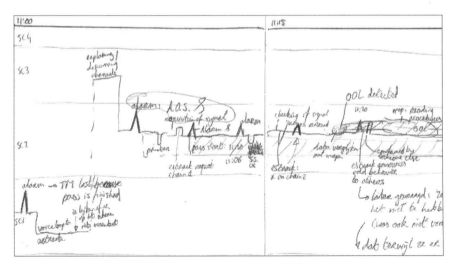

FIGURE 13.6 Original field notes of the Cluster mission control room. *Note:* Verbal data that were not transcribed during the observation session were later transcribed using audio recordings.

well (e.g., waiting for confirmations on the execution maneuvers), the purpose of this system state differs from that of the second system state. Fourth, solving problems involves activities related to unexpected messages (e.g., dealing with out of limits alarms).

13.3.4.2 Anticipation through Transitional Journey Maps

We used transitional journey maps and a state transition diagram to represent the data required to fill out the cells of Table 13.1. Anticipation of events could not be measured directly, but was inferred from spacon statements. A transition between two activity categories was attributable to a signal if the former followed shortly after the latter. Also, it was not possible to observe directly whether the situation during which a signal occurred was actionable from the spacon's perspective. The presence or absence of a transition to an activity category of problem-solving behavior was used as a proxy for the actionability of a situation. This is in line with the purpose of an alarm, which is to inform a spacon that an abnormal condition occurred, which requires a response (ANSI/ISA 18.2, 2009).

Three transitional journey maps were created. Figure 13.7 shows a 14-minute excerpt of the Cluster earth observer mission, which corresponds with a part of the data presented in Figure 13.6. The horizontal and vertical axes display time and the four activity categories, respectively. Transitions between activity categories are labeled with hexagonal boxes, which contain a reference to the mission and the order of occurrence (e.g., CL2). These lines are frequently interrupted by black circles, which represent auditory signals. A separate row for voice loop communication distinguishes between communication in the mission control room itself and communication with other rooms (e.g., engineers) or external parties (e.g., operators of the antennas, Estrack).

This excerpt illustrates the procedures between two passes. An example of anticipation can be found at the alarm at 11:05. This alarm is related to a loss of signal. The retrospective anecdote (regular font style) shows that the spacons knew this alarm would come. Additionally, anecdotes on and observations of the active situation (italic font style) suggest that this signal was used as a starting point for their activities toward the beginning of the next pass. In this case, the alarm sound did not initiate a transition toward problem-solving behavior, but functioned as a feedback signal. Another example of how spacons use auditory signals as feedback can be found at 11:12. The spacons knew that the next pass (e.g., "acquisition of signal") had started from the printer sound. This resulted in their transition (CL2) to the monitoring activity category. Finally, the alarm at 11:13 initiated a transition (CL3) to problem-solving behavior. The spacons determined that the signal-to-noise ratio of the data transmission was too low, and requested a switch to another bitrate. Their comment "Should we first try high?" indicates that they were prepared for this situation. These examples show how anticipation toward events can be derived from a transitional journey map.

13.3.4.3 Alarm Response Behavior through a State Transition Diagram

A state transition diagram was used to count and group signals that initiated problem-solving behavior. The transitional journey maps were translated into a state

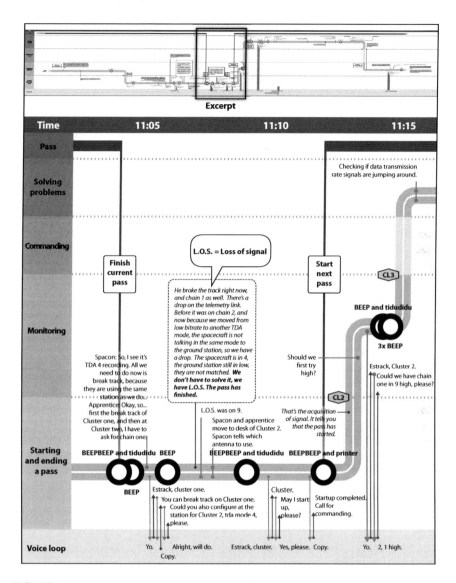

FIGURE 13.7 Excerpt (14 minutes) of a transitional journey map (100 minutes). This excerpt describes the process of ending one pass and starting the next pass.

transition diagram for an overview of transition causes and signal frequencies. In this diagram, each state corresponds with an activity category. Transitions were labeled according to the causes identified in the transitional journey maps, and the frequency at which they occurred. A total of eight transitions between activity categories were found, which are represented as solid lines in the state diagram of Figure 13.8. The transition labeled CL2 in Figure 13.7 corresponds with the orange arrow from "starting and ending a pass" to "monitoring," whereas CL3 represents one of the critical events at the arrow from "monitoring" to "solving problems." Dashed lines represent

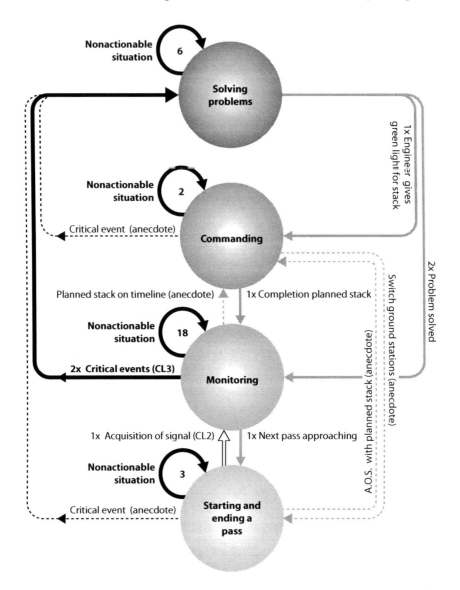

FIGURE 13.8 State transition diagram. Black and white lines represent state transitions initiated by alarm signals and printer sounds, respectively. Gray lines represent state transitions initiated by planned or finished activities. Solid lines are observed transitions. Dashed lines represent anecdotes.

transitions that were mentioned by spacons, but which were not observed. The state transition diagram shows that only two auditory signals labeled as an alarm resulted in a transition to problem-solving behavior. Statement analysis in the corresponding transitional journey maps revealed that one of these signals came as a surprise, and therefore had an alarming function. The other signal was related to an anticipated event (e.g., transition CL3 described above).

The circular arrows in the state transition diagram show that 29 auditory signals labeled as an alarm did not result in a transition to another activity category at all, which implies they were nonactionable. For 19 of these nonactionable signals, statement analysis revealed that the corresponding system events were anticipated. The role of anticipation could not be derived for the other 10 signals. They occurred in situations where asking a spacon for an explanation would have disrupted his workflow, and where a retrospective explanation was hindered by other auditory signals that had rang in the meantime.

13.3.4.4 Quantifying the Role of Anticipation with Bayes' Theorem

Bayes' theorem relates current probability to prior probability. Applied to the context of a control room, Bayes' theorem can be used to explore the causal relationship between event anticipation and situation actionability. Of particular interest in the present context is the probability that anticipation influenced spacons in deciding not to respond to a signal (e.g., the "nuisance or feedback" cell in Table 13.1). According to Westbury (2010), this study concerns the simplest form of Bayesian inference, with only two sets of mutually exclusive possibilities. Therefore, the canonical Bayesian expression can be rewritten as

$$
\begin{aligned}
&P(\text{Anticipated} \mid \text{Nonactionable})\\[6pt]
&= \frac{P(\text{Nonactionable} \mid \text{Anticipated}) \cdot P(\text{Anticipated})}{P(\text{Nonactionable})}\\[6pt]
&= \frac{P(\text{Anticipated and Nonactionable})}{P(\text{Nonactionable})}
\end{aligned}
\tag{13.1}
$$

The two signals that initiated a transition to problem-solving behavior in the state transition diagram are represented in the "Actionable" row of Table 13.2. Signals corresponding with the circular arrows in the state transition diagram are found in the "Nonactionable" row of Table 13.2. As the role of anticipation could not be

TABLE 13.2
Distribution of Alarm Sounds as a Function of Situation Actionability and Event Anticipation

Situation/Event	Anticipated	Unanticipated
Actionable	1	1
Nonactionable	$19 + (10 - x)$	x

Note: The role of anticipation could not be derived for 10 alarms. See text for more explanation. Cells are divided by the total number of alarm sounds ($n = 31$) to obtain the probabilities in Equation 13.1.

TABLE 13.3

Interpretation of Anticipated Nonactionable Signals (19)

Activity Category	Feedback Signal	Nuisance Signal	Interpretation Unclear
Solving problems	0	0	0
Commanding	1	0	0
Monitoring	2	3	11
Procedures at pass	2	0	0

Note: All signals were designed as an alarm signal.

derived for 10 alarms, the distribution of nonactionable signals in Table 13.2 is represented as a range, where $0 \leq x \leq 10$. If all signals with unknown event anticipation were in fact unanticipated (e.g., $x = 10$), then Equation 13.1 yields a probability of 66%. A probability of 100% is found if all nonactionable signals were in fact anticipated (e.g., $x = 0$).

13.3.4.5 Feedback Signals versus Nuisance Signals

The Bayesian inference showed there is a high probability that spacons were guided by anticipation in their decision not to respond to signals. This warranted investigating whether the 19 nonactionable and clearly anticipated signals were interpreted as feedback or nuisance. Statement analysis in the transitional journey maps resulted in the interpretations shown in Table 13.3. For 11 signals, the interpretation in terms of feedback value was unclear. Five signals were used to confirm the end of five unrelated processes (e.g., "loss of signal" in Figure 13.7). Therefore, these signals are marked as feedback signals. This was not the case for signals about a low temperature of the satellite. Spacons had disabled one of the heaters to save fuel, resulting in three nuisance signals.

13.3.5 Discussion

This study shows that qualitative data analysis is essential for interpreting quantitative data on alarm responses. This was facilitated by the use of transitional journey maps, by which alarm anticipation could be derived. Anticipation influenced spacons' choices to ignore auditory signals as alarms, which was confirmed through Bayesian inference. Expertise level may have induced anticipation, which allowed spacons to cope with high signal densities. This was found in all missions, which suggests that anticipation is not dependent on the type of mission (e.g., distance between satellite and earth), but rather a general strategy.

Only 2 of the 31 signals that were labeled as alarm actually corresponded with critical events. The rationalization stage in standards on alarm management (ANSI/ISA-18.2, 2009; NEN-EN-IEC 62682, 2015) prescribes that all nonactionable signals are discarded. However, five nonactionable alarms related to anticipated system events were interpreted as feedback signals. We recommend designing these signals with a lower level of perceived urgency than the alarm signals. Concluding, the identification of valuable feedback signals should be part of the alarm management life cycle. This requires distinguishing between nuisance and feedback for anticipated nonactionable signals.

Most methods for alarm reduction target alarms with the highest frequency of occurrence (e.g., "bad actors") (Izadi et al., 2010) or groups of correlated alarms (Schleburg et al., 2013; Kondaveeti et al., 2012). These quantitative approaches do not take into account the context in which alarms were triggered. Consequentially, it is not possible to determine whether related events were anticipated, or if nonactionable signals contain valuable feedback information. However, a new monitoring paradigm with novelty detection (Martínez-Heras et al., 2012) may partially address this issue, for novelty excludes anticipation by definition. It is hoped that intelligent alarm systems will eventually reduce alarm fatigue by filtering nuisance signals, while continuing to provide valuable feedback.

13.4 GENERAL DISCUSSION

The transitional journey map is a generic approach to describe workflow, connecting consecutive activities, occurring events, and verbal data as a journey along a timeline. Transitional journey maps can be used in varying contexts, ranging from rather unpredictable environments (e.g., incoming messages in operational policing) to reasonably predictable environments (e.g., anticipation in satellite control rooms). We discuss how the anatomy of a transitional journey map enables three classes of analysis. Furthermore, we discuss why the act of creating data visualizations can be viewed as a design process, which justifies *post hoc* activity categorization.

In Case Study 1, fragmentation in police work was quantified by calculating the average segment durations in each activity category (represented by horizontal arrows in Figure 13.9, left panel). This analysis demonstrates that the dynamics of operational policing are not properly represented by aggregating time expenditure per activity category over an entire shift (e.g., Anderson et al., 2005; Smith et al., 2001; Frank et al., 1997). Furthermore, differences between solo and duo patrol were examined through a comparison of their state transition diagrams. These diagrams were constructed through an inventory of all transitions between activity categories (see vertical arrows in Figure 13.9, center panel). Finally, opportunities for and limitations on human information processing were identified by clustering verbal data per activity category (see Figure 13.9, right panel). Apparently, police officers receive

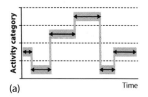 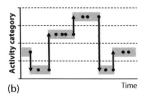 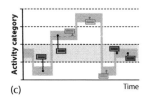

FIGURE 13.9 Analysis of a transitional journey map based on segment lengths (a), transitions (b), and statements (c).

more information than they can process, yet they still desire to be informed about everything. This warrants a reconsideration of the current information system.

In Case Study 2, auditory signals in satellite control rooms were examined. Three transitional journey maps were transformed into a state transition diagram, which showed that the majority of alarms did not result in a transition toward problem-solving behavior. The role of anticipation in decision making was identified, based on verbal data related to events that did result in a transition, and events that did not (see dots in Figure 13.9, center panel). Bayesian inference confirmed that anticipation played a role in spacons' decisions to ignore auditory signals.

13.4.1 TRANSITIONAL JOURNEY MAPS AS A RESULT

The analyses described in Figure 13.9 are possible because of the anatomy of a transitional journey map. Its main elements are the outline (e.g., *x*-axis and *y*-axis), the unit of analysis (e.g., actors, journey, and overarching system), and the interactions (e.g., observations and verbal data). In some studies, the time axis is categorized according to a predefined sequence of activities, such as customer journey maps (e.g., Trischler and Zehrer, 2012; Zomerdijk and Voss, 2010) and anesthesia procedures (e.g., Kennedy et al., 1976). However, the unpredictable aspects of police work and satellite control room operations refute predefined sequences. Therefore, no categorization is used on the *x*-axis of a transitional journey map.

A transition between two activity categories should always signify disruption. To ensure meaningful transitions with regard to workflow fragmentation, activities on the *y*-axis of a transitional journey map are categorized based on a common goal. When activities are categorized on a lower level of abstraction, there is a chance that a sequence of transitions is unintentionally associated with a high level of workflow fragmentation. For example, Zheng et al. (2010) assigned the acts of walking and talking (e.g., what) to separate categories, without reference to a goal (e.g., why). If a transition between these categories is part of the same momentary workflow goal (e.g., walking to a colleague to discuss something), it is illogical to state that the workflow was disrupted.

The combination of verbal data in a visualization helps in understanding human interactions with information systems. Previous studies have identified high levels of workflow fragmentation by showing transitions between activity categories. However, these studies stated that verbal data were required to determine the cause for these transitions (e.g., Cornell et al., 2010; Zheng et al., 2010; Potter et al., 2004). By including verbal data, transitional journey maps enable the examination of such causality. In Case Study 1, for example, incoming messages from the dispatcher made it possible to determine the cause for transitions toward the "driving to an incident" activity category. In Case Study 2, absent reactions to auditory signals were explained by verbal statements in close temporal proximity of the occurring signals. In addition, transitional journey maps incorporate the context in which statements are made. For example, the momentary communication state of the satellite control system was represented as a pass layer. The interpretation of verbal statements was facilitated by knowing that a pass had just ended. Finally, a distinction between retrospective and momentary verbal data facilitates triangulation between reported and observed events. Police officers in Case Study 1 gave retrospective accounts on differences between solo and dual patrol (e.g., police officers operate in couples in case of violence). This difference was confirmed by an event in another observation session (e.g., the motorcyclist waiting for assistance in Figure 13.3).

13.4.2 TRANSITIONAL JOURNEY MAPS AS A PROCESS

There were several iterations of data visualization during the data collection phase. This iterative process facilitated articulating insights during the analysis phase (cf. Segelström, 2009). The field notes presented in Case Studies 1 and 2 (see Figures 13.2 and 13.6) display early versions of the transitional journey map. Initially, data were collected as a set of transcriptions in the appropriate activity category. Later, arrows between the notes provided a sense of order and time, and revealed transitions between activity categories. There were several changes in the number of activity categories, and the order in which they were presented. The latter was guided by the observation that some activity categories were associated with higher levels of mental workload. This transformed a nominal categorization into a semiordinal categorization.

This iterative process shows a strong resemblance with the sketching phase of a product design process. Ferguson (1992) distinguishes three sketch types. So-called thinking sketches support the thinking process of the individual designer, whereas talking sketches support group discussion within the design team, and prescriptive sketches are used to communicate detailed information outside the design time. Goldschmidt (1991) describes thinking sketches in terms of an interpretative cycle. A designer will not "see" the entire image in his or her mind before putting a preliminary version of this image on paper. It is through the act of sketching that the image is brought into existence, both on paper (e.g., the sketch) and in the designer's mind (e.g., knowledge). The acquired knowledge may serve as an inspiration for creating another sketch. A similar interpretative cycle appears to have occurred in creating transitional journey maps. Furthermore, designers draw the same product from different angles, to explore which perspective best communicates relevant product properties. An analogy is found in the rearrangement of activity categories in Case Study 1. Finally, the original field notes comply with Buxton's (2007) characterization of sketches. In his view, sketches are suggestive and tentative, rather than descriptive and specific. Their aim is to explore, rather than to define. In sum, data visualization can be viewed as design activity in explorative field studies.

13.4.3 IMPLICATIONS

In the introduction of this chapter, a methodological question was raised about which level of abstraction results in a meaningful categorization of activities. We have argued that categorizing activities according to a common goal ensures meaningful transitions with regard to workflow fragmentation. However, in Case Study 1, the number of activity categories was extended within the same level of abstraction. Cornell et al. (2010) argue that a large set of activity categories enables one to detect subtle effects, but the downside of such a large set is the risk of losing one's overview of the larger workflow picture. In their study, previous research was used to establish a set of activity categories. An important finding of the present studies is the notion that data visualization is part of an ongoing sense-making process, which does not necessitate *a priori* categorization. Thus, in explorative field studies, the exact number of activity categories can also be determined during and after collecting the data.

Further explorations in workflow visualization should ideally collect data at multiple levels of abstraction. We envision a tool that reveals different workflow patterns by zooming in or out to a given level of abstraction, and by dynamically redefining the number of activity categories within that level. Because lower levels of abstraction require a higher sample resolution, an automatic logging system may help in gathering more detailed work patterns. We have shown that qualitative data are essential to understand quantified behavior. The challenge, then, lies in automatically capturing verbal data. While time-consuming, the presence of a researcher offers the opportunity to ask explanations about what is happening, and to clarify previous statements.

Transitional journey map visualizations combine quantitative data with qualitative data. This combination affords a better understanding of human interaction with information systems in a dynamic context. We hope this understanding will inform studies on workflow fragmentation and inspire the design of future information systems.

ACKNOWLEDGMENTS

Case Study 1 was partially funded by Staf Korpsleiding/Directie Facility Management, Nationale Politie. Case Study 2 was partially funded by the European Space Agency (project code AO 1–7223/12/F/MOS). The authors wish to thank Sacha Silvester for his contribution to Case Study 1, and all police officers and spacons who volunteered for their cooperation.

REFERENCES

Anderson, G.S., A. Courtney, D. Plecas, and C. Chamberlin. 2005. Multi-tasking behaviors of general duty police officers. *Police Practice and Research* 6(1):39–48.
ANSI/ISA. 2009. Management of alarm systems for the process industries. 18.2:2009.
Baethge, A. 2013. A daily perspective on work interruptions. PhD dissertation, Johannes Gutenberg-Universität, Mainz.
Bailey, B.P., and S.T. Iqbal. 2008. Understanding changes in mental workload during execution of goal-directed tasks and its application for interruption management. *ACM Transactions on Computer-Human Interaction* 14(4):1–28.
Beyer, H., and K. Holtzblatt. 1997. *Contextual Design: Defining Customer-Centered Systems.* San Francisco: Morgan Kaufmann.
Bitan, Y., J. Meyer, D. Shinar, and E. Zmora. 2004. Nurses' reactions to alarms in a neonatal intensive care unit. *Cognition, Technology and Work* 6(4):239–246.
Borglund, E., and U. Nuldén. 2012. Personas in uniform: Police officers as users of information technology. *Transactions on Human-Computer Interaction* 4(2):92–106.
Bouwman, H., T. Haaker, and H. de Vos. 2008. Mobile applications for police officers. In *BLED 2008 Proceedings*, 78–90, Bled, Slovenia.
Buxton, W. 2007. *Sketching User Experiences: Getting the Design Right and the Right Design.* San Francisco: Morgan Kaufmann.

Card, S.K., J.D. Mackinlay, and B. Shneiderman. 1999. *Readings in Information Visualization: Using Vision to Think*. San Francisco: Morgan Kaufmann.

Cornell, P., D. Herrin-Griffith, C. Keim, S. Petschonek, A.M. Sanders, S. D'Mello, T.W. Golden, and G. Shepherd. 2010. Transforming nursing workflow, Part 1: The chaotic nature of nurse activities. *Journal of Nursing Administration* 40(9):366–373.

Cvach, M. 2012. Monitor alarm fatigue: An integrative review. *Biomedical Instrumentation and Technology* 46(4):268–277.

De Waard, D. 1996. The measurement of drivers' mental workload. PhD dissertation, University of Groningen.

Ferguson, E.S. 1992. *Engineering and the Mind's Eye*. Cambridge, MA: MIT Press.

Frank, J., S.G. Brandl, and R.C. Watkins. 1997. The content of community policing: A comparison of the daily activities of community and 'beat' officers. *Policing: An International Journal of Police Strategies and Management* 20(4):716–728.

Goldschmidt, G. 1991. The dialectics of sketching. *Creativity Research Journal* 4(2):123–143.

González, V.M., and G. Mark. 2005. Managing currents of work: Multi-tasking among multiple collaborations. In *ECSCW 2005*, ed. H. Gellersen, K. Schmidt, M. Beaudouin-Lafon, and W. Mackay, 143–162. Berlin: Springer.

Izadi, I., S.L. Shah, and C. Tongwen. 2010. Effective resource utilization for alarm management. Presented at 49th IEEE Conference on Decision and Control, December 15–17.

Jansen, R.J., R. Van Egmond, and H. De Ridder. 2015. No alarms and no surprises: How qualitative data informs Bayesian inference of anticipated alarm sounds. *Procedia Manufacturing* 3:1750–1757.

Jansen, R.J., R. Van Egmond, H. De Ridder, and S. Silvester. 2014. Transitional journey maps: Capturing the dynamics of operational policing. In *Proceedings of the Human Factors and Ergonomics Society Europe Chapter 2013 Annual Conference*, 15–27, Turin, Italy.

Kennedy, P.J., A. Feingold, E.L. Wiener, and R.S. Hosek. 1976. Analysis of tasks and human factors in anesthesia for coronary artery bypass. *Anesthesia and Analgesia* 55(3):374–377.

Keyson, D.V., A. Al Mahmud, and N. Romero. 2013. Living lab and research on sustainability: Practical approaches on sustainable interaction design. In *Ambient Intelligence*, ed. J.C. Augusto, R. Wichert, R. Collier, D.V. Keyson, A.A. Salah, and A.-H. Tan, 229–234. Berlin: Springer.

Kondaveeti, S.R., I. Izadi, S.L. Shah, T. Black, and T. Chen. 2012. Graphical tools for routine assessment of industrial alarm systems. *Computers and Chemical Engineering* 46:39–47.

Korniewicz, D.M., T. Clark, and Y. David. 2008. A national online survey on the effectiveness of clinical alarms. *American Journal of Critical Care* 17(1):36–41.

Lundin, J., and U. Nuldén. 2007. Talking about tools—investigating learning at work in police practice. *Journal of Workplace Learning* 19(4):222–239.

Martínez-Heras, J.-A., A. Donati, M.G.F. Kirsch, and F. Schmidt. 2012. New telemetry monitoring paradigm with novelty detection. In *SpaceOps 2012 Conference*, 1–9, Stockholm, Sweden.

Matthews, M.L., and P. Desmond. 2002. Task-induced fatigue states and simulated driving performance. *Quarterly Journal of Experimental Psychology* 55:659–686.

Michon, J.A. 1985. A critical view of driver behavior models: What do we know, what should we do? In *Human Behavior and Traffic Safety*, ed. L. Evans and R.C. Schwing, 485–524. New York: Plenum Press.

Monk, C.A., J.G. Trafton, and D.A. Boehm-Davis. 2008. The effect of interruption duration and demand on resuming suspended goals. *Journal of Applied Experimental Psychology* 14(4):299–313.

Morgan, J.F., and P.A. Hancock. 2011. The effect of prior task loading on mental workload: An example of hysteresis in driving. *Human Factors* 53(1):75–86.

Nardi, B. 1995. *Context and Consciousness: Activity Theory and Human-Computer Interaction.* Cambridge, MA: MIT.

NEN-EN-IEC. 2015. Management of alarm systems for the process industries. 62682:2015.

Niitamo, V.-P., S. Kulkki, M. Eriksson, and K.A. Hribernik. 2006. State-of-the-art and good practice in the field of living labs. In *Proceedings of the 12th International Conference on Concurrent Enterprising: Innovative Products and Services through Collaborative Networks*, 249–357, Milan, Italy.

Potter, P., S. Boxerman, L. Wolf, J. Marshall, D. Grayson, J. Sledge, and B. Evanoff. 2004. Mapping the nursing process: A new approach for understanding the work of nursing. *Journal of Nursing Administration* 34(2):101–109.

Randall, M.J., E.M. Roth, K.J. Vicente, and C.M. Burns. 2000. There is more to monitoring a nuclear power plant than meets the eye. *Human Factors* 42(1):36–55.

Rasmussen, J. 1983. Skills, Rules, and Knowledge; Signals, Signs, and Symbols, and Other Distinctions in Human Performance Models. *IEEE Transactions on Systems, Man, and Cybernetics* 13(3):257–266.

Rauterberg, M. 1999. Different effects of auditory feedback in man-machine interfaces. In *Human Factors in Auditory Warnings*, ed. N.A. Stanton and J. Edworthy, 225–242. Sydney: Ashgate.

Schleburg, M., L. Christiansen, N.F. Thornhill, and A. Fay. 2013. A combined analysis of plant connectivity and alarm logs to reduce the number of alerts in an automation system. *Journal of Process Control* 23(6):839–851.

Seagull, F.J., and P.M. Sanderson. 2001. Anesthesia Alarms in Context: An Observational Study. *Human Factors* 43:66–78.

Segelström, F. 2009. Communicating through visualizations: Service designers on visualizing user research. In *First Nordic Conference on Service Design and Service Innovation*, 175–185, Oslo, Norway.

Sendelbach, S., and M. Funk. 2013. Alarm fatigue: A patient safety concern. *Advanced Critical Care* 24(4):378–386.

Smith, B.W., K.J. Novak, and J. Frank. 2001. Community policing and the work routines of street-level officers. *Criminal Justice Review* 26(1):17–37.

Sørensen, C., and D. Pica. 2005. Tales from the police: Rhythms of interaction with mobile technologies. *Information and Organization* 15(2):125–149.

Spano, R., and M.D. Reisig. 2006. "Drop the clipboard and help me!": The determinants of observer behavior in police encounters with suspects. *Journal of Criminal Justice* 34(6):619–629.

Stanton, N. 1994. Alarm initiated activities. In *Human Factors in Alarm Design*, 93–117. London: Taylor & Francis.

Stanton, N.A., D.J. Harrison, K.L. Taylor-Burge, and L.J. Porter. 2000. Sorting the wheat from the chaff: A study of the detection of alarms. *Cognition, Technology and Work* 2(3):134–141.

Stol, W.Ph., Ph. van Wijk, G. Vogel, B. Foederer, and L. van Heel. 2004. Politiestraatwerk in Nederland: Noodhulp en Gebiedswerk [Police streetwork in the Netherlands]. Apeldoorn, the Netherlands: Nederlandse Politie Academie and Zeist, the Netherlands: Kerckebosch.

Straus, S., T. Bikson, E. Balkovich, and J. Pane. 2010. Mobile technology and action teams: Assessing BlackBerry use in law enforcement units. *Computer Supported Cooperative Work* 19(1):45–71.

Streefkerk, J.W., M.P. van Esch-Bussemakers, and M.A. Neerincx. 2006. Designing personal attentive user interfaces in the mobile public safety domain. *Computers in Human Behavior* 22(4):749–770.

Trischler, J., and A. Zehrer. 2012. Service design: Suggesting a qualitative multistep approach for analyzing and examining theme park experiences. *Journal of Vacation Marketing* 18(1):57–71.

Tromp, N., M. van Dijk, and P. Hekkert. 2010. Toekomstig Mobiel Werken: een visie vanuit het sociale systeem [Future mobile workplaces: A vision based on the social system]. Internal report Delft University of Technology, Delft, the Netherlands.

Vastenburg, M.H., D.V. Keyson, and H. De Ridder. 2009. Considerate home notification systems: A user study of acceptability of notifications in a living-room laboratory. *International Journal of Human-Computer Studies* 67:814–826.

Westbury, C.F. 2010. Bayes' rule for clinicians: An introduction. *Frontiers in Psychology* 1:1–7.

Woods, D.D. 1995. The alarm problem and directed attention in dynamic fault management. *Ergonomics* 38(11):2371–2393.

Zheng, K., H.M. Haftel, R.B. Hirschl, M. O'Reilly, and D.A. Hanauer. 2010. Quantifying the impact of health IT implementations on clinical workflow: A new methodological perspective. *Journal of the American Medical Informatics Association* 17(4):454–461.

Zomerdijk, L.G., and C.A. Voss. 2010. Service design for experience-centric services. *Journal of Service Research* 13(1):67–82.

14 The Missing Links in System Safety Management

Karen Klockner and Yvonne Toft

CONTENTS

14.1   INTRODUCTION

> To an outsider it might appear that industrial accidents occur because we do not know how to prevent them. In fact, they occur because we do not use the knowledge that is available. Organisations do not learn from the past … and the organisation as a whole forgets.

These words by Kletz (1993, p. 1), from his book *Lessons from Disaster: How Organisations Have No Memory and Accidents Recur*, are still relevant and powerful for those organizations that are striving to enhance their ongoing preventive safety efforts by eliminating the blunt and sharp end factors that seem to inevitably occur despite the best safety management efforts. Invariably, organizations and safety regulators often identify that there appear to be reoccurring patterns and themes to the contributing factors identified by safety occurrence investigations. The ongoing frustration is how lessons can be learned from what has already occurred and how can that information be used to identify how areas and aspects of organizational safety management systems are negatively contributing to safety occurrences.

Perhaps in an effort to solve this very problem, the notion of accident modeling came about in the early 1930s and was an attempt to give a visual and graphical understanding of how safety occurrences happen. The first model was simple, and can be thought of as a fair representation of the industrial age of work, considering the somewhat simple interactions between man and machine that occurred at that time. It was able to provide a notion that, by viewing these simple system interactions laid out in a pictorial linear fashion, the factors that contributed to a safety occurrence could be identified with a view to learning from those negative outcomes. From the safety practitioner's point of view, it provided an understanding of the trajectory of events as they occurred and pointed toward which factors need to be examined and identified as contributing to the accident. Organizations could then focus safety efforts on removing or controlling the identified issues.

The theoretical viewpoint of reviewing safety occurrences after the event and collecting information on what happened has always been to try to make sense of those events from a lessons learned perspective. This paradigm has not changed. What has changed is the complexity of systems of work in which humans find themselves today, and a better appreciation is now had that safety occurrences are caused by contributions of both technical and sociocultural factors, including what are well-accepted human factors, that is, ergonomics, fatigue, etc. Safety researchers, in acknowledging this, have called for accident models that better represent these modern-day complex systems and better display the relationship between what is now a much wider list of contributing factors.

This ongoing appreciation of complex safety system theory is now pointing organizations to view and understand their safety management systems as a network of factors that operate as an integrated whole (system) rather than a stand-alone, separate parts model. This has led to the development and testing of new methodologies that result in accident models that are now truly reflective of the current complex sociotechnical system (network) theories. This type of modern-day accident modeling will be covered here, starting with a review of the development of accident modeling as a safety management tool.

14.2 SEQUENTIAL ACCIDENT MODELING

The concept of accident modeling developed in an attempt to further the understanding of why accidents came about and how to prevent them from occurring in the future. The desire for this understanding resulted in a need to explore accident trajectories looking at causal and contributing factors and a further desire to control accidents through preventive barriers that strive to eliminate the escalation path of the safety occurrence contributing factors.

A historic timeline of how accident models have developed can clearly be established, and it is evident that they have certainly moved through various reiterations since 1931, when the first accident model was proposed. The history of accident modeling itself can be traced back to the original work by Herbert. W. Heinrich, whose book *Industrial Accident Prevention* in 1931 became the first major work on understanding accidents and preventing them. One of the first sequential types of accident models to be espoused was that of the domino effect or domino theory (Heinrich, 1931). This model proposed that certain accident factors could be thought of as being able to be stacked in a line like dominos, and represented the earliest notion that there is a temporal sequence to an accident that progresses like a chain of connected events or situations, one happening before the other, which in effect compounds and leads to a negative outcome.

Heinrich (1931) gave the five domino factors labels, ranging from the social environment, to fault of the person, to the unsafe acts or conditions that lead to the accident and then the injury itself. Heinrich held that "the occurrence of a preventable injury is the natural culmination of a series of events or circumstances, which invariably occur in a fixed or logical order ... an accident is merely a link in the chain" (p. 14). When one of the dominos (contributing factors) falls, it has an ongoing knockon effect, which ultimately results in an accident. Accident prevention was therefore understood to be the result of the removal of one of the factors because, if "this series is interrupted by the elimination of even one of the several factors that comprise it, the injury cannot possibly occur" (p. 15).

Heinrich believed that "the unsafe act and mechanical hazard constitute the central factor in the accident sequence," and furthermore stated that "the removal of the central factor makes the action of the preceding factors ineffective." as shown in Figure 14.1.

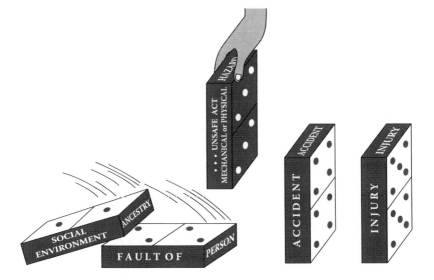

FIGURE 14.1 Heinrich: the removal of the central factor.

Heinrich's key tenet, which still remains valid in terms of risk management theory today, is that the removal of the hazard would prevent the accident or injury from occurring. This long-held view goes to the heart of preventive safety measures in that safety professionals still strive to identify risks and hazards in an attempt to control or remove them, and hence prevent safety occurrences from happening.

Sequential linear models therefore offered an easy visual representation of the "path" that accidents took in their progression leading to an accident event. Accident modeling in this fashion did not escape from the widely accepted linear time aspect of events, which is tied into the "western cultural world-view of past, present and future as being part of everyday logic, prediction and linear causation" (Buzsákl, 2006, p. 8).

Bird and Germain (1985) later acknowledged that the domino model had been widely used to convey the principles of accident prevention and loss control, with Heinrich's original domino sequence being a classic in safety thinking for more than 30 years. They were therefore happy to develop an "updated domino model to reflect the direct management relationship with the causes and effects of accident loss ... whereby arrows were incorporated to show the *multilinear interactions* of the cause and effect sequence" (p. 34). This new model became known as the Loss Causation Model. Bird and Germain recognized the need for management to prevent and control accidents in what were fast becoming highly complex situations attributed to advances in technology. Their model was viewed as being able to enable "the user to understand and retain the critical few facts important to the control of the vast majority of accidents and management problems" (p. 22).

The first call for major change around the understanding of safety occurrences (at that time called events) can be attributed to Floyd Allport, a social psychologist, as early as 1954 (Allport, 1954). He can perhaps now be considered, looking back at his writing at that time, the father of complex system thinking in the area of accident modeling. He was possibly the first to outline the need for accident models to be shown as events in a cyclical structure that were "connected by continuous ongoings ... but extend ... from an indefinite past into an indefinite future" (p. 297). His work most closely aligned with the concept of how events are now thought to be patterns of interconnectedness, and importantly influenced by the relationships of constituent parts.

14.3 SYSTEMIC ACCIDENT MODELS

By 1984, papers presented at the International Seminar on Occupational Accident Research (Benner, 1984) had identified that previous accident models did not reflect any realism as to the true nature of the observed accident phenomenon. Benner stated that "one element of realism was non-linearity ... models had to accommodate non-linear events. Based on these observations, a realistic accident model must reflect both a sequential and concurrent non-linear course of events, and reflect event interactions over time" (p. 177). Rasmussen's (1990) discussion of causal accident factors acknowledged that the identification of events and objects in an accident are not isolated but "depend on the context of human needs and experience in which they occur and by definition ... therefore will be circular" (p. 451). Systemic accident model development was therefore commenced to examine in greater detail the idea that system failure, rather than just human failure, was a major contributor to accidents. These models began to address these issues and recognized that events do not happen in isolation from the systemic environment in which they occur.

Rasmussen (1990) also wrote extensively on the problem of causality in the analysis of accidents and explored the struggle to decompose real-world events and objects, and explain them in a causal path. This resulted in the notion that causes can be found upstream from the actual accident, where these latent effects from earlier events or acts may be lying dormant. He recognized that sociotechnical systems were both complex and not stable, and that any attempt to discuss a flow of events does not take into account "closed loops of interaction among events and conditions at a higher level of individual and organizational adaption ... the causal tree found by an accident analysis is only a record of one past case, not a model of the involved relational structure" (p. 454). His call was for a new approach to the analysis of causal connections found in accident reports. It heralded in a realization that there was a growing complexity in attempting to both graphically display accidents (for ease of use and reader understanding) and capture the temporal and complex system and events that surrounded their initial manifestation.

James Reason also accepted that accidents were not solely due to individual operator error (active errors), but lay in the wider systemic organizational factors (latent conditions) in the upper levels of the organization. Reason's (1990) model is commonly known as the Swiss Cheese Model, but is also known as the Model of Organizational Accidents.

Unlike the early modeling work of Heinrich, and Bird and Germain, Reason did not attempt to specify what the holes represented or what the various layers of cheese represented. The model left the safety practitioner to his or her own investigations as to what factors within the organization these might be.

One of the most forward-thinking researchers in the area of accident modeling and the understanding of causal factors is Erik Hollnagel. Hollnagel's (2004) book on barriers and accident prevention saw a number of challenges put forward in relation to the current way of thinking about accident modeling. Indeed, in this book, Hollnagel introduces the reader to the concept of a three-dimensional way of thinking about accidents in what are now known to be highly complex and tightly coupled sociotechnical systems in which people work.

Hollnagel (2004) originally proposed a new model called the Functional Resonance Accident Model (FRAM); however, his later book on FRAM (Hollnagel, 2012) acknowledged that FRAM was to become a methodology that resulted in a model, and while keeping the acronym FRAM, he changed the name to Functional Resonance Analysis Method. This model is the first attempt to place accident modeling into a three-dimensional framework, as Hollnagel (2012, p. 171) recognized that "forces (being humans, technology, latent conditions, barriers) do not simply combine linearly thereby leading to an incident or accident."

Hollnagel's originally proposed systemic FRAM began by looking at how different functions within an organization were linked or coupled to other functions. This was done with a view to understanding the variability of each of the functions and how that variability could be both interpreted and managed in terms of an accident model. Variability in one function would also affect and be affected by the variability of other functions, as would be expected in system thinking.

14.4   METHODS-SINE-MODEL VERSUS MODEL-CUM-METHOD

The understanding of how things happen, in this case safety occurrence investigation, has recently seen a major shift in paradigm thinking around the issue of whether investigative models produced the investigation methods (which are required to fit that model), or if investigation methods can produce results that can be understood in the resultant model produced by using a methodology. Hollnagel (2012) has now eloquently discussed this issue in relation to his latest work:

> In most approaches to accident analysis, the underlying model defines or describes a set of relations while the associated method provides a way to interpret events in terms of those relations.... Since the models provide a clearly structured representation of the 'world', the methods are typically linear with either single or multiple cause-effects paths. In these approaches, the methods in practice impose an a priori interpretative structure on the event. The value of the results of an analysis therefore depends on the correctness of the model.... In everyday practice, which means in the short-term, the advantage of an articulated model is the efficiency of the associated method—even if the model is incorrect ... commonly used methods try to describe relations derived from the model, and hence represent a model-cum-method approach. (pp. 131, 132)

Hollnagel's FRAM method is a breakaway from the traditional way of using a model to dictate a method of accident investigation, which then needs to fit into the model structure, be it one of hierarchy, layers, parts, or components, or any other simplified representation of how accidents happen (Hollnagel, 2012). Hollnagel has proposed that his new method (FRAM) results in a model, and that by undertaking the methodology of FRAM, there is no presumed organization of the system under investigation or about the cause-and-effect relationship. The system is revealed by the resultant model, and while the method may produce similar models of systems under examination, this would be "simply because the cases or situations are similar" (p. 132). Therefore, Hollnagel has moved to a method-sine-model approach whereby the method is determining the output results in a model form. Hollnagel, as one of the most widely recognized safety scientists stepping into this way of rethinking accident modeling, has opened the door for this new thinking to be accepted and followed within the safety sciences.

This latest insight into another way of developing accident models via the "method first" trajectory has been timely in solving a major issue for this research. While one of the aims of this research was to develop a new model for four types of major railway accidents under review, the way of getting to the necessary models can now be the result of the methodology used for this research (i.e., network analysis), rather than trying to fit a method into a model; it will now be acceptable for the method to develop the model in the method-sine-model rationale.

14.5   ACCIDENT MODELING SUMMARY

It is perhaps difficult to say that accident modeling has had a rich history, but it certainly has had a history that can be traced back to the 1930s and the zenith of the industrial age. The industrial age and its technological developments saw the birth of the concept of management efforts to prevent and control occupational injury and illness (Bird and Germain, 1985), and accident modeling has continued to develop since that time. The question of how it can now be undertaken to truly reflect the complex sociotechnical systems of work in today's modern world is the endeavor this research seeks to fulfill.

It may appear that progress in this area has been slow, but concepts of complex systems and system theory have also been developing for some time. Leplat (1984) summarized that "an accident is a phenomenon resulting from the intervention of a set of variables for which where is no simple model ... their study will often necessitate the combination of several models and the concepts of systems theory" (p. 87).

These ideas continue to be supported by the latest expert opinion that "accidents are complex processes involving the entire socio-technical system. Traditional event-chain models cannot describe this process adequately" (Leveson, 2011, p. 31), and that view fully supports the accident modeling research and methodology being developed here.

14.6   IDENTIFICATION OF CONTRIBUTING FACTORS

The identification of causal and contributing factors related to a safety occurrence is explicably tied to the accident model favored at the time. Accident investigations are usually based on a model of some type. Typically in today's thinking based around systemic models, investigators collect data about individual or team actions at the sharp end, and organizational influences and technical issues at the blunt ends. While human error is still a prominent area of interest, the movement toward understanding how organizational factors enable and promote human error is becoming much more mainstream.

There are numerous tools for use within the area of safety science to identify contributing factors. Since Van Vuuren's review in 1999, there has been a growth in the number of tools developed to capture systemic contributing factors and, in particular, the human factors and upstream organizational factors in accident causation.

14.6.1   CONTRIBUTING FACTORS FRAMEWORK FOR THE RAIL INDUSTRY

The Contributing Factors Framework (CFF) was developed for use within the rail industry in Australia and was launched in the industry in early 2009 throughout the various state Rail Safety Regulation jurisdictional offices. The CFF is primarily used as a tool, rather than an investigation method, but does assist in this endeavor by providing accident investigators within the rail industry a specific set of criteria for the identification of contributing factors to safety occurrences. A contributing factor has been defined as "any element of an event that, if removed from the sequence of the events leading to the occurrence, could have prevented the occurrence or reduced the severity of the consequences of the occurrence" (Grey et al., 2011, p. 1).

Rail Safety Regulators in each jurisdiction in Australia collect data on rail safety occurrences using criteria specified within a standard that was known as the Occurrence Notification Standard One (ON-S1) from 2004 to 2008 and the Occurrence Classification—Guideline One (OC-G1) Version 1(1) after 2008 (Grey et al., 2011).

OC-G1 is the coding system used to identify the different types of railway safety occurrences under investigation in this research. While the use of OC-G1 has provided the rail industry with a standard safety occurrence type of coding system for various types of safety occurrences (derailments, signals passed at danger, etc.), the development of the CFF allowed the coding of why the safety occurrence occurred by providing a tool for identifying the contributing factors.

14.7   REOCCURRING SAFETY OCCURRENCE THEMES

Some safety researchers (Maurino, 1999) have stated that "accidents in sociotechnical systems rarely repeat themselves, in part because the exact conditions and environment surrounding an accident seldom (if ever) recur" (p. 417). This appears not to be the case for the rail industry, and what is evident from a review of major incidents in this research ($n = 429$) is that reoccurring themes can be identified and, in fact, types of incidents can be routinely categorized by railway reporting codes for the various types of safety occurrences. Regularity in terms of causal relationships can be found between types of events and subtypes of events, and event causal connections therefore depend on categories that are identified by typical examples and prototypes, with the very nature of causal explanations shaping the analysis of accidents (Rasmussen, 1990; Rosch, 1978).

The rail industry is not the only one to have identified that incidents have trends, and it is suspected that most industries could identify reoccurring themes in their safety occurrences, as discussed by Perrow (1984) for the nuclear industry: "Two-thirds of the problems discussed in this issue are strikingly similar to ones previously reported in *Nuclear Safety* ... in the hope and expectation that we will all be able to learn from the experiences of others ... operators should take particular note of these occurrences so that they can more readily avoid similar happenings in their own plants" (as cited in Perrow, p. 48, quoting Castro, 1972).

Benner (2009), in discussing road vehicle accidents, has acknowledged that the word *retrocursors* has been given to such events and has been defined as "accidents that replicate the behaviour of prior occurrences" (p. 5). Benner has advanced the view that these events often demonstrate that lessons that could have been learned from prior mishaps have yet to be recognized, and that "retrocursors occur often enough that they should have captured the attention of system safety practitioners" (p. 5).

Benner's rationale as to why this happens is perhaps the crux of the research being conducted here, namely, that up until now, there have been opportunities to prevent accidents that have occurred, but this has not happened for a number of reasons: (1) a lack of sufficient robust data are captured that indicate a system's dynamic performance, (2) there is little cross-referencing of either obvious or subtle similarities across systems, (3) historical investigation data are not captured to allow for easy retrieval by key players who need those data, and (4) lessons to be learned are seldom, if ever, itemized (through visual modeling) to provide a road map for improving a system's operational efficiency. His overall call was for system safety analysts to "demand improved investigation lessons learned quality, dissemination and access" to information on accidents (p. 7).

This research attempted to not only meet those demands for the rail industry, but also offer the industry more. What this research points toward is a new way of examining safety occurrences and capturing the information about what has contributed to them on a larger scale, but one that is nevertheless simple enough to be used by safety analysts. What is required is a way to view safety occurrences in systems as visual patterns, so that reoccurrences can be seen and preventive action taken on those factors that contribute to the repetitious nature of these events. This leads to support for a new way of accident modeling that meets these demands.

While developments in accident modeling appear to have been slow, the assumption that safety occurrences can be understood by looking at a linear chain of events is now widely implied to be outdated (Leveson, 2011). Researchers have continued to acknowledge that accident models need to be more reflective of the interaction of parts within the complex systems in which they occur. This new challenge calls for the understanding that "accidents are complex processes involving the entire sociotechnical system ... and we need to understand how the whole system, including organisational and social components, operating together led to the loss" (Leveson, 2011, p. 31). Leveson goes on to call for a method that can expand the investigation beyond the proximate event and its individual factors, to a wider view of "the most important factors in terms of preventing future accidents" (p. 33).

Essentially, the call was for how current-day accident modeling should be done in a way that allows a sound representation of safety occurrences in the modern complex sociotechnical systems in which humans now work.

The starting point is perhaps to understand that different types of safety occurrences should be represented by different accident models because types (meaning the same category of event) of safety occurrences have their own uniqueness. The older style generic accident model, characterized as one model fits all, is no longer applicable or valid. While each safety occurrence may be unique, this is not to say that reoccurring types of occurrences do not have their own similarities, and these are usually around the contributing factors identified as being present at the time of the occurrence. Patterns therefore can and do emerge where the same types of events happen over and over again, and rail safety occurrences are no exception to this.

The development of accident models that visually demonstrate a pattern of reoccurring contributing factors reflective of the whole operating system is the key in being able to fully understand the big picture system complexity. Aspects of safety management systems can then be examined, and by understanding which factors are present for reoccurring types of safety occurrences over a period of time, it is therefore possible to fully grasp the lessons that need to be learned.

14.8 SOLVING THE CALL FOR NEW ACCIDENT MODELS

In complex systems where safety events occur, the traditional notion of pure linear causality begins to crumble and safety science researchers are beginning to recognize that (1) events are not clearly delineated or independent of the context in which they occur, (2) there is complexity in identifying the flow of influence from one contributing factor to the next in that they may (but often don't) contribute in a linear way, (3) small changes might lead to big effects as flow on effects compound, and

(4) two or more factors alone may not be harmful, but when combined they interact to become dangerous (e.g., fatigued workers performing safety-critical tasks).

Complex systems in this context are not implied to mean complicated, but the implication is toward a nonlinear relationship between the interacting and relational system components. These complex systems are also termed open, and while they appear stable, they are in constant modes of change; this is seen as a defining feature compared to "closed" systems that remain in balance, stable, and hard to perturb. Complex systems are characterized as also having constituent parts that tend to become adaptive to change, being multileveled and hierarchical, and often far from operating in a state of equilibrium (Buzsáki, 2006).

While accident investigation and the identification of causal or contributing factors is said to have moved to a system approach (based on system theory, whereby one has to understand the components that work separately, as well as how the components work within the larger system in which they operate), the linear accident trajectory model has not caught up with the concept of nonlinear system thinking.

At the very least, new models that give a clearer picture and can examine and explain the interconnectedness of aspects within the system, and the relationship of constituent parts, are a good starting point. A way of modeling accidents where the traditional cause-and-effect rules are challenged to allow the joining and linking of contributing factors, previously unconsidered in this way, was the drive behind this research.

14.9 INFORMATION VISUALIZATION

The research method here finds its pedigree in the area of information visualization. This field of research is less than 15 years old, but is now extending into being a far-reaching, interdisciplinary field (Chen, 2006). Network analysis builds upon graph theory to study the visualization of these graphs. A network can therefore be thought of, "in its simplest form, as a collection of points joined together in pairs by lines" (Newman, 2010, p. 109). This research was interested in both the mathematical outputs in relation to the graphs under investigation, with these outputs (models) being examined as measures of centrality, and the actual visual graphs displayed by using network analysis software.

Newman (2010) identifies various focuses to any study of networks, but three main reasons prevail: (1) to review the nature of the individual components, (2) to study the nature of the connections or interactions, and perhaps most importantly, (3) to see and understand the pattern of connections between the components. For researchers wanting to use network analysis as a research method, modern software programs based on Social Network Analysis (SNA) offer the ability to transform data to enable it to be visualized both as a visual network and in a statistical form suitable for network analysis, predominantly including measures of centrality.

14.10 LOGICAL USE OF SOCIAL NETWORK ANALYSIS

The history of SNA dates back to 1934, with the work of Jacob Moreno, who is recognized as doing the first work in the field of sociometry, or SNA as it is now known

(Newman, 2010). The use of information visualization using SNA for this research made sense given that one of the aims of the research was to examine the interactions of the contributing factors as a network of connections in a larger system with a view to having a model as a resulting outcome.

Social networks traditionally represent the connections between people, with the *vertices* representing people or groups of people and the *edges* representing some form of interaction (usually social) between the people or actors in a system. SNA, however, lends itself to the study of many different types of interactions, with connections able to represent any type of connection, from friendships to exchange of money, professional relationships, or connections between actors and things.

The supposition used to conceptualize and operationalize the translation of the research presented here into the SNA realm was that the contributing factors to safety occurrences could also be thought of as people or actors, and that these actors (factors), when identified as having been part of a safety occurrence (as a contributing factor), would have met each other, as if they had attended a party, but instead they attended a safety occurrence. The contributing factors could therefore be thought of as individuals who knew and met each other every time they showed up at a safety occurrence (the event). If they were present at the event, then they met all the other factors that also showed up at that event. The data on which factors showed up at each event, and hence knew and met each of the other factors at that same event, could be investigated using the tools of SNA. This would enable an examination of the networking of the factors for each type of event under study. Information visualization using SNA as a method both offered a resulting accident model and aided the understanding of the actual contribution to the safety occurrence as identified through the examination of centrality measures in SNA.

What was therefore performed was a conversion of the traditional data collected under the individual headings of the CFF into a relationship format, where one factor was linked to another factor.

14.10.1 Social Network Analysis Centrality Measures

Result outcomes in SNA are examined in terms of patterns of connections around an individual (or other entity), in this case a contributing factor, which results in a visualization displayed as a network accident model.

Statistical results are also available that are centered on network metrics in the form of centrality measures, which determine the relative position of the contributing factors within the larger network. Network analysts use centrality measures to determine the importance or most central factors in the network. In a typical social network problem, these central factors might be key people who are influencers. Of interest in the present research is which of the contributing factors act as the major influencers in the different types of railway safety occurrences.

There are many common formally established measures of centrality, such as degree centrality, closeness centrality, betweenness centrality, eigenvector centrality, and Katz centrality. The objectives of the research generally guide which measures will be used. The Safety and Failure Event Network (SAFE-Net) models displayed in this chapter show betweenness centrality.

Betweenness centrality is of particular interest because it measures the relative importance of a node by measuring not how well connected it is, but instead where it falls between other nodes in the network. It is, in effect, an information broker to the other nodes. The nodes with the highest betweenness centrality derive a lot of power from their position in the network, and their removal from the network will disrupt communications to other nodes (Newman, 2010).

Betweenness centrality is also a measure of how much removing a factor would disrupt the connections between one factor and another factor. Betweenness centrality may be important in accident modeling, because—theoretically—the effective disruption of a network of contributing factors may prevent a future safety occurrence from manifesting. This premise lies at the heart of preventive risk management safety strategies, which expect that the control or removal of identified risk factors will lead to the prevention of safety occurrences.

14.11   RESEARCH METHOD STEPS

This research required several steps to be taken. This included the steps of data collection by using the CFF to code the contributing factors found by examining major railway safety reports, traditional number counts, and data analysis, and then the use of data modeling using SNA, with further data analysis based on the model outputs.

One of the aims of this research was to collect information on the contributing factors to major types of railway accidents. Major incident reports were therefore analyzed using the CFF tool, and data were collected for the 5-year period from 2006 to 2010. Overall, 429 major rail incident reports were analyzed using the CFF. The data set for this research came from the accident investigation reports for major rail safety occurrences submitted from rail transport operators to the office of the Rail Safety Regulator in Queensland between the years 2006 and 2010. These major incident reports were read and coded using the CFF to identify the contributing factors for each type of safety occurrence under review. This coding was completed by staff from the Rail Safety Regulation Branch who had been trained in the use of the CFF tool to identify the contributing factors to major railway safety occurrences in Queensland.

14.12   SAFETY AND FAILURE EVENT NETWORK METHODOLOGY

The new method developed as part of this research to allow the modeling of safety occurrences using a network methodology, in particular SNA, has been called the SAFE-Net method. This method in effect demonstrates a way of taking traditional safety occurrence data and using an SNA platform to enhance the understanding of the relationship between contributing factors for various types of railway occurrences, and also to provide a new way of modeling railway safety occurrences that better reflects the complex sociotechnical systems of today's modern railways.

Discussed and demonstrated below is how accident modeling can be done through the use of the SAFE-Net method, providing the reader with an overview of just some of the models available to be examined using this method.

14.12.1   SAFE-Net Method and Modeling Options

To begin the process, the original data obtained on the contributing factors for rail safety occurrences were required to be converted into relational data, as the use of an SNA method requires that there is a connection between one contributing factor and another contributing factor, a mutual (social or other relationship) connectedness.

In order to obtain a relational set of data, the SAFE-Net method was therefore operationalized on the premise that for every safety occurrence where two or more contributing factors were identified, they could be thought of as having a relationship such that they met each other at the time of the occurrence. As mentioned above, this can be theoretically conceived as viewing each safety occurrence like a party and each factor like a person who attended that party. At each party (safety occurrence), all the factors (contributing factors) that attended (were identified) met each other and therefore formed a relationship.

Traditional data were therefore converted into a social format by linking together any two factors that were identified as attending the same safety occurrence (party). A simple example is provided below for three factors that attend one safety occurrence. The three factors are fatigue, business planning, and training. The relationship for the three factors meeting each other would be as shown below, where fatigue meets business planning, fatigue meets training, and business planning meets training:

Fatigue ↔ Business planning
Fatigue ↔ Training
Business planning ↔ Training

Fatigue has a mutual or reciprocal (undirected) relationship with both of the other factors: business planning and training. Business planning and training also have the same type of relationship with each other. All factors that were identified as being present at the time of the occurrence met each other.

The method then involves making a list of these relationships for each safety occurrence. The safety occurrences can then be viewed for each type or subtype of railway safety occurrence as identified by the OC-G1. This converts the data into a social network format ready to be accepted by SNA programs.

Once the data were converted into relational data, they could be entered into an SNA program to enable the program to both calculate centrality measures and show models of the safety occurrence under review. While a few different programs were used, the main software program used for modeling of the safety occurrences was TouchGraph Navigator 2 (TouchGraph), with NetMiner being used for obtaining some of the centrality measures. Once the data are entered into the SNA program, many variations of the data can then be examined and modeled, depending on the area of interest.

SAFE-Net modeling has also been shown to provide models that individually reflect the safety occurrence type under investigation, and highlight those factors that can be deemed of greatest interest based on their contribution to the model. These models can be manipulated by adding filters to measurements of interest, and also color coded to allow further analysis of research around subtypes of contributing factors. Furthermore, the models can be reflective of many similar types of safety incident reoccurrences over a period of time, or provide valuable data on a single occurrence requiring analysis to direct future preventive safety actions.

The SAFE-Net method also delivers on the requirement stated by Hollnagel (2012) that systems are revealed using a methodology that produces a model, rather than the other way around.

The below results obtained using a traditional data analysis method are directly compared to the results obtained using the SAFE-Net methodology to not only allow a direct comparison, but also show an understanding of what value can be had by looking at the new resultant accident models through a new lens.

14.12.2 Comparing Results of Traditional Data Analysis versus SAFE-Net

Here, one subtype of railway safety occurrences is directly compared, looking at the distinction between those factors that were identified as having the highest number counts using traditional data analysis, and then reviewing those factors with the highest betweenness centrality measures for the same types of incidents. This allows a direct comparison of the results from the two types of data analysis to be seen.

Safety occurrence data have traditionally been limited to nominal data analysis methods, being number counts and frequency counts. This is primarily due to the fact that a safety occurrence (accident or incident) either did or did not occur. Of course, the counts are reflective of only the events that actually did occur.

The data collected for this research on the contributing factors to safety occurrences are in this category, meaning that if the contributing factor was identified as having occurred at the time of the event, it can be counted and grouped together to give a number or percentage of some total number. Analysis, then, is traditionally limited to the understanding of what can be gleaned by standard data and charts reflecting number counts.

The data were reviewed across the three factor category headings of the CFF: Individual or Team actions, Technical Failures, and Local Conditions and Organizational factors. The CFF tool was modeled on a linear context of accident phenomenology, meaning that the model hypothesizes that the organizational factors that contributed to the accident occurred at the blunt end, farthest away from the accident. At the sharp end, individual or team factors, as well as technical factors, have a close relationship to the eventual accident. This places contributing factors into three individual streams for analysis and data collection, under these separate category headings.

14.13 DERAILMENTS

One subtype of derailment safety occurrences will be discussed, running line derailments, with comparisons made between the understandings provided by the two methods for safety occurrence data analysis: traditional number counts and the SAFE-Net method using measures of centrality from an SNA perspective. A running line derailment is defined by the OC-G1 as "any derailment that affects the safe operation of a running line" (Queensland Transport, 2008, p. 21). Traditional number

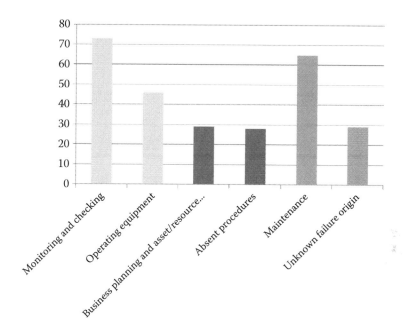

FIGURE 14.2   Running line derailments top six factors by number count grouped by CFF categories.

count bar charts are initially shown and discussed, followed by the SAFE-Net model analysis for these types of events.

14.13.1   DERAILMENTS: RUNNING LINE DERAILMENTS

Traditional number counts of the factors identified in the traditional data of running line derailments are presented in Figure 14.2.

The bar chart shows two factors from each of the three CFF categories, with monitoring and checking and maintenance having the top number count of being identified 73 and 65 times, respectively, for these types of safety occurrences across the years 2006–2010. What is unknown by looking at these data in the traditional number count view is how the factors are related across the three CFF categories of individual or team actions, technical failures, and local conditions and organizational factors; this means that they can only be viewed as factors within their stand-alone groups. To assist in a quick visual presentation of the alternative SAFE-Net method, the model of these top six factors based on betweenness centrality is presented in Figure 14.3.

Here it is possible to see that the top six factors based on betweenness centrality (where betweenness centrality has been filtered to only show those factors with this measurement greater than 112 and reduce the model to six factors) relate only to Individual or Team Actions (shown in light gray) and Technical Failures (shown in dark gray), and no Local Conditions and Organizational Factors become of interest.

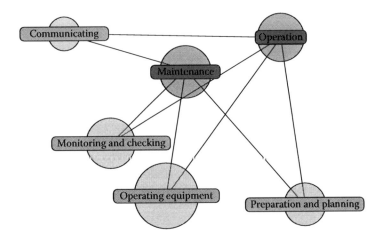

FIGURE 14.3 Running line derailments model showing top six factors for betweenness centrality.

What becomes evident is that using the SAFE-Net method would give the safety analyst a different view or focus of which contributing factors warrant further investigation. It is clear here that all the individual or team action factors are linked to the two Technical Failure factors, which become apparent for this type of safety occurrence. This appears to indicate that the human actions involved are related to technical issues, or vice versa, as the direction of the relationship is unknown in this undirected model, as previously discussed. Digging a little deeper with the SAFE-Net model, it is possible to bring in the Local Conditions and Organizational Factors (shown in medium gray) to see which of those contribute to the resultant accident model.

By adjusting the filter requirements, the model can be expanded to include these factors, as shown in Figure 14.4.

In order for any Local Conditions and Organizational Factors to appear in the model, more Individual or Team Actions and Technical Failure factors are also brought into the model, as they have higher betweenness centrality measures than the Local Conditions and Organizational Factors.

The safety analyst can now review which factors are networked or linked to which other factors in the system, and resolve to put attention on those areas that have the greatest influence for these types of safety occurrences.

What is evident is that the SAFE-Net model gives a richness of information and shows from a system perspective how the factors are influencing each other and what role they play in this complex system. By filtering the model and looking at various measures of centrality, it is possible to focus on the most important aspects of the factors in the model.

14.14   RESULTS

The traditional results from data analysis, being limited to the count of numbers of contributing factors, are clearly restricted in providing the ability to view these

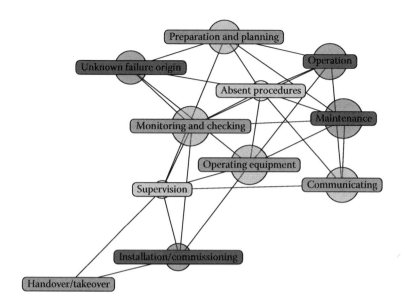

FIGURE 14.4 Running line derailments model of betweenness centrality greater than 70.

data as anything more than bar charts, percentages, and other simple mathematical displays. The CFF data are limited to being viewed under the three CFF categories, and no relationships between the contributing factors can be identified. Each factor stands on its own merits, and there is no understanding able to be gained about how the factors are interconnected. If the examination of causation and contributing factors is to catch up with complex system thinking, it cannot be done via this method. The next step is one that takes account of how the constituent parts in the system are related to each other. This is the benefit of the SAFE-Net method, which has demonstrated that this is possible.

In reviewing the SAFE-Net method models, each type of safety occurrence appears to have its own contributing factor footprint, and each SAFE-Net model for the various types of safety occurrences reveals a different set of relationships between the factors. The resultant accident model for each type of occurrence provides a specific and unique model of those factors that are contributing for each type of occurrence based on their relationship to the other factors in the model. What does become apparent is that this supports complex system theory where "the whole is based upon cooperation and competition among its parts, and in the process certain constituents gain dominance over the others" (Buzsáki, 2006, p. 14).

Not only are the relationships revealed, but also the contribution of each identified contributing factor can be determined to understand its dominance over the others. While dominance in the traditional data analysis was limited to those with the highest number counts, dominance can now be calculated using centrality measures. These measures provide information on various types of dominance, and while betweenness centrality has been the measure of choice here, many others can be examined that can provide a wealth of information about contributing factor influence and value. The ability to now display and explore this information is much greater and can ultimately provide several lenses through which to look at accident modeling options.

It has been demonstrated that the SAFE-Net models can be expanded or reduced (filtered) to reveal more or less factors, displaying a more interesting and comprehensive view on which factors are the most important and offering a deeper understanding of the complex nature of the interrelationships between the factors for the various types of safety occurrences under investigation. The SAFE-Net models, which originate from industry-wide accident reports, provide an insight that the Rail Safety Regulator can use to provide information to the wider railway industry and community on what factors should be of interest in the prevention of railway safety occurrences.

Provided here is a quick overview of the comparisons that can be made between traditional number counting of contributing factors and the information and models that can be obtained by looking at centrality measures. The information obtained by looking at the SAFE-Net models allows a rich and deeper understanding of how the contributing factors are actually contributing to the safety occurrence. The key factors across the three main categories of the CFF can now be identified and preventive safety work can be focused on those factors that most influence the safety occurrence based on centrality measures of interest.

Consideration has also been given during the development of this research that the method needs to be transferable to other industries to allow for its uptake and use outside of the rail industry. The SAFE-Net method can be used for any data collected on the contributing factors to safety occurrences. While individual organizations will need to determine their own set of conversion "rules," the methodology appears sound for transferability.

14.14.1 Representation of Sociotechnical Systems

The SAFE-Net models presented in this research appear to better represent accident models of the complex sociotechnical systems they represent for a number of reasons.

First, complexity exists in modern-day systems and comes in many forms, including "interactive complexity (related to interaction among system components) ... and non-linear complexity (where cause and effect are not related in a direct or obvious way)" (Leveson, 2011, p. 4). These models deliver on both the notion of interaction and nonlinear relationships.

Sociotechnical systems by their definition are interrelationships between people and technology and the system in which they exist.

The primary contribution is the proof of a safety analysis methodology that elicits advanced information on the relationship between contributing factors for railway safety occurrences that allows previously unknown information to be obtained. The methodology developed here will enable accident modeling to move into the modern era and out of the limited linear era in which it has been stuck for some time. The ability to review safety occurrences as a relationship among the constituent parts and to better represent what are now termed complex sociotechnical systems in the accident investigation world should be a welcome avenue of exploration for the safety analyst.

In examining contributing factors for safety occurrences, the goal has and will always be to prevent similar events from reoccurring in the future. Safety professionals strive to achieve this goal and have always been challenged by how to best know what has happened so that it can be thwarted in the future. At the heart of accident investigation is to know the why of serious negative safety events. But knowing the why really only delivers the who or what in terms of the critical factors (causal or contributed); it is the how that becomes important. How did those factors come together to create that event? How did one factor interact with the other? How was the combination of factors influential in creating and allowing the event to occur? These are the questions that safety professionals can now answer through examination of the relationships evident using the SAFE-Net method.

The SAFE-Net methodology contribution is that it offers a solution that bridges the identified gap between theory and practice, offering a relatively easy pathway for enhancing knowledge and determining appropriate preventive actions in the real world.

14.15 CONCLUSION

This research has contributed to the body of knowledge in the human factors, safety science, and rail safety regulation disciplines by providing a methodology that allows safety occurrences to be modeled based on the contemporary understanding of the complex sociotechnical systems in which they occur. The methodology, which has been called the SAFE-Net method, allows the jurisdictional Rail Safety Regulation Branches to examine quantitatively and graphically the linkages between contributing factors at both a micro and a macro level, depending on the focus of the examination or level of interest. It provides answers on which factors are contributing the most for types of safety occurrences and reduces knowledge doubts and misdirection of safety actions.

In particular, this research has

- Established a method for modeling types of railway safety occurrences
- Provided a methodology that allows the Rail Safety Regulator or other safety analyst to visualize the relationships of factors, human or social, technical, and organizational, across the headings of the CFF tool currently in use within the rail industry
- Provided a complete end-to-end example of the methodology's use and value, which can be replicated in other organizations and industries

The importance of these contributions should not be treated lightly. The reoccurrence of major rail safety occurrences is a serious commercially and socially expensive reality. The SAFE-Net methodology offers a relatively straightforward means to enhance regulatory and organizational understanding and reduce risk. Its contribution to the understanding of complex system functioning by the recognition of the interrelationships of parts of the system provides a collective knowledge where this was once not possible or available. Collective knowledge on contributing factors can now be captured, visualized, filtered, and modeled to demonstrate how best to proactively manage these elements. It is hoped that this research will lay the foundations for many others to follow and for accident modeling to move into the future with a new lens based on complex sociotechnical system thinking.

REFERENCES

Allport, F. H. (1954). The structuring of events: Outline of a general theory with applications to psychology. *Psychological Review*, 61(5), 281–303.

Benner, L. (1984). Accident models: How underlying differences affect workplace safety. *Journal of Occupational Accidents*, 6(1–3), 127.

Benner, L. (2009). Outside the lines: The curse of the retros. *Journal of System Safety*, 45(4).

Bird, F. E. J., and Germain, G. L. (1985). *Practical Loss Control Leadership*. Loganville, GA: International Loss Control Institute.

Buzsáki, G. (2006). *Rhythms of the Brain*. New York: Oxford University Press.

Chen, C. (2006). *Information Visualization: Beyond the Horizon*. London: Springer.

Grey, E., Klampfer, B., Read, G., and Doncaster, N. (2011). Learning from accidents: Developing a contributing factors framework (CFF) for the rail industry. Paper presented at HFESA 47th Annual Conference 2011, Crows Nest, New South Wales, Australia.

Heinrich, H. W. (1931). *Industrial Accident Prevention: A Scientific Approach*. New York: McGraw-Hill.

Hollnagel, E. (2004). *Barriers and Accident Prevention*. Aldershot, UK: Ashgate.

Hollnagel, E. (2012). *FRAM: The Functional Resonance Analysis Method, Modelling Complex Socio-Technical Systems*. Surrey, UK: Ashgate.

Kletz, T. (1993). *Lessons from Disasters: How Organisations Have No Memory and Accidents Recur*. Warwickshire UK: Institution of Chemical Engineers.

Leplat, J. (1984). Occupational accident research and systems approach. *Journal of Occupational Accidents*, 6, 77–89.

Leveson, N. (2011). *Engineering a Safer World: Systems Thinking Applied to Safety*. Cambridge, MA: MIT Press.

Maurino, M. (1999). Safety prejudices, training practices, and CRM: A midpoint perspective. *International Journal of Aviation Psychology*, 9(4), 413–422.

Moreno, J. L. (1934). *Who Shall Survive?* Beacon House, Beacon, NY.

Newman, M. E. J. (2010). *Networks: An Introduction*. Oxford: Oxford University Press.

Perrow, C. (1984). *Normal Accidents: Living with High-Risk Technologies*. New York: Basic Books.

Queensland Transport. (2008). *Guideline for the Top Event Classification of Notifiable Occurrences: Occurrence Classification—Guideline One (OC-G1)*.

Rasmussen, J. (1989). *Human error and the problem of causality in analaysis of accidents*. Unpublished invited paper for Royal Society meeting on Human Factors in High Risk Situations, 28–29 June, 1989, London, UK.

Rasmussen, J. (1990). *Human error and the problem of causality in analaysis of accidents*. Paper presented at the Human Factors in Hazardous Situations. Proceedings of a Royal Society Discussion Meeting. London.

Reason, J. T. (1990). *Managing the Risks of Organisational Accidents*. Aldershot: Ashgate.

Rosch, E. (1978). Principles of categorization. In E. Rosch and B. B. Lloyd (eds.), *Cognition and Categorization* (pp. 28–49). Hillsdale, NJ: Erlbaum.

TouchGraph. TouchGraph Navigator 2. http://www.touchgraph.com (accessed June 1, 2013).

Van Vuuren, W. (1999). Organisational failure: Lessons from industry applied in the medical domain. *Safety Science*, 33, 13–29.

15 Prediction of High Risk of Drowsy Driving by a Bayesian Estimation Method

An Attempt to Prevent Traffic Accidents due to Drowsy Driving

Atsuo Murata

CONTENTS

The aim of this study was to predict in advance drivers' drowsy states with a high risk of encountering a traffic accident and prevent drivers from continuing to drive under drowsy states. While the participants were required to carry out a simulated driving task, electroencephalography (EEG) (*EEG-MPF* and *EEG-α/β*), electrocardiography (ECG) (*RRV3*), tracking error, and subjective rating of drowsiness were measured. On the basis of such measurements, we made an attempt to predict in advance the point in time with a high risk of a crash (a state of a remarkably decreased arousal level with a high risk of a crash) using Bayesian estimation. As a result of applying the proposed method to the data points of each participant, it was verified that the proposed method could predict in advance the point in time with a high risk of a virtual crash before the point in time of a virtual accident when the participant would surely have encountered a serious accident with a high probability.

15.1 INTRODUCTION

According to statistics by the Japanese National Police Agency, the percentage of fatal crashes due to drowsy driving during 2015 is the highest to date and equals 17.9% of all fatal crashes (http://www.e-stat.go.jp/SG1/estat/List. do?lid=000001132129). The American Automobile Association (AAA) Foundation for Traffic Safety reported that drowsy driving plays a significant role in an average of 328,000 crashes annually. This total includes 109,000 crashes that result in injuries and 6,400 fatal crashes (http://school.sleepeducation.com/drowsydrivingstats. aspx). According to the Fatality Analysis Reporting System (FARS) of the U.S. Department of Transportation, 416,000 accidents (fatal, injury, and property damage only [PDO]) occurred during the 5 years from 2005 to 2009 (http://www-nrd. nhtsa.dot.gov/Pubs/811449.pdf).

More and more attention has been paid to the importance of monitoring drowsiness, predicting a risky state, and warning drivers of such a state before falling asleep during driving. The development of the system that can monitor drivers' arousal level and warn drivers of a risk of falling asleep is essential for assuring safety during driving. Advanced vehicle control systems (AVCSs) such as longitudinal and lateral collision avoidance systems or (Shladover, 1995; Kuroda et al., 2014) lane-keeping systems have been developed to enhance safety. However, these systems do not always work for the prevention of collisions or deviations from a lane as expected. Doubled safety measures would compensate for such a shortcoming and further enhance the safety. Therefore, such systems (AVCSs) will further contribute to enhanced traffic safety if they are used together with drowsiness prediction and warning systems.

Many studies used measures such as blink, electroencephalography (EEG), saccade, and heart rate to assess drowsiness induced by fatigue (Fukui and Morioka, 1971; McGregor and Stern, 1996; Milosevic, 1978; Piccoli et al., 2001; Sharma, 2006; Tejero and Choliz, 2002; Yoshida, 1998). McGregor and Stern (1996) suggested that caution is necessary in interpreting saccade velocity change as an index of drowsiness induced by fatigue since most of the reduction in average saccade velocity might be secondary to the increase in blink frequency.

Brookhuis and Waard (1993) carried out an on-road experiment to assess driver status using measures such as EEG and electrocardiography (ECG). Here, the driver status was changed by administration of alcohol or increase of drowsiness due to long-hour engagement in driving task, and was measured by standard deviation of lateral position or steering wheel movement. They found that changes in EEG and ECG reflected changes in driver status. Kecklund and Akersted (1993) recorded EEG continuously during a night or evening drive for 18 truck drivers. They showed that during a night drive, a significant intraindividual correlation was observed between subjective sleepiness and the EEG alpha burst activity. Subjective drowsiness (sleepiness) and the EEG alpha burst activity at the end of drive were significantly correlated with total work hours. As a result of a regression analysis, total work hours and total break time predicted about 66% of the variance of EEG alpha burst activity during the end of the drive. Galley (1993) overcame a few disadvantages of electrooculography (EOG) in the measurement of gaze behavior by using online computer identification of saccades. As EOG, especially saccades and blinks, is regarded as a useful measure to evaluate drivers' drowsiness, such an improvement might be useful for detecting the low arousal state of drivers. Wright and McGown (2001) investigated sleepiness in an aircrew during long-haul flights, and showed that EEG and EOG are potentially promising measures on which to base an alarm system. Skipper and Wierwillie (1986) attempted to detect drowsiness of drivers using discrimination analysis, and showed that a false alarm or miss would occur.

Murata and Hiramatsu (2008) and Murata and Nishijima (2008) attempted to evaluate the drowsiness of drivers objectively using EEG or heart rate variability (HRV) measures. Murata and Hiramatsu (2008) and Murata and Nishijima (2008) succeeded in clarifying the decrease of *EEG-MPF* (mean power frequency of EEG power spectrum) or the increase of *RRV3* (representative of parasympathetic nervous system) when the participant's arousal level was low. However, it was not possible to predict in advance the risky state due to increased drowsiness on the basis of the time series of *EEG-MPF* or *RRV3*.

Only the assessment of a tendency of a reduced arousal level is insufficient to prevent drivers from driving under a drowsy state that causes a crash. It is necessary to identify accurately the point in time when the drowsy state certainly leads to a crash if a driver continues to drive further. In other words, detecting the decreased arousal level with a high risk of a crash automatically and warning drivers of the risky state is the ultimate goal in the studies of developing a preventive safety system that contributes to decreasing crashes caused by drowsy driving. Although the above studies can assess the driver status using a variety of physiological and behavioral measures, it is impossible to detect the point in time of a risky state in advance before the occurrence of the drowsy driving that induces a crucial crash at the worst case. Therefore, these studies must be further enhanced so that the method and technique for predicting the point in time of a risky state in advance before a crash occurs can be developed.

Murata et al. (2012) attempted to predict in advance the risky state caused by drowsy driving using a Bayesian estimation algorithm. Although the relationship between the posterior probability $P(H_1|x_j)$ (H_1: drowsy, x_j: measured data, such as *EEG-MPF*, *EEG-α/β* [ratio of the α-wave power to the β-wave power], or *RRV3*) and the subjective rating of drowsiness was identified, this study could not identify the point in time of a risky state in advance before the genuinely risky state occurred in a simulated driving task.

Therefore, an attempt was made to identify the point in time with a high risk of a crash in advance before the point in time of a virtual crash by means of the integrated posterior probability, the subjective rating on drowsiness, and the tracking error in a simulated driving task. First, the EEG and ECG during a simulated driving task were measured, and it was investigated how these measures changed under the low arousal (drowsy) state. Based on the measurements of the EEG, ECG, and a tracking error during a simulated driving task, a method was proposed to identify and predict the point in time with a high risk of a virtual crash in advance before the point in time of a virtual accident occurring, using the posterior probability $P(H_1|x_j)$ above obtained by a Bayesian estimation method (Murata et al., 2012), a tracking error, and a subjective rating of drowsiness. It was explored whether the proposed method can predict in advance the point in time with a high risk of a virtual crash before the point in time of a virtual accident to verify the effectiveness of the proposed method.

15.2 METHOD

15.2.1 PARTICIPANTS

Thirteen male graduates or undergraduates (from 21 to 26 years old) participated in the experiment. They were all healthy and had no orthopedic or neurological diseases. All participants had held a driver's license for 3–4 years and were required to stay up all night and visit our laboratory early in the morning (at about 5:00). They were not permitted to drink anything containing caffeine or exercise excessively during the sleep deprivation. Screen time, such as spending time interacting with a PC or a smart phone, was also confined to less than 1 h during the sleep deprivation. We judged that such a condition would make participants readily feel sleepy or very drowsy. The visual acuity of the participants was matched and more than 20/20. All signed the informed consent after receiving a brief explanation about the experiment.

15.2.2 APPARATUS

EEG and ECG activities were acquired with an A/D instrument PowerLab 8/30 and a bioamplifier ML132. Surface EEGs were recorded using A/D instrument silver/silver chloride surface electrodes (MLAWBT9), and sampled with a sampling frequency of 1 kHz. According to the International 10–20 standard, EEGs were led from O_1 and O_2. The ECG was led from V_5 using BiolaoDL-2000 (S&ME). The outline of the display of a driving simulator, a steering wheel, and switches for evaluating subjective drowsiness is depicted in Figure 15.1.

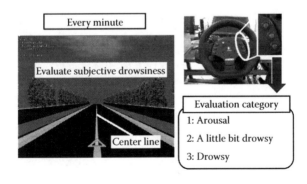

FIGURE 15.1 Outline of display of driving simulator, steering wheel, and switch for evaluating subjective drowsiness.

15.2.3 TASK, DESIGN, AND PROCEDURE

The participants sat on an automobile seat and were required to carry out a simulated driving task using a steering wheel, shown in Figure 15.1. The inside lane in Figure 15.1 is displayed on the screen 3.2 m in front of a participant via a projector (EPSON, EB-S12H). Participants were required to steer a steering wheel and keep their own vehicle at the center line in Figure 15.1 as much as they could. The driving simulator consisted of three inside lanes in one direction, and the participants were required to run the second lane. The width of each lane inside the simulated driving task was assumed to be 3.6 m. In a simulated driving task, the tracking error between the center line and the present location of their own vehicle (an arrow in Figure 15.1) was recorded every second. The mean tracking error every minute was calculated. This was used to identify the point in time of a virtual accident, described in Section 15.2.5.

The participants were also asked to evaluate their arousal level every minute according to the following categories: 1, arousal, 2, a little bit drowsy, and 3, drowsy. It is possible to use a more detailed category, such as a 7-point (e.g., Stanford Sleepiness Scale [SSS]; Hoddes et al., 1973) or a 9-point (e.g., Karolinska Sleepiness Scale [KSS]; Åkerstedt and Gillberg, 1990) category. Such a detailed categorization has the following advantages and disadvantages. While a finer categorization enables us to conduct a finer evaluation of drowsiness, the disadvantage of such a categorization is that the exact evaluation according to finer categories is difficult and suffers from more frequent false (ambiguous) evaluations at the boundary of two arbitrary categories. Moreover, it is pointed out that the more choice alternatives there are, the less effective the choice is (Schwartz, 2005). In this study, the participant must report his subjective rating on drowsiness every minute. The greater number of alternatives force the participants to feel it is difficult to evaluate their feeling, and it is possible that a subjective evaluation cannot be carried out appropriately. The subjective rating of drowsiness was not a principal means for identifying the point in time with a high risk of potential crash. Therefore, we adopted a 3-point categorization of drowsiness. The psychological evaluation of drowsiness was incorporated into the experimental procedure to help the identification of point in time with a high risk of a virtual crash.

EEG (*EEG-MPF*, *EEG-α/β*) and HRV (*RRV3*) during a simulated driving task were derived to evaluate drowsiness. The relation between these measurements and drowsiness was analyzed. As well as the physiological measures above, the tracking error during a simulated driving task was recorded. The psychological rating of drowsiness reported every minute was used to grasp the change of drowsiness over time and detect a high risk of a virtual crash, together with the tracking error and the posterior probability $P(H_1|x_j)$ (H_1: drowsy, x_j: values of evaluation measures) using the Bayesian estimation stated in Section 15.2.4. The duration of each experiment ranged from 420 to 2400 s since the measurement was continued until the experimenter judged that the participant was unable to carry out an experimental task any more. Although the participants were required to stay up all night and the experiment began early in the morning (at about 5:00), there were individual differences in the extent of induced drowsiness. Therefore, the duration of experimental measurement differed among participants.

15.2.4 DATA ANALYSIS

An FFT (Fast Fourier Transform) program was carried out every 1024 data points (1.024 s) for EEG. Before the EEG time series were entered into an FFT program, they were passed through a cosine taper window. Based on such an FFT analysis, the mean power frequency (*EEG-MPF*) and *EEG*-α/β were calculated. *EEG-MPF* was calculated as follows:

$$EEG - MPF = \frac{\sum\limits_{i=1}^{30} f_i \cdot pow_i}{\sum\limits_{i=1}^{30} pow_i} \tag{15.1}$$

where f_i and pow_i correspond to the frequency and the corresponding power, respectively. The values of f_1 and f_{30} were set to about 1 and 30 Hz, respectively (the frequency band was assumed to range from 1 to 30 Hz). If the value of *EEG-MPF* is low, this indicates that the arousal level is decreased. *EEG*-α/β can be calculated as the ratio of the sum of the α-band (8–12 Hz) power and θ-band (4–7 Hz) power to the power of the β-band (13–30 Hz). The higher the value, the lower the arousal level becomes.

R-R intervals (interbeat intervals) were obtained by detecting R-waves from an ECG waveform. HRV measure *RRV3* was derived as follows. The moving average per three interbeat intervals was calculated. The variance of the past three interbeat intervals was calculated as *RRV3*, which is regarded to represent the state of the parasympathetic nervous system. The reason why *RRV3* was used to evaluate the state of parasympathetic nervous system is shown in Figure 15.2. Here, the

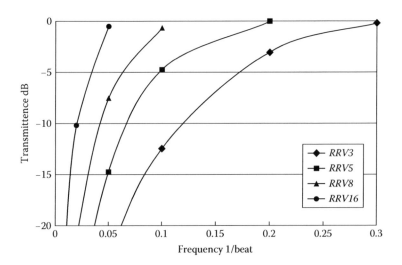

FIGURE 15.2 Filter characteristics of *RRV3* (transmittance for moving averaged time series of 3, 5, 8, and 16 interbeat intervals is plotted as a function of frequency [1/beat]).

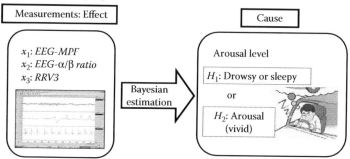

FIGURE 15.3   Explanation of Bayesian estimation and its relevance to this study.

transmittance for the moving average of 3, 5, 8, and 16 interbeat intervals is plotted as a function of the frequency (1/beat). The transmittance of *RRV3* for the frequency corresponding to the respiration frequency (more than 0.25 1/beat) is nearly equal to zero. Therefore, *RRV3* can be regarded as reflecting the state of parasympathetic nervous system (respiratory sinus arrhythmia). A value of *RRV3* increases under a low arousal condition owing to the dominance of the parasympathetic nervous system.

The basic concept of the Bayesian theorem (Swinburne, 2002) is depicted in Figure 15.3. The prior probabilities of events H_1 (drowsy) and H_2 (arousal) are given by $P(H_1)$ and $P(H_2)$. The conditional probabilities (likelihood) $P(x_j|H_1)$ and $P(x_j|H_2)$ represent the probabilities of observing the data x_j given that the hypothesis H_i is true. Based on evaluation measures such as x_1: *EEG-MPF*, x_2: *EEG-α/β*, and x_3: *RRV3*, the probability of event (in this study, H_1: drowsy, H_2: arousal) given that the measurements x_j are obtained (in this study, $P(H_1|x_1)$, $P(H_1|x_2)$, $P(H_1|x_3)$, $P(H_2|x_1)$, $P(H_2|x_2)$, and $P(H_2|x_3)$) is estimated using the Bayesian estimation method below. The three measures x_1, x_2, and x_3 were plotted as an X-bar control chart (Stamatis, 2003). An X-bar control chart of *EEG-MPF* is demonstrated in Figure 15.4. Using an X-bar control chart, the judgment (assessment) of drowsiness of each participant was carried out according to the procedure stated below.

First, the procedure for calculating the likelihood is stated (see Figure 15.4). In order to apply a Bayesian estimation method, the likelihood $P(x_j|H_i)$ ($i = 1, 2; j = 1, 2, 3$) was calculated as the ratio of the number of judgments (assessment) as H_i to the total number of judgments (30 judgments) using an X-bar chart of each measurement. An interval of 30 s was selected. In case of *EEG-MPF*, it was assessed that the arousal level was low when more than 10% of data were below the threshold (lower control limit) $(CL - \sigma)$. In case of *EEG-α/β* and *RRV3*, it was assessed that the arousal level was low when more than 10% of data were above the threshold (upper control limit) $(CL + \sigma)$. The 30 s interval was moved forward by 1 s, and the judgment (assessment) of arousal level was carried out for the whole analysis interval. In case of EEG-based measures such as *EEG-MPF* and *EEG-α/β*, 30 and 1 s exactly

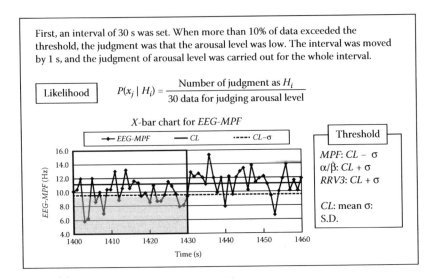

First, an interval of 30 s was set. When more than 10% of data exceeded the threshold, the judgment was that the arousal level was low. The interval was moved by 1 s, and the judgment of arousal level was carried out for the whole interval.

Likelihood $\quad P(x_j \mid H_i) = \dfrac{\text{Number of judgment as } H_i}{\text{30 data for judging arousal level}}$

FIGURE 15.4 Explanation for calculating likelihood necessary for Bayesian estimation.

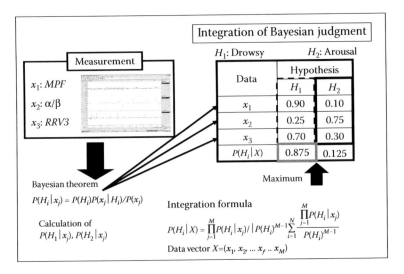

FIGURE 15.5 Explanation for calculating $P(H_1|x_j)$ and $P(H_2|x_j)$.

corresponded to 30.72 and 1.024 s, respectively. As for *RRV3*, 30 and 1 s exactly corresponded to 30 beats and 1 beat, respectively.

Using the following Bayesian theorem and the likelihood $P(x_j|H_i)$ ($i = 1, 2; j = 1, 2, 3$) mentioned above, the posterior probabilities $P(H_1(\text{drowsy})|x_j)$ and $P(H_2(\text{arousal})|x_j)$ ($j = 1, 2, 3$) can be calculated using the following formula (see Figure 15.5):

$$P(H_i \mid x_j) = P(H_i)P(x_j \mid H_i) / P(x_j) \tag{15.2}$$

An initial value of prior probability is usually set to $P(H_1) = P(H_2) = 1/2$, and $P(x_j)$ is the probability of observing x_j. According to the procedure shown in Figure 15.5, the calculation of $P(H_i|X)$ (estimation of integrated posterior probability of drowsiness) by a Bayesian estimation method (Hershman, 1971) was carried out. The integration formula is given by Equation 15.3. X corresponds to a vector that integrated x_1, x_2, and x_3. As the three measures *EEG-MPF*, *EEG-α/β*, and *RRV3* are not highly correlated, it is possible that the integration of three measures corrects for the redundancy of *a priori* probability in each measure, leading to the enhanced reliability of drowsiness prediction.

$$P(H_i \mid X) = \prod_{j=1}^{M} P(H_i \mid x_j) \prod_{j=1}^{M} P(x_j) / \left\{ P(H_i)^{M-1} \sum_{i=1}^{N} P(H_i) \prod_{j=1}^{M} \frac{P(H_i \mid x_j)P(x_j)}{P(H_i)} \right\}$$

$$= \prod_{j=1}^{M} P(H_i \mid x_j) / \left\{ P(H_i)^{M-1} \sum_{i=1}^{N} \frac{\prod_{j=1}^{M} P(x_j \mid H_i)}{P(H_i)^{M-1}} \right\}$$

(15.3)

Here, $X = (x_1, x_2, ..., x_M)$. In this study, M and N are equal to 3 and 2, respectively.

Figure 15.5 demonstrates the calculated probabilities $P(H_1|x_1)$, $P(H_1|x_2)$, $P(H_1|x_3)$, $P(H_2|x_1)$, $P(H_2|x_2)$, and $P(H_2|x_3)$, which are used to calculate the integrated posterior probabilities $P(H_1|X)$ and $P(H_2|X)$. In the example in Figure 15.5, the value of $P(H_1|X)$ is by far larger than that of $P(H_2|X)$. Here, H_1 is estimated as true. These values were calculated every second (*EEG-MPF* and *EEG-α/β*) or one interbeat interval (*RRV3*).

Bayesian estimation remains unbiased with sample size due to exact estimation. Therefore, one can judge that Bayesian estimation can benefit with a limited sample size. Moreover, the proposed method mentioned in Chapter 3 is applied to each participant in order to predict the point in time with a high risk of a virtual crash of each individual in advance. With the elapse of experiment, the data usable for Bayesian inference are accumulated. Therefore, it can be regarded that a sample size of 13 is not so inadequate.

Although initial values of prior probability are usually set to $P(H_1) = P(H_2) = 0.5$, it is necessary to renew the prior probabilities $P(H_1)$ and $P(H_2)$ so that the estimation accuracy is enhanced. Therefore, the prior probabilities $P(H_1)$ and $P(H_2)$ were renewed every second or one interbeat interval according to the procedure in Figure 15.6. In this manner, the integrated posterior probabilities $P(H_1|X)$ were calculated to identify the point in time with a high risk of a virtual crash, and this value was used to warn drivers of a risky state before the point in time of a virtual accident, and prevent them from encountering a crash.

15.2.5 Definition of a Virtual Crash

The virtual crash was defined as follows, and the point in time of the virtual accident was obtained according to the following procedure (see Figure 15.6). One can judge

Renewal of prior probability

Using mean tracking error every 1 min, prior probability
was renewed every 1 min.

Mean tracking error	$P(H_1)$(drowsy)	$P(H_2)$(arousal)
0 m ~ 0.9 m	0.1	0.9
0.9 m ~ 1.8 m	0.3	0.7
1.8 m ~ 5.4 m	0.7	0.3
5.4 m ~	0.9	0.1

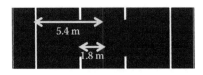

FIGURE 15.6 Explanation for renewing prior probability.

that the participant surely would have been encountered a serious accident with a
high probability if he continued driving when the following two conditions were
simultaneously satisfied:

1. The mean tracking error per 1 min is more than 1.8 m.
2. The participant could not report subjective drowsiness using a switch.

The check of the identification of a virtual crash was carried out by the two
experimenters. Only when the rating (evaluation) on conditions 1 and 2 above by
the two experimenters completely coincided was this regarded as the point in time
of a virtual accident. Before carrying out the experiment, the two experimenters
were trained so that the interrater (experimenter) reliability of decision making was
ensured. As shown in Figure 15.6, the tracking error of 1.8 m corresponds to half of
the lane width and shows that the vertical location is dispersive to a larger extent (this
cannot be judged to be driving normally).

15.2.6 IDENTIFICATION PROCEDURE OF A RISKY STATE

The aim of this study is to verify for each participant whether it is possible to predict
the point in time with a high risk of a virtual crash (the risky state of encountering a
virtual crash) in advance before the point in time of a virtual accident by means of
the integrated posterior probability $P(H_1|X)$, the subjective rating of drowsiness, and
the tracking error in a simulated driving task.

The following algorithm was proposed for identifying the point in time with a high
risk of a virtual crash. When conditions 1–3 below were simultaneously satisfied, the
corresponding point in time was judged to have a high risk of a virtual crash if the
participant was further required to continue carrying out the simulated driving task.

1. The point in time when the posterior probability $P(H_1|X)$ is above x (= 0.8)
   occupied more than $y\%$ (= 60%) of the z-second interval (= 30 s).

2. The point in time when the mean tracking error is more than 1.8 m occupied more than $v\%$ (= 20%) of the w-second interval (= 60 s).
3. The rating of drowsiness corresponded to 3 (very drowsy) or the missing of response switch pressing. In addition to this, the ratings of drowsiness 1 min before and after this point in time also correspond to 3 (very drowsy) or the missing of response switch pressing.

This procedure was applied to all participants, and it was explored whether the point in time with a high risk of a virtual crash could be identified using the procedure above.

15.3 RESULTS

As a result of checking the data of all participants according to the rule described in Section 15.2.5, the point in time with a high risk of a virtual accident was detected in 8 out of 13 participants. As for five participants, no definite virtual accident could be identified. Using the proposed method, it was explored whether the proposed method can predict the point in time with a high risk of a virtual crash in advance before the point in time of a virtual accident.

Examples of the change of tracking error and the posterior probability $P(H_1|X)$ are depicted in Figures 15.7 (Participant E), 15.8 (Participant F), and 15.9 (Participant H). In Figure 15.7, the point in time of a virtual accident was detected. When the tracking error increased, the probability $P(H_1|X)$ increased and kept high values until the point in time of a virtual accident. It seems reasonable to regard the point in time when $P(H_1|X)$ is more than 0.8 for more than 1 min as a risky state. In this case, a

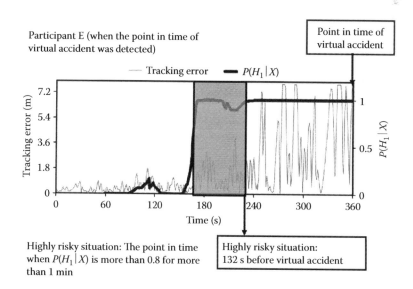

FIGURE 15.7 Change of tracking error and $P(H_1|X)$ with time (Participant E). The point in time of a virtual accident was detected.

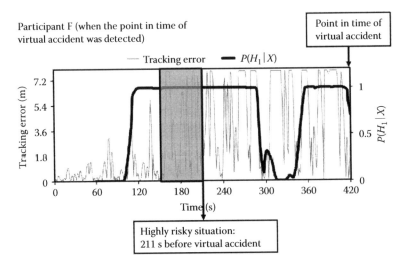

FIGURE 15.8 Change of tracking error and $P(H_1|X)$ with time (Participant F). The point in time of a virtual accident was detected.

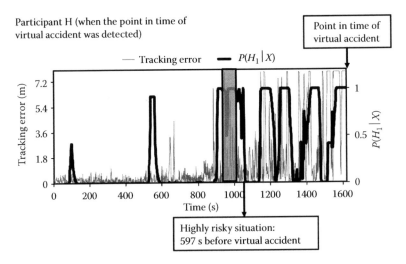

FIGURE 15.9 Change of tracking error and $P(H_1|X)$ with time (Participant H). The point in time of a virtual accident was detected.

risky state was detected 132 s before a virtual accident occurred. This indicates that the virtual accident can be prevented by providing the participant with some alarm signal or forcing him to stop the simulated driving task at the identified point in time of a risky state. In Figure 15.8, a similar change of tracking error and $P(H_1|X)$ with time (Participant F) is shown. Even in this case, the risky state was detected 211 s before a virtual accident occurred. In Figure 15.9, a similar change of tracking error and $P(H_1|X)$ over time (Participant H) is shown. Even in this case, a risky state was detected 597 s before a virtual accident occurred.

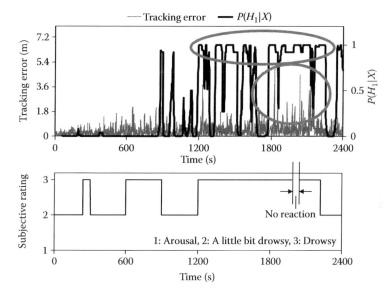

FIGURE 15.10   Change of tracking error and $P(H_1|X)$ with time (Participant A). The point in time of a virtual accident was not detected.

Examples of the change of tracking error, $P(H_1|X)$, and subjective rating of drowsiness over time and the identification of the point in time with a high risk of a virtual accident are shown in Figures 15.10 (Participant A), 15.11 (Participant D), 15.12 (Participant K), 15.13 (Participant G), 15.14 (Participant M), and 15.15 (Participant M). In Figure 15.10, the change of tracking error, $P(H_1|X)$, and subjective rating of drowsiness over time is shown (Participant A). This corresponds to the case when the point in time of a virtual accident was not detected. Although the virtual accident was not detected, the subjective drowsiness and the tracking error increased over time. At the same time, the high value of $P(H_1|X)$ continued. Even when the virtual accident could not be detected, the synthetic evaluation of $P(H_1|X)$, the tracking error, and the subjective rating of drowsiness might enable us to identify the point in time with a high risk of driving under a drowsy state.

15.4   DISCUSSION

Although Murata et al. (2012) proposed a method to calculate $P(H_1|X)$, they did not show a procedure for predicting the point in time with a high risk of a virtual crash that leads to a crash if the participant continues performing a simulated driving task. They also did not define and identify the point in time of a virtual accident. This study made such an attempt to predict the point in time with a high risk of a virtual crash before a virtual accident occurs in the simulated driving task and verified the effectiveness of the proposed method for the purpose of contributing to the development of drowsiness prediction and a warning system.

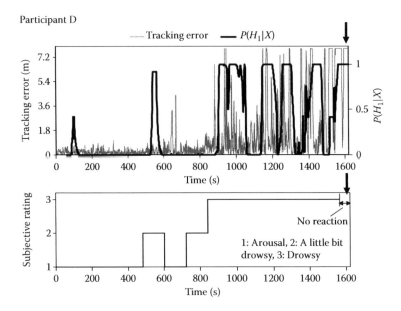

FIGURE 15.11 Change of tracking error, $P(H_1|X)$, and subjective rating of drowsiness with time (Participant D). The point in time of a virtual accident was detected.

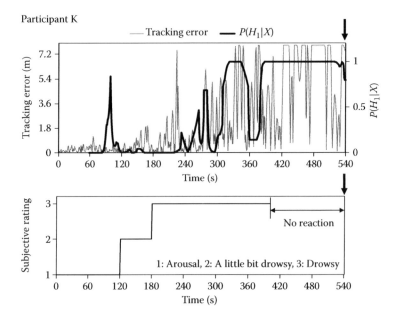

FIGURE 15.12 Change of tracking error, $P(H_1|X)$, and subjective rating of drowsiness with time (Participant K). The point in time of a virtual accident was detected.

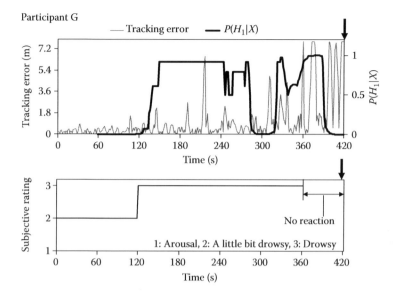

FIGURE 15.13   Change of tracking error, $P(H_1|X)$, and subjective rating of drowsiness with time (Participant G). The point in time of a virtual accident was detected.

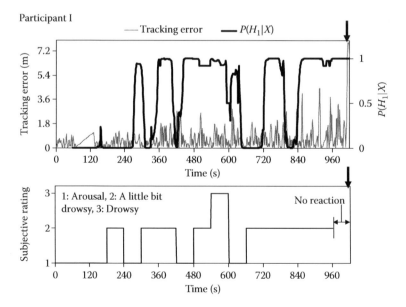

FIGURE 15.14   Change of tracking error, $P(H_1|X)$, and subjective rating of drowsiness with time (Participant I). The point in time of a virtual accident was detected.

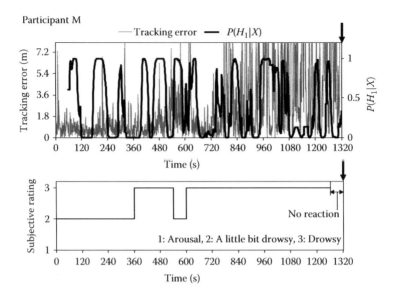

FIGURE 15.15 Change of tracking error, $P(H_1|X)$, and subjective rating of drowsiness with time (Participant M). The point in time of a virtual accident was detected.

The risky state was detected using the proposed procedure mentioned in Chapter 3 for all eight participants for whom the point in time of a virtual accident was detected according to the procedure mentioned in Section 15.2.5. On the other hand, the point in time with a high risk of a virtual crash was not detected for all five participants for whom the point in time of a virtual accident was not identified. The mean percentages of drowsiness evaluation (rating 3) or missing reaction were 64.66% and 18.72% for the identified (Participants B, D, E, F, G, I, K, and M) and nonidentified (Participants A, C, H, J, and L) groups of a virtual accident, respectively. This indicates that the degree of drowsiness is larger for the participants whose point in time of a virtual accident was identified than for the participants for whom the point in time of a virtual accident was not identified. Therefore, it can be judged that the procedure for identifying the point in time of a virtual accident is appropriate.

Although McDonald et al. (2012) showed that steering wheel angle could be used to predict drowsiness-related lane departures 6 s before they occurred, they only showed the relationship between lane departures and drowsiness and could not predict the point in time with a high risk of a crash. Brookhuis and Waard (1993) showed that the driver status, such as drowsiness due to long-hour engagement in driving, was measured by standard deviation of lateral position or steering wheel movement and concluded that lane departure measures can be used effectively for the assessment of drowsiness. From Figures 15.7 through 15.9, it is clear that the tracking error is dispersive and variable throughout the experiment. On the other hand, the posterior probability $P(H_1|X)$ was less variable and dispersive than the tracking error. It must be noted that one cannot visually predict the risky state in advance on the basis of only lane departure measures such as a tracking error, although this measure is to some extent effective for assessing the drowsy state. Therefore, it was judged that only the tracking error could not be used practically and effectively for predicting the risky state (the point in time with a high risk of a virtual crash) under which a driver is likely to encounter a crash. This measure must be used in combination with other more stable and reliable measures for predicting the risky state in advance. Therefore, as stated in Section 15.2.6, an attempt was made to predict the risky state in advance before a virtual crash occurred in the simulated driving by integrating $P(H_1|X)$, the tracking error, and the subjective rating on drowsiness.

As the aim of this study was to propose a method to predict in advance the point in time with a high risk of a crash (risky state) before the point in time of a virtual accident using the proposed procedure, we applied the proposed procedure to all participants so that the effectiveness of the proposed procedure can be demonstrated. The risky state was not identified for the participants for whom the point in time of a virtual accident was not identified. The risky state was identified for only the participants whose point in time of a virtual accident was identified. This verifies the validity of the procedure for identifying and predicting in advance the risky state. The validity of the proposed procedure should be further verified in future work.

The risky state indicates that the participant would be likely to encounter a crash due to drowsy driving if no countermeasures, such as providing drivers with some alarm, were taken. The risky state could be identified effectively before a virtual accident occurred. This indicates that the virtual accident can be prevented by predicting the risky state (the point in time with a high risk of a virtual crash) in advance and providing the participant with some alarm, or forcing him to stop the simulated driving task. The identification technique of the risky state before the virtual crash by means of the integration of the Bayesian estimation, the tracking error, and the subjective rating on drowsiness was found to be effective for preventing drowsy driving.

The parameters x ($= 0.8$), y ($= 60\%$), z ($= 30$ s), v ($= 20\%$), and w ($= 60$ s) in Section 15.2.6 were empirically determined. Future work must explore how parameters x, y, z, v, and w should be systematically determined. As the participants consisted of a sample of male university students, this might limit the validity of this study. Future research should verify whether the proposed method can effectively predict the risky and drowsy state in advance for a population such as long-haul truck drivers or lorry drivers. The driving simulator in this study corresponds to a low-fidelity one (see Figure 15.1). It is clear that the cognitive workload might be different between low- and high-fidelity simulators. The aim of this study was not to evaluate the cognitive load, but to predict in advance the risky state (where the participant would be likely to encounter a crash because of drowsy driving if one leaves such a state and takes no countermeasures) for the sleep deprivation participants who were required to stay up all night and visit our laboratory early in the morning (at about 5:00). Therefore, it can be judged that the fidelity of the simulator does not have a major influence on the findings.

15.5   CONCLUSIONS

The aim of this study was to predict in advance drivers' risky state due to drowsiness and prevent drivers from driving under such a drowsy state and eventually encountering a crash in order to prepare against crashes. While the participants were required to carry out a simulated driving task, EEG (*MPF* and α/β ratio), ECG (*RRV3*), tracking error, and subjective rating of drowsiness were measured. Using these measurements, a method to identify and predict the point in time with a high risk of a crash (a risky state due to drowsiness) in advance was proposed.

The point in time of a virtual accident was defined, and it was explored whether this point in time was observed for each participant. On the basis of such measurements, using a Bayesian estimation, a tracking error, and a subjective rating of drowsiness, an attempt was made to identify and predict the point in time with a high risk of a virtual crash (a state where the participant would be likely to encounter a crash owing to drowsy driving if one leaves such a state and takes no countermeasures) in advance before the point in time of virtual accident.

The proposed procedure could effectively predict the risky state in advance for only the participants for whom the point in time of a virtual accident was identified. The risky state was not identified for the participants for whom the proposed procedure did not detect the point in time of a virtual accident. In this manner, the validity of the proposed method for predicting the risky state in advance before the point in time of a virtual accident was demonstrated.

Future research should further verify the effectiveness of the proposed method in more real-world environments. Using physiological measures such as *EEG-MPF*, *EEG*-α/β ratio, and *RRV3* is not practical and feasible due to the high price (cost) of such physiological measurement apparatus. Thus, it should be explored whether the application of the proposed method to only the behavioral measures (Murata and Koriyama, 2012; Murata et al., 2013a, 2013b) can effectively identify the risky state.

REFERENCES

Åkerstedt, T., and Gillberg, M. 1990. Subjective and objective sleepiness in the active individual. *International Journal of Neuroscience*, 52: 29–37.

Brookhuis, K.A., and Waard, D. 1993. The use of psychophysiology to assess driver status. *Ergonomics*, 36: 1099–1110.

Fukui, T., and Morioka, T. 1971. The blink method as an assessment of fatigue. *Ergonomics*, 14: 23–30.

Galley, N. 1993. The evaluation of the electrooculogram as a psycho-physiological measuring instrument in the driver study of driver behavior. *Ergonomics*, 36: 1063–1070.

Hershman, R.L. 1971. A rule for the integration of Bayesian options. *Human Factors*, 13: 255–259.

Hoddes, E., Zarcone, V., Smythe, H. Phillips, R., and Dement, W.C. 1973. Quantification of sleepiness: A new approach. *Psychophysiology*, 10: 431–436.

Kecklund, G., and Akersted, T. 1993. Sleepiness in long distance truck driving: An ambulatory EEG study of night driving. *Ergonomics*, 36: 1007–1017.

Kuroda, H., Yokoyama, A., Tanimichi, T., and Ohtsuka, Y. 2014. Advanced vehicle safety control system. *Hitachi Review*, 63: 61–66.

McDonald, A.D., Schwartz, C., Lee, J.D., and Brown, T.L. 2012. Real-time detection of drowsiness related lane departures using steering wheel angle. In *Proceedings of the Human Factors and Ergonomic Society 56th Annual Meeting*, Boston, MA, 2201–2205.

McGregor, D.K., and Stern, J.A. 1996. Time on task and blink effects on saccade duration. *Ergonomics*, 39: 649–660.

Milosevic, S. 1978. Vigilance performance and amplitude of EEG activity. *Ergonomics*, 21: 887–894.

Murata, A., and Hiramatsu, Y. 2008. Evaluation of drowsiness by HRV measures—basic study for drowsy driver detection. In *Proceedings of IWCIA2008*, Hiroshima, Japan 99–102.

Murata, A., and Koriyama, T. 2012. Basic study on the prevention of drowsy driving using the change of neck bending angle and the sitting pressure distribution. In *Proceedings of SICE2012*, Akita, Japan, 274–279.

Murata, A., Koriyama, T., Ohkubo, Y., and Moriwaka, M. 2013a. Verification of physiological or behavioral evaluation measures suitable for predicting drivers' drowsiness. In *Proceedings of SICE2013*, Nagoya, Japan, 1766–1771.

Murata, A., Matsuda, Y., and Moriwaka, M. 2012. An attempt to predict drowsiness by Bayesian estimation. In *Proceedings of AHFE2012*, San Francisco, CA, 7435–7444.

Murata, A., Nakatsuka, A., and Moriwaka, M. 2013b. Effectiveness of back and foot pressures for assessing drowsiness of drivers. In *Proceedings of SICE2013*, Nagoya, Japan, 1754–1759.

Murata, A., and Nishijima, K. 2008. Evaluation of drowsiness by EEG analysis—basic study on ITS development for the prevention of drowsy driving. In *Proceedings of IWCIA2008*, Hiroshima, Japan, 95–98.

Piccoli, B., D'orso, M., Zambelli, P.L., Troiano, P., and Assint, R. 2001. Observation distance and blinking rate measurement during on-site investigation: New electronic equipment. *Ergonomics*, 44: 668–676.

Schwartz, B. 2005. *The Paradox of Choice: Why More Is Less*. New York: Harper Perennial.

Sharma, S. 2006. Linear temporal characteristics of heart interbeat interval as an index of the pilot's perceived risk. *Ergonomics*, 49: 874–884.

Shladover, S.E. 1995. Review of the state of development of advanced vehicle control systems (AVCS). *Vehicle System Dynamics*, 24: 551–595.

Skipper, J.H., and Wierwillie, W. 1986. Drowsy driver detection using discrimination analysis. *Human Factors*, 28: 527–540.

Stamatis, D.H. 2003. *Six Sigma and Beyond—Design for Six Sigma*. Boca Raton, FL: St. Lucie Press.

Swinburne, R. 2002. *Bayes's Theorem*. New York: Oxford University Press.

Tejero, P., and Choliz, M. 2002. Driving on the motorway: The effect of alternating speed on driver's activation level and mental effort. *Ergonomics*, 45: 605–618.

Wright, N., and McGown, A. 2001. Vigilance on the civil flight deck: Incidence of sleepiness and sleep during long-haul flights and associated changes in physiological parameters. *Ergonomics*, 44: 82–106.

Yoshida, F. 1998. Frontal midline theta rhythm and eyeblinking activity during a VDT task and a video game; useful tools for psychophysiology in ergonomics. *Ergonomics*, 41: 678–688.

16 Space Missions as a Safety Model

Irene Lia Schlacht

CONTENTS

16.1 INTRODUCTION

From the International Space Station to extreme environments on Earth, people are risking their lives every day to enhance scientific progress and perform their work. To do so, they work in very dangerous environments under extreme working conditions that seriously affect their safety and performance. Safety is defined in the *Oxford Dictionary* as "the condition of being protected from or unlikely to cause danger, risk, or injury" (Oxford Dictionary 2015). "Safety first"—it is well known that safety is the most important factor that is always given top priority in every project, especially in recent decades. In addressing particularly dangerous contexts, such as space missions, safety requirements need even more attention. These requirements need to be considered in the design approach of the overall system structure and when testing the system with dedicated simulations.

This chapter addresses the improvement of safety in working and living conditions using space missions as a model and starting from the results presented at the Applied Human Factors and Ergonomics (AHFE) 2015 conference (Schlacht et al. 2015c). It presents research performed regarding safety optimization in space missions using

- The Integrated Design Process (IDP)
- A sustainable system
- A mission simulation

The final section describes why the transfer of research and design solutions can be valuable in terms of addressing safety problems faced in extreme or dangerous environments on Earth.

16.2 INTEGRATED DESIGN PROCESS VERSUS SAFETY

The space scenario was selected as being characterized by the most life-threatening challenges. Indeed, this context includes the most extreme and adverse factors for human life, such as radiation, absence of pressure and oxygen, physical adaptation to microgravity, social isolation, and spatial confinement (Schlacht 2011, 2012). A very specific and small range of users also characterizes this context: astronauts. On Earth, life-threatening challenges are not rare either, but they affect a wide range of users in comparison to space missions. For instance, the literature on accidents such as Chernobyl, Bhopal, and Deepwater Horizon highlights the catastrophic consequences that those accidents brought with them. In contrast, in space every user has a fundamental role, and his or her safety is strictly related to the safety of the entire mission. This is why in space, the approach used to support the safety of each individual needs to be optimal. User-centered design is the foundation for this.

One option for optimizing safety in space missions is to use computer simulations. "To prepare for a potential disaster, decision-support systems based on computer simulations can enable safety managers to determine mitigation projects, and better understand the different risks associated with operations. For example, if toxic gases are released, there is a need to predict where the gas plume will go, how far it will extend, the expected concentration of toxins, and the health and safety consequences" (Rabelo et al. 2005, p.1). However, the space context is very particular; it can be compared to an aquarium where each element has a strong influence on the others and where it can be very difficult to analyze with a computer simulation each element separately in order to verify the whole system. For these reasons, all of the human factors interacting with the system need to be part of the simulation. So, if toxic gases are released in a space station, the astronauts cannot just escape to the outside. This may also interact with other safety elements that need to be simulated, such as psychological factors. With computer simulations, it is difficult to consider all the factors and their interactions at the same time, but it is easy to consider each factor separately. However, as Aristotle said, "The whole is greater than the sum of its parts." Applied to our case, this means that the whole needs to be simulated at the same time in order to achieve a better result than through the simulation of individual factors. In other words, a holistic approach needs to be used (covering all the aspects together as a whole; *holos* = all) (Schlacht 2012; Bandini Buti 2011).

To achieve this, the simulation for the space environment is done by simulating physically or virtually (with virtual simulation) each interaction factor within the IDP. The IDP is a design model developed by the author to approach this particular context of design. The IDP combines user-centered design, a holistic approach, and human factors needs to achieve an enjoyable, comfortable, and safe environment for users who need to perform under extreme and life-threatening conditions.

Specifically, the IDP integrates operational, physical, environmental, psychological, and socio-cultural factors. It is based on a human-centered approach and a holistic methodology to support the human side of the project, such as cultural dimensions within the technological interaction. The human-centered design focuses on three techniques: designing the experience of the user (user experience), designing together with the user (participatory design), and designing by identifying oneself with the user (empathetic design). The holistic methodology aims to support the user in relation to the system and is composed of the interrelations among three mainly quality-oriented methods: a multidisciplinary team (integrating humanities), concurrent design (concurrently working together with all the disciplines on each phase of the project life cycle), and support through dedicated development of human–machine–environment interactions. The application of the design model in respect to the current methodology increases safety and productivity because it supports usability, livability, and flexibility, which are very important elements in extreme and emergency contexts (Schlacht et al. 2012b) (Figure 16.1).

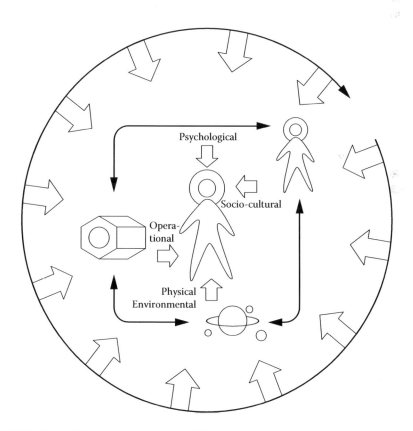

FIGURE 16.1 IDP design model. In the IDP graphical representation, the circular arrow is the concurrent design, the arrows in the circle are the different disciplines, the human-machine-environment interrelation is represented inside the circle, and the human-centered design is represented with the human in the middle surrounded by the five human factors: operational, environmental, physiological, psychological, and socio-cultural. (Copyright Schlacht.)

16.3   SUSTAINABLE SYSTEM VERSUS SAFETY

"Space stations are working places operating in extreme and isolated environments. In isolation, having no access to resources, these places need to be self-sufficient and sustainable and be able to reuse their resources" (Schlacht et al. 2015a, p. 1). Sustainability of the system is a key factor for safety (Figure 16.2).

Sustainability is related to safety in two complementary ways: the first is operational and technical and related to achieving a "closed-loop system" that optimizes its performance through precise management of system resources and operations (e.g., recycling of goods and *in situ* energy production); the second is sociopsychological and related to the "slow design" approach, which uses sustainability to create a user experience that increases psychological well-being and user reliability in isolated environments (e.g., by supporting direct production and consumption of goods as a form of qualitative user experience) (Schlacht et al. 2015a).

The integration of safety and sustainability objectives is necessary to achieve optimal safety and performance, as well as mission success.

From an operational and technical perspective, a closed-loop system that is autonomous, self-sufficient, and regenerative is necessary in remote locations with limited or absent infrastructure, transportation, and resources (Bannova and Bell 2011). A closed-loop or self-sufficient system requires no additional input, as it employs recycling of goods and *in situ* energy production to perform system operations. In creating a closed-loop system for extreme environments such as space, the objectives of sustainability can complement and even enhance user safety. Through a comprehensive life cycle analysis, the system structure can be designed to sustain an

FIGURE 16.2   Mars Desert Research Station (MDRS) isolated in the middle of the "Martian desert."

equilibrium that supports safety and optimizes the use and reuse of resources for the duration of the mission (Takata and Kimura 2003). Using this holistic approach toward all system elements and their interactions, the closed-loop system is optimized and preserved. Consumed resources once regarded as waste products become inputs for other operations; by-products of production and consumption processes are harvested as resources as well. Methods of decontamination and recycling of materials for reuse are devised to optimize product recovery and minimize user hazard (e.g., use of bacteria on mycellium to break down toxic materials). Efficient *in situ* energy increases reserves for periods when additional energy is required, such as during system failure. The identification of relationships between safety and sustainability motivates efforts to find new and improved organization, planning, and design solutions for optimal performance and user safety, and supports the design of closed-loop systems for isolated environments, from space to Earth (Schlacht et al. 2015a) (Figure 16.3).

From a sociopsychological perspective, the slow design supports sustainability addressing human needs that improve and sustain the psychological health of users. This is imperative for safety during situations with limited human interaction, confined environments, and conditions of isolation (Bannova 2014). "A human who is not reliable psychologically may make mistakes and disrupt the small and fragile closed-loop system of a Space station" (Schlacht et al. 2015a). Research on the effect of isolation and confinement on humans has broadened to encompass states of consciousness, stress, health, small group dynamics, personnel selection, crew training, and environmental engineering (Harrison et al. 1990). Improving and sustaining the psychological well-being of users can improve human–machine–environment interaction and increase the probability of user safety and survival. A design approach that supports the user's quality of life is strictly related to the equilibrium of the system (Ceppi 2012). The slow design approach enforces quality in the design approach by supporting sustainability and equilibrium in the system (Schlacht et al. 2015a). "Slow Design means cultivating quality: linking products and their producers to their places of production and to their end-users who, by taking part in the

FIGURE 16.3 Technical system for recycling water at MDRS. (Copyright Schlacht 2010.)

FIGURE 16.4 Experiencing sustainable food production at MDRS. (Copyright Schlacht 2010.)

production chain in different ways, become themselves coproducers" (Capatti et al. 2006). Principles of slow design can facilitate the design of a system that reduces stress, improves intellectual processing, and boosts overall morale (Schlacht et al. 2015a). By encouraging user interaction with the production and consumption of goods, slow design enables communication, control, and feedback. This improves overall mental health and perception and enables successful execution of operational tasks and emergency response, thus supporting overall safety (Figure 16.4).

16.4   MISSION SIMULATION VERSUS SAFETY

On the basis of the IDP, three mission simulation scenarios were investigated from the safety perspective in order to increase safety in particularly life-threatening environments:

1. MDRS (2-week simulation): The case of the Mars Desert Research Station (MDRS) in Utah, where a crew of six members took part in a real simulation of a space mission, testing safety, user interfaces, and procedures
2. ExoLab (1-day simulation): The International Lunar Exploration Working Group (ILEWG) mission, where a four-member crew in a basic container tested safety procedures for extra vehicular activity (EVA) on the Moon and their potentiality for application on Earth
3. V-ERAS (virtual simulation): The Mars Society mission, where a crew of four members took part in a virtual simulation of a Mars mission, testing safety, user interfaces, procedures, and reduced gravity

A safety factor being specifically studied in terms of human interaction is habitability, which is defined as "the usability of the environment" (Blume Novak 2000). During selected missions, safety and performance have been investigated with the help of the "habitability debriefing" developed by the author. The habitability debriefing is a new instrument for the analysis of safety and performance based on the IDP presented here (Hendrikse et al. 2011; Schlacht et al. 2012a). During each simulation mission, the debriefing was performed by all crew members together reporting the result as a group and anonymously. In order to increase the overall system safety and performance, the methodological aim was to let the group of users collectively discuss each possible problem and problem solution covering all the different human factors aspects. To cover all the human factors aspects, the discussion was guided in particular to operational, psychological, socio-cultural, environmental, and physiological factors. As a part of the IDP, a holistic approach was used (covering all the aspects together as a whole) (Schlacht 2012; Bandini Buti 2011). This approach is quite different from the traditional approach, where each crew member is questioned individually and each factor is studied separately. For example, operational problems are traditionally investigated after the mission, with each user facing a team of experts, while in the habitability debriefing, operational problems are investigated in relation to and in parallel with psychological, socio-cultural, environmental, and physiological factors, and the investigation is carried out jointly by all crew members during the mission (again, "the whole is greater than the sum of its parts").

16.4.1 MDRS

At the MDRS in the Roswell desert in Utah, every year a rotating crew of six members simulates a mission of the Moon–Mars scenario, testing safety as well as factors such as human interaction and procedures (Foing et al. 2010, 2011; Schlacht et al. 2010; Voute 2010; Mangeot et al. 2012). Research performed at the MDRS is optimal for testing and optimizing mission safety, in particular considering the specific desert surrounding the station, which is a perfect analog of the Mars environment: the natural reserve of the San Rafael Swell, a red-colored desert in Utah (Mars Society 2014) (Figures 16.5 through 16.7).

FIGURE 16.5 Working and living inside MDRS. (Copyright Schlacht 2010.)

FIGURE 16.6 Equipment safety check before EVA at the MDRS. (Copyright Schlacht 2010.)

FIGURE 16.7 EVA in the Martian desert during mission simulation at the MDRS. (Copyright Schlacht 2010.)

To create the optimal environment for the simulation, each individual factor needs to be planned and organized, such as the environment and the architecture of the space station; the system, which needs to be as sustainable as possible; the equipment; the behavioral procedure; and the hierarchy, selection, and training of the crew. Indeed, the organization of a simulation campaign is a complex procedure related to several factors and has to take into account specific rules.

At the MDRS, the system has been developed to give the greatest possible feeling of an autonomous system: the energy is provided by batteries, the communication connection is also autonomous, and the water is recycled for use in flushing the toilet. This is close to the concept of sustainability and autonomy. However, in space, the recycling rate of the water is much higher and reached 90% in the International Space Station.

To increase the credibility of the simulation, each subject needs to mimic a real space mission. One of the most important restrictions is the confinement in the habitat, with no possibility to escape without authorized EVA. The EVAs themselves need to be performed taking into account and applying all the regulations and equipment as in a real mission.

To give a better understanding of the MDRS simulation, the main points of the structure are described below.

- Crew composition: Usually mixed gender (men and women).
- Crew selection: Based on motivation and profiles.
- Crew structures and hierarchies: The crew has a sound structure with fixed tasks. There are six main roles that need to be covered: commander, executive officer, crew engineer, health and safety officer, journalist, and crew scientists (e.g., human factors researcher, geologist, and biologist). Extra roles outside the crew are campaign director, mission support, and project scientists.
- Training: Around 6 months before the mission, crew meetings are organized via remote conference calls in an attempt to accommodate the different goals and instruct the members to follow the strict safety rules and ethical restrictions of the station.
- Mission schedule: During the 2 weeks of the mission, each crew member carries out planned tasks, including scientific research, social activities, and station maintenance, in accordance with the simulation requirements (Figure 16.8).

    The isolation in a space analog environment such as the Utah desert, the strict procedures, and the crew hierarchy are some of the constraints that make this mission simulation an optimal scenario for verifying, testing, and increasing safety in extreme contexts. The methodology used to achieve a deep understanding of safety and performance problems concerns the application of the habitability debriefing. The debriefing is performed the day before the end of the mission. In complete privacy, the crew is guided for 90 min by a strict procedure regarding multidisciplinary analysis and collective discussion of the overall mission. The main mission problems and possible solutions are discussed from the perspectives of safety,

FIGURE 16.8 Soil measurement during MDRS mission simulation. In the image, the gloves are removed to interact with the instrument. Dedicated instruments are needed to perform a safe and successful mission.

performance, and comfort by the crew alone. Regarding the results during the 2014 MDRS mission, six crew members consisting of male and female members with international identities were able to spend 2 weeks simulating life on Mars. The human factors discipline was integrated and evaluated during the simulation to find problems and solutions, as well as propose implementation recommendations to increase the overall system performance.

- Find problems and solutions: Socio-cultural, psychological, operational, environmental, and physiological aspects were investigated. Operational aspects emerged as the most frequently discussed problem; in particular, communication was the most frequently recurring topic associated with this problem (Table 16.1). The main problems and solutions referred to increasing the quality of
  - Communication (as operational factors)
  - Equipment and structure (as operational, psychological, and environmental factors)
  - As a solution, to increase the overall system performance the crew proposed to improve the design of the equipment (particularly regarding EVA, toilet, and habitat structure) and the communication (particularly regarding manual and guideline)

In conclusion, in particular extreme and isolated contexts, safety, performance, and comfort are elements that are strongly correlated. A very uncomfortable scenario in a Mars mission will influence the performance and, as a consequence, also impact the safety of the crew.

TABLE 16.1

Problems and Solutions Voted as Most Important and Discussed by the Crew during the MDRS Mission (November 2014)

Problem (P)	Problem	Solution	Field	Crew Vote
P1: EVA equipment	Space suit fatigue, CO_2 buildup, poor air circulation, helmet fogging	Better design of air distribution; sensors; water cooling system; antifog system	Psychological— EVA Operational—EVA	6/6 6/6
P2: Toilet smell	Toilet smell	Increase ventilation; difficult to clean the room (new design); closable trash; more frequent flushing (recycling water)	Psychological— IVA Environmental— IVA	6/6 5/6
P3: Mission control communication	Lack of transparency and knowledge transfer	Manual, guideline improvement	Operational—IVA	6/6
P4: Station incomplete structure	Fake tunnel "breaking" simulation	Finish the tunnel and roof over the porch of the engineering airlock	Operational—IVA	6/6
P5: Communication on maintenance	Limited flexibility to make easy fixes; unclear what maintenance requires mission approval	Manual and guidelines improvement	Operational—IVA	6/6

16.4.2 EXOLAB

The ExoLab mission simulation consisted of testing the safety and performance during a specific scenario related to a communication breakdown during EVAs. In the following, the mission results will be described as presented during the International Astronautical Conference (Schlacht et al. 2015b) (Figures 16.9 and 16.10).

This simulation addressed the context of building a minimum autonomous modular architecture for the Moon and extreme environments on Earth. The simulation was also performed to investigate the potential use for art- and science-related applications. More specifically, ExoHab and ExoLab have been set up as technical mock-ups at the European Space Research and Technology Centre (ESTEC) of the European Space Agency (ESA) in the Netherlands for the purpose of multidisciplinary mission simulation (Schlacht 2011, 2012; Schlacht et al. 2015b).

The structural project is based on particular restrictions in order to be applicable in extreme environments. These restrictions were to organize a living space for two scientists inside an International Organization for Standardization (ISO) 20 container—about 15 m² in size. In the project, the space is multifunctional and convertible; the different areas (working station, kitchen, and lounge) are mostly open and common, but guarantee privacy when convenient (Cenini et al. 2015).

FIGURE 16.9 ExoLab container for mission simulation at ESA-ESTEC. (Copyright Schlacht 2014.)

In May 2015, the ExoLab was structured and equipped as a technological mock-up to perform a space mission simulation. A team of nine members was invited by the ILEWG to address specific tasks as part of the safety simulation in order to verify how persons from different humanities and technical fields could make both cultural and technical contributions.

The crew had a classical task and hierarchy structure, but consisted of members from different humanities and scientific fields, divided between

- Remote support: Campaign director, commander, mission support
- ExoLab and ExoHab: Executive officer, crew engineer, health and safety officer, crew biologist
- ATV observatory: Crew astronomy specialist, crew scientist

Appropriate procedures are one of the most important things to create the feeling of simulation and make the results reliable. The procedure used referred to a basic space mission configuration.

FIGURE 16.10 ExoLab EVA communication check. (Copyright Schlacht 2014.)

Ordinary equipment was used to simulate professional equipment for extreme environments in order to perform the simulation at this first stage and learn and understand which equipment development would need to have priority in the next step of the project.

During the simulation, the crew performed research on life-forms living on rocks during EVAs, while the crew astronomer and the biologist worked in their fields of research, also getting inspirations for cultural applications. During the EVA, a communication breakdown was planned and two astronauts in EVAs performed a safety emergency procedure, while the crew biologist and the crew health and safety officer were left alone, each in one of the modules, to try and reconnect the communication. After the communication breakdown with ExoLab, the crew decided to get in contact with ExoHab. The crew member in ExoHab communicated to them the way to reach the ExoHab module, while the member in ExoLab was left alone with no communication connection. The crew members left alone experienced psychological reactions related to the feeling of isolation.

The complete EVA took about 60 min, and the ExoLab crew member left alone experienced isolation.

TABLE 16.2

Problems and Solutions Voted as Most Important and Discussed by the Crew during the ExoLab Mission (May 2015)

Problem	Problem	Solution	Field	Crew Vote
Instrument	Communication	Dedicated equipment	Operational—EVA/IVA	3/4
	Nothing was working in ExoLab		Operational—IVA	1/4
	Could not use trackpad with gloves		Operational—EVA	2/4
	During the day, the UV instrument had no darkness setting to do proper work and identify the right sample			
Panic	I got bored and panicked because of the boredom	Social care	Socio-cultural and psychological—IVA	2/4
	Panic; no equipment was working; I focused on sensory perception; I was scared and panicked; I tried to calm down by remembering past socio-cultural experiences from my life			

The simulation was performed successfully; each task was addressed appropriately.

After the EVA, the simulation concluded with a debriefing, performed in accordance with the habitability debriefing procedure described above, which holistically considers all human factors involved in the mission to learn how to improve safety and performance in the mission. The debriefing results are reported here anonymously and divided according to the factors used in the procedure (Table 16.2 and Figure 16.11).

FIGURE 16.11 ExoLab isolated during communication breakdown. (Copyright Schlacht 2014.)

1. Find problems and solutions: Socio-cultural, psychological, operational, environmental, and physiological aspects were investigated. Operational aspects emerged as the most frequently discussed problem; communication, in particular, was the key problem. Other problems were the breakdown of the equipment, the difficulties of using the trackpad with gloves, and using the ultraviolet (UV) instrument with daylight for sampling. Also, the cold emerged as a key problem related to physical and environmental factors. Boredom and panic emerged as socio-cultural and psychological problems.

2. Propose implementation recommendations to increase the overall system performance: Improvement of social care emerged as a keyword. Moreover, the development of dedicated equipment for communication technology, sampling technology, and gloves was suggested.

3. Also, successful achievements were discussed by the crew. The crew recorded these as short descriptions of positive personal experiences related mostly to psychological factors. Some examples include performing the sampling successfully, learning how to handle the feeling of isolation, learning to avoid boredom, experiencing a slow perception of time, sharing hands with the others when using instruments, and having an interesting experience with regard to waiting time.

Finally, a comparison between the scenarios was discussed. The aim was to discover the applicability and use of simulation in disaster contexts, such as the Fukushima radioactive scenario on Earth.

The crew had different comments about this. For example, they said both are hostile environments; however, there is a different psychological approach: in a radioactive environment on Earth, performing a mission is not something one wishes to do, while in space it is (Figure 16.12).

FIGURE 16.12 ExoLab communication breakdown during EVA. In the image, the gloves are removed to interact with the instrument; dedicated instruments are needed to perform a safe and successful mission. (Copyright Schlacht 2014.)

16.4.3 V-ERAS

Another scenario used to test safety and performance in relation to procedures, equipment, mission structure, and system is a simulation in virtual reality. In order to effectively test such a particular extreme environment, equipment needs to be developed to properly simulate the effect of specific factors, such as the different gravities or the absence of oxygen during EVA. In this case, a simulation realized by the Italian Mars Society is presented; specifically, the V-ERAS mission carried out in December 2014 in Italy is used as a case study. The virtual simulation is composed of a complex infrastructure and team structure (Figures 16.13 and 16.14).

The infrastructure is mainly based on four virtual stations characterized by the following key elements:

- Immersive virtual simulations on the Blender Game Engine (BGE) with three-dimensional (3D) virtual reality headset (Oculus Rift).
- Full-body tracking via a Kinect device.
- Main component: Four motivity omnidirectional treadmills (also called stations), which are specific structures where the user can visualize and interact with the virtual environment and which with a modified version of Motivity called Motigravity, also simulate the difference in gravity. These stations are linked via dedicated multiplayer support capable of synchronizing the events happening at the four simulation nodes.

FIGURE 16.13 Motivity V-ERAS. (Copyright Schlacht 2014.)

FIGURE 16.14   V-ERAS astronaut avatar on the Martian surface. (Copyright Schlacht 2014.)

- Mumble voice chat software is used to ensure the overall voice communication infrastructure.

The people involved are assigned specific roles and tasks:

- The team is composed of a mission director, science officer, technical support team, outreach communication team (Earth-based), and the crew that is performing the mission simulation (Mars-based).
- The crew is composed of a commander, executive officer, crew engineer, and a health and safety officer.
- The team supports the crew regarding the performance of the following experiments: habitat design and station design review, communication test, health monitoring, simulation of telemedical support session, ATV vehicle review, EVA missions review, simulating Martian reduced gravity (motigravity omnidirectional treadmill), test of the analog space suit during the simulation, human performance in teleoperation, and human factors analysis.
- The mission director is responsible for the overall mission operation, coordinating all necessary actions with the team. As far as we know, this virtual mission simulation was carried out for the first time with this configuration of equipment and experiments, developed specifically to achieve the most reliable conditions to simulate the main factors related to a Mars mission, from the difference in gravity to the difficulties in performing activities during EVA.

To test safety and performance using the IDP approach, the habitability debriefing was applied both in the virtual reality context (in intravehicular activity [IVA] and EVA conditions) and in the real context outside. The crew debriefing allowed learning from the crew how to improve the overall safety, performance, and comfort of the mission. Regarding the results obtained during the Italian Mars Society mission, four members of the crew with international and mixed-gender identity were tracked in virtual reality and were able to interact through an avatar with different field tasks on the Martian surface. The human factors discipline was integrated and evaluated during the simulation to find problems and solutions, as well as to propose implementation recommendations to increase the overall system performance, as described here:

1. Find problems and solutions: Socio-cultural, psychological, operational, environmental, and physiological aspects were investigated. Operational aspects emerged as the most frequently discussed problem; in particular, *motigravity* was the most frequently recurring word associated with *uncomfortable*. The main problems and solutions referred to increasing the quality of
    a. The system (test the system before the mission and increase the number of team members)
    b. The tasks (increase the margin among tasks to avoid overload; ensure free time for the crew, in particular after dinner, and physical training)
    c. The equipment (increase the comfort of motivity and the quality of the navigation)
2. Propose implementation recommendations to increase the overall system performance: It was proposed to implement the system for different user typologies and anthropometrics, tracking users with the anthropometrics of a 2-year-old child, using an extremely small human size to verify the performance of the system in abnormal situations; to implement the interior design and interface with movement data from tests in Martian gravity; and finally, to provide the possibility of interaction among crew members in VR (virtual reality). Social aspects did not emerge as a problem; however, late work and short periods of free time led to dissatisfaction, which was not approached by the team. In conclusion, it was verified again (as in the MDRS mission) that safety is strictly correlated with performance and comfort (Figure 16.15).

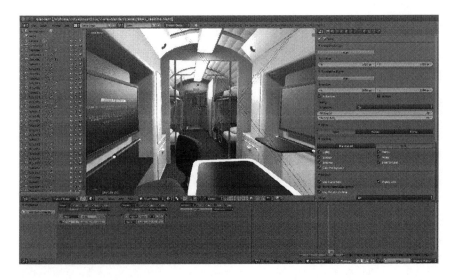

FIGURE 16.15 Interior of the V-ERAS Martian space station. (Copyright Schlacht 2014.)

16.5   SAFETY FROM SPACE TO EARTH

To support innovation in safety procedures for extreme, dangerous, and isolated environments, research related to the space environment has been presented here. Space was selected as the most extreme environmental condition that incorporates all the factors that characterize other extreme environments. For this reason, space missions can be used as a source for learning how to increase safety and improve user–system interaction in other extreme contexts on Earth. The transfer of research and design solutions from space can be valuable in terms of addressing problems faced in extreme environments, as well as in providing the base setting for transferring solutions to practical problems, both for today and for the future.

This chapter can be used to inspire specialists to use the space missions described in the scenarios as a model for realizing possible safety procedure implementations in all life-threatening, isolated, and extreme contexts. Although these environments are quite different, they share many of the problems regarding the support of human life in them. For example, considering the context of hospitals, the architecture needs to be user-centered in order to create an environment where it is enjoyable to spend time. As mentioned in the abstract of the International Conference Cluster in Design of Health Facilities, "The design of healthcare facilities requires a sensitive approach to minimise the sense of alienation and offer a welcoming and comfortable place for people for improving health through healing environments. Nowadays, architectures for health are suffering not only from a lack of resources but also from a whole vision of hospital's needs. Therefore, the attention to topics like safety, environmental sustainability, comfort and well-being, requires a typological, structural and managerial re-organization of the design process: for this reason, it is important to involve hospital planners and design teams for boosting innovative design strategies through highly experienced professionals and stimulating debate on these issues" (SItI 2015), and maybe get inspired by space missions.

Space is a domain characterized by a huge amount of complexity in terms of technical details as well as operational processes. The interconnections among the agents involved in this complex system add to the risk-proneness. The user astronauts, who are the social components of these sociotechnical systems, play a vital role in overall safety. Thus, the right operator or astronaut must be selected for this job, and training is needed to provide the necessary skills to ensure smooth as well as safe operation in these contexts.

It has been verified that training in an immersive environment allows increasing the safety of the equipment and the operator during an accident scenario. As described here with three different simulation approaches, it has been verified both in virtual reality and in analog environment simulations that safety is strictly correlated with performance and comfort, involving the optimization of human factors. This is because in the space context, the user is involved in an extremely dangerous context during a period of several months, which includes not only working time, but also living time (Schlacht 2012). This is why it is important to have an environment that can simulate all the factors that may impact safety, performance, and comfort. With the holistic approach, socio-cultural, psychological, operational, environmental, and physiological factors are investigated together by all the crew members to show the interconnections related to safety, performance, and comfort. To understand why all these human factors are so important for the overall safety in extreme contexts, one can easily imagine that in a space mission, the user cannot be replaced (just like a hivernaut on Antarctica, an astronaut cannot easily return and be replaced either). This is the reason why he needs to be supported with a holistic perspective in order to keep up a high level of performance and reliability. However, with respect to isolated environments such as space or Antarctica, the overall risk involved in a dangerous scenario on Earth may be much higher, as the consequences for the surrounding population and the environment of a production site, for example, could be considerable (e.g., in the case of an explosion at a chemical plant). The above-mentioned factors need to be predicted and recognized in time, which is why it is rather important to adopt a holistic approach, in particular during measures aimed at prevention, such as operator training. Finally, in order to prevent accidents in particularly dangerous environments, simulation and training in scenarios that are as similar as possible to the real condition can be accomplished in an analog or 3D immersive virtual environment.

16.6   CONCLUSIONS AND FURTHER DEVELOPMENTS

In conclusion, we can summarize that in all contexts presented in this chapter, the users played a vital role regarding overall safety and need to be approached holistically. Indeed, in space as on Earth, the most important variable is the user, the operator, or better still, the "unpredictable human." This is why whenever there is any human interaction, it is more important to test the elements not only in isolation, but also holistically, as an overall system simulation, in order to predict possible user interactions. Especially during an extended system simulation, the variables interact with each other, which leads to much more reliable results regarding the increase of user and system safety. To repeat the words of Aristotle, "The whole is greater than the sum of its parts."

Further investigation is needed regarding the application of

- IDP as a methodology for developing projects based on human factors, user-centered design, and a holistic approach
- A sustainable system with the benefit of technical system autonomy, as well as user experience and reliability
- Overall human–machine system simulations and IDP-based debriefings
- Crossover benefits to increase safety in extreme contexts both on Earth and in space

ACKNOWLEDGMENTS

I would like to acknowledge the crew members and teams involved in the mission simulations, in particular Alexandre Mangeot, Antonio Del Mastro (director) and the V-ERAS team from the Mars Society, the MDRS crew and team, and the ExoLab crew and team. I also have the pleasure to acknowledge CMIC (PSE-Lab) and the design department at Politecnico di Milano, the Mars Society, the ILEWG, the Dipartimento di Scienze della Vita e Biologia dei Sistemi at Università di Torino, and the Mensch-Maschine-Fachgebiet at Technische Universität Berlin. Finally, I am grateful to Manjula Singh, researcher in sustainable architecture, ecology, and psychology, for her contribution to Section 16.3 on sustainable system versus safety.

REFERENCES

AA.VV. 2015. Healing architecture: Designing innovations in architectures for health abstract. Presented at VI International Conference Cluster in Design of Health Facilities, Milan, Italy, October 14.

Bandini Buti, L. 2011. *Ergonomia olistica. Il progetto per la variabilità umana.* Milano, Italy: Franco Angeli.

Bannova, O. 2014. Extreme environments—design and human factors considerations. Doctoral dissertation, Chalmers University of Technology. http://publications.lib. chalmers.se/records/fulltext/204341/204341.pdf.

Bannova, O., Bell, L. 2011. Designing from minimum to optimum functionality. *Acta Astronautica*, 68(7–8), 760–769.

Blume Novak, J. 2000. Summary of current issues regarding space flight habitability. *Aviation, Space, and Environmental Medicine*, 71(9), A131–A132.

Capatti, A., Ceppi, G., Colonetti, A., et al. 2006. Slow + design l manifesto l. Presented at International Seminar on Slow Approach to Distributed Economy and Sustainable Sensoriality, Milan, Italy.

Cenini, E., Desole, E., Facchinetti, et al. 2015. Moon habitat module: New ways of living in extreme spaces. Presented at 66th International Astronautical Congress 2015.

Ceppi, G. 2012. Awareness design. Dodici Edizioni. http://www.awarenessdesign.it/.

Foing, B., 2010. ExoHab pilot project and field tests for Moon-Mars human laboratories. Presented at EGU General Assembly 2010, Vienna, Austria, May 2–7, 2010. http://adsabs.harvard.edu/abs/2010EGUGA..1213688.

Foing, B.H., Stoker, C., Ehrenfreund, P., et al. 2011. Astrobiology field research in Moon/Mars analogue environments: Instruments and methods. *International Journal of Astrobiology*, 10(03), 141–160.

Harrison, A.A., Clearwater, Y.A., McKay, C.P. 1990. *From Antarctica to Outer* Space. New York: Springer-Verlag.

Hendrikse, J., Ono, A., Schlacht, I.L., et al. 2011. Human and robotic partnerships from EuroMoonMars analogue missions 2011. Presented at 62nd International Astronautical Congress 2011, Cape Town, South Africa.

IAF International Astronautical Congress, Jerusalem, Israel, 12–16 October 2015.

Mangeot, A., Hendrikse, J., Schlacht, I.L., et al. 2012. Human-robotic partnerships and performance: Lessons learned from ILEWG EuroMoonMars Campaigns 2012 and 2011. Presented at 63rd International Astronautical Congress 2012, Naples, Italy.

Mars Society. 2014. Mars Desert Research Station. Lakewood, CO: Mars Society. http://desert.marssociety.org/home/about-mdrsMars.

Oxford Dictionary. 2015. Safety. Oxford: Oxford University Press. http://www.oxforddictionaries.com/definition/english/safety.

Rabelo, L., Sepulveda, J., Compton, J. et al. 2005. Simulation of range safety for the NASA space shuttle. *Aircraft Engineering and Aerospace Technology*, 78(2), 98–106. Online: http://ntrs.nasa.gov/archive/nasa/casi.ntrs.nasa.gov/20110024186.pdf

Schlacht, I.L. (ed.) 2011. Space extreme design. *Personal and Ubiquitous Computing*, 15(5), 487–526.

Schlacht, I.L., 2012. Space habitability: Integrating human factors into the design process to enhance habitability in long duration missions. Doctoral dissertation, Technische Universität Berlin. https://depositonce.tu-berlin.de/handle/11303/3390.

Schlacht, I.L., Ceppi, G., Nazir, S. 2015a. Sustainable quality: From space stations to everyday contexts on earth. Presented at Conference NES 2015. Nordic Ergonomics and Human Factors Society annual conference Lillehammer, Norway 1-4/11/2015. www.nes2015.no.

Schlacht, I.L., Foing, B., Ceppi, G. 2015b. ExoHab1 development: Spin-in/out from space habitat to disaster management facility. Presented at 66th International Astronautical Congress 2015. IAF International Astronautical Congress, Jerusalem, Israel, 12–16 October 2015.

Schlacht, I.L., Hendrikse, J., Hunter, J., et al. 2012a. MDRS 2012 ILEWG campaign: Testing habitability and performance at an analogue moon base infrastructure outpost on earth. Presented at 63rd International Astronautical Congress 2012, October 1–5 Naples, Italy.

Schlacht, I.L., Nazir, S., Manca, D. 2015c. Space vs. chemical domains: Virtual and real simulation to increase safety in extreme contexts. Presented at 6th International Conference on Applied Human Factors and Ergonomics and the Affiliated Conferences, 26–30 July 2015 Las Vegas, United States AHFE 2015.

Schlacht, I.L., Roetting, M., Masali, et al. 2012b. Human factors in the space station design process. Presented at 63rd International Astronautical Congress 2012, Naples, Italy, October 1–5.

Schlacht, I.L., Voute, S., Irwin, S., et al. 2010. Moon-Mars analogue mission at the MDRS. EuroMoonMars-1 Mission. Presented at GLUC 2010 Congress, Beijing.

Takata, S., Kimura, T. 2003. Life cycle simulation for life cycle process planning. *CIRP Annals—Manufacturing Technology*, 52(1), 37–40.

Voute, S. 2010. Testing planetary geological sampling procedures in a Mars analogue site. Presented at 38th COSPAR Scientific Assembly, Bremen, Germany, July 18–25.

17 Categorization of Effective Safety Leadership Facets

Sari Tappura and Noora Nenonen

CONTENTS

17.1 INTRODUCTION

In the safety literature, generally, managers' critical role in developing occupational health and safety (OHS) is addressed. Managers are commonly considered key factors in safety improvement and implementing safety management systems because they have the capacity and power to make decisions about safety investments and can influence the safety culture of an organization (e.g., Dedobbeleer and Béland, 1998; DeJoy et al., 2004; Fernández-Muñiz et al., 2009; Flin, 2003; Flin et al., 2000; Flin and Yule, 2004; Guldenmund, 2000; Hale et al., 2010; Hofmann and Stetzer, 1996, 2003; Zohar, 1980). For establishing successful organizational OHS policies and procedures, managers' resources, competence, and commitment are crucial (Conchie et al., 2013; Fruhen at al., 2013; Hale et al., 2010; Hardison et al., 2014; Tappura and Hämäläinen, 2012). In particular, safety leadership is considered important for developing safety culture, climate, and performance, and it has been studied actively in recent years (e.g., Barling et al., 2002; Conchie et al., 2013; Eid et al., 2012; Hoffmeister et al., 2014; Hofmann and Morgeson, 2004; Kapp, 2012;

Künzle et al., 2010; Lu and Yang, 2010; O'Dea and Flin, 2001; Wu et al., 2008; Zohar, 2010). In addition, novel OHS risks such as psychosocial risks have emerged, and the mental demands of work have increased (Leka et al., 2011; Siegrist et al., 2004). Therefore, managers should possess different types of leadership competencies for ensuring employees' health and safety at work.

Safety management and leadership are part of managing other business activities and should be integrated closely in general business management within organizations (Bluff, 2003; EU-OSHA, 2010). In an exploratory study involving 10 cases, Veltri et al. (2013) conjoined the safety and organizational contexts. They suggested that rather than trading off safety against operational practices, operational excellence should be pursued simultaneously with safety excellence. Moreover, excellent organizations blur the boundaries between safety management and operations management or fully incorporate safety management into general management. In this study, however, leadership is reviewed from the safety perspective.

Organization-specific safety procedures and safety culture may be supported by management practices, at least to fulfill regulatory requirements related to the physical and psychosocial well-being of all employees (Tappura and Hämäläinen, 2011). To promote safety performance, leadership is a key factor for motivating both safety participation and compliance of employees (Kapp, 2012). Moreover, safety leadership may affect the productivity of an organization via employee motivation and commitment; work fluency; and costs related to accidents, absences, or quality (e.g., Biron and Bamberger, 2012; Sievänen et al., 2013; Tappura et al., 2013). Thus, a lack of leadership skills may impede overall improvement actions and safety performance (Tappura and Hämäläinen, 2012).

When striving for safety performance, several possible weaknesses in management skills and supervisory processes might hinder safety rather than promoting it. According to Bentley and Haslam (2001), supervisors' influence on safety results from both their attitudes and actions. Zohar (2002b) found that training supervisors to enhance their monitoring and rewarding of workers' safety behaviors had a positive effect on injury rates and safety climate scores. Thus, undefined supervisory processes, and lack of management training and education of managers and supervisors might impede improvement actions (Carder and Ragan, 2005).

Various leadership behaviors suitable for improving safety performance have been suggested in the literature (e.g., Barling et al., 2002; Clarke, 2013; Eid et al., 2012; Hale et al., 2010; Kapp, 2012), as presented in Section 17.6. Törner (2011) suggests the following mechanisms for promoting occupational safety within an organization: a leadership style that promotes cooperation, inspires, fosters group goals, provides individualized support, and empowers workers may intrinsically be expected to comprise rich and open communication, thus supporting the development of high-quality manager–employee interaction. Such interaction and communication may promote the development of mutual trust and a good workgroup climate. Trust, in turn, may further promote communication and interaction. Mutual trust, high-quality relationships, and a strong group climate may promote worker motivation and intention to contribute toward organizational goals. Managers successful in demonstrating honest and consistent prioritization of worker safety may promote the development of workers' trust in the notion that safety is a prime organizational goal.

This may motivate workers to behave safely. Trustful relationships characterized by empowerment and participation are then likely to support the transition of safety intentions into safe behaviors.

Despite awareness of the importance of safety leadership, managers tend to have little safety training and limited understanding of their important roles (Tappura and Hämäläinen, 2012). Moreover, managers' safety competence requirements are often unclear (Hardison et al., 2014; Tappura and Hämäläinen, 2012). Hardison et al. (2014) identified supervisors' knowledge-based safety competencies, and Tappura and Hämäläinen (2011, 2012) defined an outline of managers' OHS competence areas and training requirements. Biggs and Biggs (2013) developed a construction safety competency framework that included identification of the knowledge, skills, and behaviors required for safety management tasks. However, there is a lack of studies pertaining to managers' safety competence, especially of studies related to effective safety leadership facets.

In this chapter, effective safety leadership facets are discussed as a part of managers' leadership competencies. The objective is to suggest a categorization scheme and examples of effective safety leadership facets. Relevant safety leadership facets are identified from the existing literature on safety and empirical findings, and structured according to the transactional and transformational leadership theory (Bass, 1985). Moreover, the importance of developing a manager's safety leadership competence in an organization is discussed.

17.2 RELATIONSHIP BETWEEN SAFETY LEADERSHIP AND SAFETY PERFORMANCE

According to Yukl (2008), organizational effectiveness consists of an organization's ability to survive, perform its mission, and maintain favorable earnings, financial resources, and asset values. Organizational effectiveness depends on performance determinants, namely, the efficiency of internal processes and the ability to adapt to the external environment. In addition to the type of industry and turbulence in the external environment, leaders' actions and decisions influence these determinants. Leaders can improve their performance by using specific leadership behaviors, and their leadership skills influence their ability to use favorable behaviors. According to Bass and Avolio (1990), effective leadership is based on transactional leadership, and transformational leadership builds on this by broadening the leader's effect on performance.

This study focuses on leadership behaviors when striving for safety performance improvements, which may be considered a subsystem of organizational performance (Wu et al., 2008). Leadership behavior and leader–member exchange relationships influence subordinates' performance and outcomes (e.g., Bass and Avolio, 1990; Michael et al., 2006; Stinglhamber and Vandenberghe, 2003; Yukl, 2008). Safety performance refers to the concept of safety-related actions and behaviors that workers exhibit in almost all kinds of work to promote the safety and health of themselves or others (Burke et al., 2010). Good safety performance influences, for example, efficiency through reduced accident costs or improved productivity (Sievänen et al., 2013; Tappura et al., 2015). Moreover, it may affect an organization's ability to adapt to changing customer needs and preferences.

According to Hofmann and Morgenson (1999), employees' safety performance improves when they have a clear understanding of safe procedures and the consequences of unsafe behaviors, and when their safety behaviors are supported by their supervisors. Both safety coaching and control have been found to be important elements of safety leadership (Blair, 2003; Cooper, 1998; Williams, 2002; Wu et al., 2008). They affect employees' safety compliance and safety participation (Griffin and Neal, 2000), resulting in compliance with safety rules and procedures, and improving workplace safety (Kapp, 2012). Similarly, transactional leadership (Bass, 1985) impacts safety compliance, and transformational leadership (Bass, 1985) impacts safety participation and the overall safety performance of employees (Kapp, 2012).

A study by Köper et al. (2009) links OHS to overall business issues (performance and competitiveness). The results of their study suggest a correlation between health-related issues and organizational performance, and they found that adverse work conditions influence organizational performance negatively. Improving health, job satisfaction, and motivation have positive effects on performance, and these factors may be influenced by a transformational leadership style (Bass and Avolio, 1990). Moreover, developing a positive safety climate requires that managers regularly and visibly demonstrate their commitment and actions toward safety (Wu et al., 2008).

According to Zohar (2002a), leadership and safety climate play important roles in occupational safety where traditional approaches of accident prevention, focusing on better engineering, and work site monitoring ignore the role of line managers. This may be the missing element in accident reduction schemes, which have not improved worker safety beyond a certain level (Shannon et al., 1997). In their studies, Wu et al. (2008) and Clarke (2013) suggest that safety leadership and safety climate are important predictors of safety performance. Moreover, leadership has been identified as a major factor in safety climate (Barling et al., 2002; Zohar, 2010). Similarly, Blair (2003) argued that both safety climate and safety leadership must be improved considering safety performance; thus, the quality of leadership influences safety performance in two ways (see Figure 17.1).

A safety performance evaluation is mostly based on accident rates, climate scores, audit scores, and expert judgment (e.g., Bigelow and Robson, 2005; Chang and Liang, 2009; Hale et al., 2010). In the scientific evaluation of successful safety

FIGURE 17.1 Relationship between safety leadership, safety climate, and safety performance. (From Wu, T.-C., et al., *Journal of Loss Prevention in the Process Industries* 21, 307–318, 2008.)

interventions by Hale et al. (2010), safety performance measurement was based mainly on output indicators (e.g., accident rate and lost days) and expert judgment. Intermediate indicators (e.g., safety climate scores, dangerous situation reports, and observation rounds) were used when available. In a study by Hoffmeister et al. (2014), safety climate scores were considered the most important safety performance indicators. From the viewpoint of this study, safety climate is a relevant indicator because it is influenced by leaders' actions and leadership behavior (e.g., Eid et al., 2012; Hoffmeister et al., 2014; Kapp, 2012; Wu et al., 2008).

17.3 LEADERSHIP COMPETENCE DEVELOPMENT

Leadership is one component of management competence (e.g., Garavan and McGuire, 2001; Viitala, 2005; Yukl, 2008). In the broadest sense, competence refers to the specific knowledge, experience, abilities, skills, traits, values, attitudes, and behaviors necessary for achieving a required level of performance (e.g., Boyatzis, 1982; Königová et al., 2012; Pickett, 1998). Essential leadership competencies are, for example, resourcefulness, change management, interaction, building relationships, communication, learning from difficult situations, being open to new ideas, composure, team leadership, integrity, and trust (Dainty et al., 2004; De Meuse et al., 2011; White et al., 1996).

Leadership skills are difficult to develop because they involve a complex mix of behavioral, cognitive, and social skills (Lord and Hall, 2005; Yukl, 2008). Moreover, they are usually connected to a manager's personal traits and personal growth (Garavan and McGuire, 2001; Viitala, 2005). Managers' competencies are generally developed through formal and on-the-job training, induction, work experience, and experiential learning (e.g., Fong and Chan, 2004; Pfeffer, 1998; Tappura and Hämäläinen, 2011; Fruhen et al., 2013). According to Schoonenboom et al. (2007), competence development typically involves formal and informal learning, and is not necessarily related to specific types of learning activities.

Several management development studies have suggested that improving self-knowledge is the basis for all true management development (e.g., Pedler et al., 1986; Viitala, 2005). According to Yukl (2008), leadership competence development includes (1) maintaining self-awareness of needs, emotions, abilities, and behavior; (2) continuous learning and self-development of relevant skills; (3) remembering that a strength can become a weakness when a situation changes; and (4) compensating for weaknesses, for example, by delegating or sharing responsibilities. Thus, the manager's desire for competence development and participation in related activities is important (Schoonenboom et al., 2007).

17.4 THEORETICAL FRAMEWORK FOR SAFETY LEADERSHIP

Safety leadership research leans on the leadership theory. Transactional and transformational leadership styles (Bass, 1985) have raised interest among safety researchers (e.g., Barling et al., 2002; Christian et al., 2009; Conchie and Donald, 2009; Kapp, 2012; Kelloway et al., 2006; Michael et al., 2006; Zohar, 2002a). However, the focus has been more on transformational than transactional leadership (Clarke, 2013). In their study, Eid et al. (2012) suggested that authentic leadership (Gardner et al., 2005) is a suitable construct for safety-focused leadership owing to its explicit emphasis on personal and social identification processes, role modeling, and value-based leadership (Avolio and Gardner, 2005). Furthermore, the leader–member exchange theory (Dansereau et al., 1975) has been used to explain the influence of leadership on safety outcomes (e.g., Hofmann and Morgeson, 1999; Hofmann et al., 2003; Michael et al., 2006). In this study, owing to their demonstration of positive safety impacts on employees' safety compliance and participation, as well as the safety climate (Barling et al., 2002; Clarke, 2013; Griffin and Hu, 2013; Hoffmeister et al., 2014; Kapp, 2012), the transactional and transformational leadership theories are applied to structure our findings.

In transactional leadership, the leader establishes goals (e.g., safety-related goals), actively monitors employee performance with regard to these goals, and provides rewarding or corrective feedback about employee performance (e.g., safe behavior). Transformational leadership achieves results by increasing employees' acceptance of set goals, for example, in safety-related behavior. Leaders serve as role models, inspire commitment toward achieving set goals, show active interest in individual employees, and challenge employees to overcome obstacles that prevent

TABLE 17.1
Characteristics of Transactional and Transformational Leadership Facets

Transactional Leadership	Transformational Leadership
Contingent reward: Providing appropriate rewards and recognition for positive behaviors (Bass, 1985). Clearly communicating desired behaviors and reward contingencies to employees, and actually recognizing accomplishments for reinforcing desired behaviors (Bass, 1985, 1990).	*Idealized influence*: Instilling pride, evoking integrity, trust, and respect in employees (Bass, 1990; Bass and Riggio, 2006), who ultimately view leaders as role models (Bass and Riggio, 2006).
	Individualized consideration: This involves giving personal attention to employees (Bass, 1990), addressing the differences in employees' needs individually, and coaching and mentoring employees to help them reach their full potential (Avolio, 1999; Bass and Riggio, 2006).
Management by exception: Discouraging negative behavior. Active management by exception is proactive and focused on prevention (Bass, 1985). Employee performance is monitored actively to detect deviations from rules and standards, and take corrective action. Passive management by exception involves reactive intervening, only when standards are not met (Bass, 1985, 1990).	*Inspirational motivation*: Leader's clear articulation of a compelling vision and the need for employees to work toward this mission, thus resulting in more inspired employees. Encouraging employees to strive for something beyond their individual goals (Bass, 1985).
	Intellectual stimulation: Promoting intelligence, rationality, and careful problem solving (Bass, 1990). Reflects the extent to which a leader solicits employees' perspectives on problems and considers a wide variety of opinions in making decisions. Inspiring employees to think creatively and innovatively (Bass, 1985).

them from achieving said goals (Barling et al., 2002; Bass, 1985; Kapp, 2012). Both the transactional and transformational leadership styles are related to effective leadership, with the best leaders demonstrating both styles (Bass, 1985; Hoffmeister et al., 2014). The transactional and transformational leadership styles consist of theoretically distinct multidimensional constructs and can be divided into more specific leadership facets (Bass, 1985), which may affect safety in different ways and for different reasons (Hoffmeister et al., 2014). The characteristics of major leadership facets related to the transactional and transformational leadership styles are summarized in Table 17.1.

17.5 METHODS

A literature review was carried out using electronic databases of scientific journals (e.g., Science Direct Elsevier), and the main search terms used were related to safety leadership. Studies related to safety leadership were browsed, and those associated with safety performance measures were included in the review. Moreover, safety management literature was reviewed to determine the interconnections between the good practices of safety management and safety leadership. A couple of major reviews (Hale and Hovden, 1998, as cited in Hale et al., 2010; Shannon et al., 1997, as cited in Hale et al., 2010), which have identified organizational factors affecting safety management and performance, were our major sources. A framework of competencies was built based on appropriate literature. Thematic interviews were carried out in a Finnish expert organization to empirically supplement the framework; 17 line managers were interviewed. The interviewees were mostly senior, experienced managers, and they were asked about their considerations of effective safety leadership. The results of the literature review and interviews were compared and structured according to the transactional and transformational facets of leadership (see Table 17.1).

17.6 RESULTS

17.6.1 TRANSACTIONAL LEADERSHIP

The interview results did not include leadership facets related to transactional leadership. In the safety literature, the following leadership facets related to transactional leadership were discussed:

- *Contingent reward:* Having a reward or incentive system (Hale and Hovden, 1998, as cited in Hale et al., 2010) and rewarding employees' safety behaviors (Lu and Yang, 2010; Zohar, 2002a)
- *Management by exception:* Monitoring employees' safe and unsafe behaviors (Griffin and Hu, 2013; Shannon et al., 1997 as cited in Hale et al., 2010; Zohar, 2002a; Zohar and Luria, 2003), correcting employee behaviors (Lu and Yang, 2010), enforcing employees' observance of safety regulations (Wu et al., 2008), and sanctioning rule violations (Hale and Hovden, 1998 as cited in Hale et al., 2010)

Both of the leadership facets, that is, contingent reward and management by exception, were linked to lower injury rates (Hale and Hovden, 1998 as cited in Hale et al., 2010; Zohar, 2002a) and better safety climate scores (Zohar, 2002a; Zohar and Luria, 2003). Moreover, both facets were positively associated with employee safety behaviors, such as compliance (Griffin and Hu, 2013; Lu and Yang, 2010), participation (Lu and Yang, 2010), housekeeping, and use of protective equipment (Zohar and Luria, 2003).

17.6.2 Transformational Leadership

17.6.2.1 Idealized Influence

Related to idealized influence, the interviewees mentioned the following leadership facets:

- Speaking respectfully about employees
- Treating all employees well and even-handedly
- Complying with organizational procedures and rules
- Being present
- Having an open-door policy to enable subordinates to discuss relevant issues when necessary
- Believing in employees' expertise
- Broaching discussion on conflicting issues and working out problems

In the literature, the following facets related to idealized influence were found:

- Stressing the importance of safety (Hale and Hovden, 1998, as cited in Hale et al., 2010; Lu and Yang, 2010), being a role model for safety (Lu and Yang, 2010), and demonstrating true and consistent priority of employee safety (Törner, 2011)
- Managers' commitment (Hale and Hovden, 1998, as cited in Hale et al., 2010; Hoffman and Morgeson, 1999) and personal commitment (Hale and Hovden, 1998, as cited in Hale et al., 2010)
- Managers' active role (Shannon et al., 1997 as cited in Hale et al., 2010), participative leadership style (Hale and Hovden, 1998, as cited in Hale et al., 2010), and the amount of energy and creativity injected by managers (Hale et al., 2010)
- Informal organization (Hale and Hovden, 1998, as cited in Hale et al., 2010) and good (Hoffman and Morgeson, 1999; Michael et al., 2006; Shannon et al., 1997, as cited in Hale et al., 2010) and trusting (Hale and Hovden, 1998, as cited in Hale et al., 2010; Kelloway et al., 2012; Törner, 2011; Zacharatos et al., 2005) relationships between the management and workforce, thus promoting cooperation (Törner, 2011)
- Interpersonal and group communication (Hale and Hovden, 1998, as cited in Hale et al., 2010) and constructive dialogue (Hale et al., 2010) between managers and the workforce
- Availability, openness to criticism, and work as a source of pride (Hale and Hovden, 1998, as cited in Hale et al., 2010)

Demonstrating true safety concerns, managers' commitment and active roles, and high-quality relationships through constructive dialogue have all been linked to lower injury rates (Hale and Hovden, 1998, as cited in Hale et al., 2010; Shannon et al., 1997, as cited in Hale et al., 2010) and safety incidents (Michael et al., 2006; Zacharatos et al., 2005). These types of leadership behaviors support trust and a position of safety as the prime organizational goal (Törner, 2011), and support employees' reporting of safety concerns (Hoffman and Morgeson, 1999). Trusting relationships (Törner, 2011) support the realization of safety behaviors (Lu and Yang, 2010). The level of trust in managers mediates personal safety orientations (i.e., safety knowledge, safety motivation, safety compliance, and safety initiative) and has a positive relationship with employees' psychological well-being (Kelloway et al., 2012). Constructive dialogue between the shop floor staff and line management has been identified as a key factor for successful safety interventions with improvements in safety performance (a combination of several measures, e.g., accidents, unsafe behavior, dangerous situations, and safety climate) (Hale et al., 2010).

17.6.2.2 Individual Consideration

In the interviews, the respondents mentioned the following leadership facets related to individualized consideration:

- Asking how they feel
- Offering help proactively
- Accepting differences in personalities
- Accepting different kinds of expressions
- Creating prerequisites for working efficiently

Respectively, the following facets found in the literature were classified as relating to individualized leadership facets:

- A culture of caring (Hale and Hovden, 1998) and providing individualized support (Törner, 2011), reflecting care and concern for the well-being of employees (Mearns and Reader, 2008)
- Human resources planning (Hale and Hovden, 1998, as cited in Hale et al., 2010) and modified work provision after accidents (Shannon et al., 1997, as cited in Hale et al., 2010)

Support considering individual needs promotes employee safety behavior (Mearns and Reader, 2008), and therefore their contribution toward organizational goals (Törner, 2011), such as lower accident rates (Hale and Hovden, 1998, as cited in Hale et al., 2010).

17.6.2.3 Inspirational Motivation

In the interview results, no leadership facet could be linked directly to inspirational motivation. The findings from the literature, however, are as follows:

- Promoting safety (Hale and Hovden, 1998, as cited in Hale et al., 2010) and motivating and inspiring safety (Griffin and Hu, 2013; Hale and Hovden,

1998, as cited in Hale et al., 2010; Shannon et al., 1997, as cited in Hale et al., 2010; Törner, 2011)

For example, by using inspirational appeals (using emotional language to emphasize the importance of a new task and arouse enthusiasm) (Clarke and Ward, 2006), empowering leader behavior (Martínez-Córcoles et al., 2011), and encouraging the workforce toward long-term commitment (Shannon et al., 1997, as cited in Hale et al., 2010)

- Having goals, standards, and resources defined and used (Hale and Hovden, 1998, as cited in Hale et al., 2010; Lu and Yang, 2008), fostering group goals (Törner, 2011), and communicating safety (Hoffman and Morgeson, 1999; Michael et al., 2006)

Promoting safety and motivating employees to engage in safety behaviors result in lower accident rates (Hale and Hovden, 1998, as cited in Hale et al., 2010; Shannon et al., 1997, as cited in Hale et al., 2010) through an improved safety climate (Clarke and Ward, 2006; Martínez-Córcoles et al., 2011; Törner, 2011) and increased employee safety participation (Griffin and Hu, 2013). Proper declaration and fostering of safety goals also support the formation of better relationships in a group climate, and they can be linked to decreased safety-related events and lower accident rates (Hale and Hovden, 1998, as cited in Hale et al., 2010). Communicating safety can help employees feel freer to raise safety concerns (Hoffman and Morgeson, 1999) and can be linked to the occurrence of fewer safety events (Michael et al., 2006) and accidents (Hoffman and Morgeson, 1999).

17.6.2.4 Intellectual Stimulation

The following themes, which were mainly linked to intellectual stimulation, came up in the interviews:

1. Encouraging employees to contemplate solutions along with their supervisor or colleagues
2. Asking them for their interpretations

In the literature, such issues were mentioned in relation to intellectual stimulation:

- Coordination, centralization (Hale and Hovden, 1998, as cited in Hale et al., 2010), and delegation of safety activities (Shannon et al., 1997, as cited in Hale et al., 2010)
- Empowering (Hale and Hovden, 1998, as cited in Hale et al., 2010; Shannon et al., 1997, as cited in Hale et al., 2010) and consulting (Clarke and Ward, 2006) with employees; using a coalition, that is, coworkers to create pressure to comply (Clarke and Wards, 2006)
- Adopting a problem-solving (Hale and Hovden, 1998, as cited in Hale et al., 2010) and learning (Griffin and Hu, 2013) approach to safety. Using logical arguments and factual evidence (rational persuasion) to motivate safety (Clarke and Ward, 2006)

According to Clarke and Ward (2006), leadership behaviors such as coalition, consultation, and rational persuasion influence employee safety participation. Empowering the workforce in different ways contributes to safety performance through an improved climate (Clarke and Ward, 2006; Törner, 2011), trust, and relationships between employees and leaders (Törner, 2011). According to Griffin and Hu (2013), safety monitoring positively influences safety participation when the leader encourages safety-related learning. Both a problem-solving approach and employee empowerment are associated with lower accident rates (Hale and Hovden, 1998, as cited in Hale et al., 2010; Shannon et al., 1997, as cited in Hale et al., 2010).

17.7 DISCUSSION

Because safety and related demands are increasingly an integrated part of business (e.g., EU-OSHA, 2012; Veltri et al., 2013), managers' safety leadership competence should be developed accordingly. Current work life presumes managers have different types of safety management and leadership competencies (e.g., Conchie et al., 2013; Hale et al., 2010; Leka et al., 2011) to ensure employee health and safety.

Managers need safety leadership competencies when motivating employee safety participation and compliance (Kapp, 2012), as well as improving safety performance (e.g., Barling et al., 2002; Clarke, 2013; Hale et al., 2010; Hoffmeister et al., 2014; Törner, 2011). Based on the current research, certain safety leadership competencies are vital when striving for good safety performance. By developing these competencies, organizations may improve their effectiveness through better safety performance.

In this study, a categorization scheme for effective safety leadership facets was suggested based on the leadership theory (Bass, 1985; Bass and Avolio, 1990). Moreover, the relationship between safety leadership and safety performance, and the importance of developing safety leadership were discussed. The categorization of safety leadership facets and related examples provides information for safety leadership competence development.

According to the results of this study, safety leadership facets linked to safety performance in the literature were found to be related to all of the studied leadership facets. This indicates that each of the facets is important with regard to safety performance. The facet of idealized influence leadership was emphasized both in the literature and in the interview findings. This is in line with the study of Hoffmeister et al. (2014), who found that idealized attributes and behaviors were the most important leadership facets explaining the studied safety outcomes (safety climate, safety behaviors, injuries, and pain). Many of the literature findings were also related to inspirational motivation, intellectual stimulation, and management by exception. The findings in the literature were less often related to individual consideration and contingent rewards.

In the interviews, idealized influence, individual consideration, and intellectual stimulation were emphasized. The major difference in the results was that individual consideration was emphasized to a lesser extent in the literature than in the interviews. It is possible that individual consideration is less studied in the safety literature. Nevertheless, Hoffmeister et al. (2014) found that individualized consideration was less important with regard to safety performance. That the interviews were carried out in an expert organization may explain the fact that individual consideration and intellectual stimulation were highlighted in the interviews. In addition, the fact that the interviews were conducted in an expert organization may explain the finding that there was no support for transactional leadership facets in the interviews. Contrary to the findings of Hoffmeister et al. (2014), there was support for management by exception leadership in the literature. Additionally, Clarke (2013) argued that active management by exception has rarely been featured in safety studies, and that it should be emphasized when encouraging safety participation.

It should be noted that the classification of safety leadership practices into leadership facets is subjective. In addition, there exist many interconnections among various practices and leadership facets. Here, each practice was classified under a facet to which the characteristics of the said practice were mainly related. Moreover, in many studies, the influence of some leadership styles with safety outcomes is studied, but there is less research on the relationships between specific leadership practices and different leadership styles (e.g., Christian et al., 2009).

Based on the current study, all traditional leadership facets of transactional and transformational leadership are relevant to safety leadership. Moreover, several previous studies have suggested that transformational and transactional leadership are suitable constructs for safety leadership (e.g., Barling et al., 2002; Clarke, 2013; Kapp, 2012; Michael et al., 2006). In her meta-analysis, Clarke (2013) found that a combination of transformational and active transactional leadership styles is the most effective for managing workplace safety. Thus, effective interventions to improve safety leadership require both the transactional and transformational facets of leadership. Most of the previous interventions have focused on transformational leadership, and leaders could benefit from a wider range of safety leadership styles, and a more situational approach (Clarke, 2013). According to Bass and Avolio (1990), general leadership training programs are often based on transactional leadership, and many aspects of effective leadership are missing when the transformational aspect is undervalued. However, both the transactional and transformational leadership styles are important from the training, education, and development viewpoints. Safety leadership competencies may be evaluated and developed as a part of the general competence development of managers. According to Kapp (2012), safety-specific transformational leadership training may improve safety performance, resulting in improved safety participation by employees. Thus, information about managers' safety leadership competence requirements is valuable for developing their competence, as well as for developing safety training programs for managers.

Safety leadership is often studied separately from safety management. However, as the results of this study show, safety management and related studies may also include elements of safety leadership. For example, employee participation is considered one of the key elements for effective safety management. Nevertheless, the extent to which a leader solicits employees' perspectives on problems and considers a wide variety of opinions in making decisions is part of the intellectual stimulation leadership style. Hence, in many companies, safety management practices could provide easy ways to incorporate safety leadership competencies into existing practices.

Further research is needed to better define the contextual factors and situational flexibility of leadership styles, as well as efficient leadership practices. Authentic leadership is another interesting construct for safety leadership (Eid et al., 2012), and it should be further implemented in safety research in the future.

ACKNOWLEDGMENTS

The authors sincerely wish to thank the Finnish Work Environment Fund and the participating organizations for funding this study. In addition, the authors are thankful to the interviewees for their contributions to this study.

REFERENCES

Avolio, B.J. (1999). *Full Leadership Development: Building the Vital Forces in Organizations.* Thousand Oaks, CA: Sage.

Avolio, B.J., Gardner, W.L. (2005). Authentic leadership development: Getting to the root of positive forms of leadership. *Leadership Quarterly* 16, 315–338.

Barling, J., Loughlin, C., Kelloway, E. (2002). Development and test of a model linking safety-specific transformational leadership and occupational safety. *Journal of Applied Psychology* 87, 488–496.

Bass, B.M. (1985). *Leadership and Performance Beyond Expectation.* New York: Free Press.

Bass, B.M. (1990). From transactional to transformational leadership: Learning to share the vision. *Organizational Dynamics* 18(3), 19–31.

Bass, B.M., Avolio, B.J. (1990). Developing transformational leadership: 1992 and beyond. *Journal of European Industrial Training* 14(5), 21–27.

Bass, B.M., Riggio, R.E. (2006). *Transformational Leadership.* Mahwah, NJ: Lawrence Erlbaum.

Bentley, T.A., Haslam, R.A. (2001). A comparison of safety practices used by managers of high and low accident rate postal delivery offices. *Safety Science* 37(1), 19–37.

Bigelow, P., Robson, L. (2005). Occupational health and safety management audit instruments: A literature review. Toronto: Toronto Institute for Work and Health. http://www.iwh.on.ca/sys-reviews/occupational-health-and-safety-management-audit-instruments-a-literature-review (accessed December 2014).

Biggs, H.C., Biggs, S.E. (2013). Interlocked projects in safety competency and safety effectiveness indicators in the construction sector. *Safety Science* 52, 37–42.

Biron, M., Bamberger, P. (2012). Aversive workplace conditions and absenteeism: Taking referent group norms and supervisor support into account. *Journal of Applied Psychology* 97(4), 901–912.

Blair, E. (2003). Culture and leadership: Seven key points for improved safety performance. *Professional Safety* 48(6), 18–22.

Bluff, L. (2003). Systematic management of occupational health and safety. Working Paper 20. Canberra: National Centre for OHS Regulation, Australian National University. http://asiapacific02.cap.anu.edu.au/sites/default/files/WorkingPaper_20.pdf (accessed June 2014).

Boyatzis, A.R. (1982). *The Competent Manager: A Model for Effective Performance.* New York: Wiley.

Burke, M.J., Sloane M., Signal, S.M. (2010). Workplace safety: A multilevel, interdisciplinary perspective. *Research in Personnel and Human Resources Management* 29, 1–47.

Carder, B., Ragan, P. (2005). *Measurement Matters. How Effective Assessment Drives Business and Safety Performance.* Milwaukee, WI: ASQ Quality Press.

Chang, J.I., Liang, C.-L. (2009). Performance evaluations of process safety management systems of paint manufacturing facilities. *Journal of Loss Prevention in the Process Industries* 22, 398–402.

Christian, M.S., Bradley, J.C., Wallace, J.C., Burke, M.J. (2009). Workplace safety: A meta-analysis of the roles of person and situation factors. *Journal of Applied Psychology* 94, 1103–1127.

Clarke, S. (2013). Safety leadership: A meta-analytic review of transformational and transactional leadership styles as antecedents of safety behaviours. *Journal of Occupational and Organizational Psychology* 86, 22–49.

Clarke, S., Ward, K. (2006). The role of leader influence tactics and safety climate in engaging employees' safety participation. *Risk Analysis* 26, 1175–1185.

Conchie, S.M., Donald, I.J. (2009). The moderating role of safety-specific trust on the relation between safety-specific leadership and safety citizenship behaviors. *Journal of Occupational Health Psychology* 14, 137–147.

Conchie, S.M., Moon, S., Duncan, M. (2013). Supervisors' engagement in safety leadership: Factors that help and hinder. *Safety Science* 51, 109–117.

Cooper, D. (1998). *Improving Safety Culture: A Practical Guide.* Chichester, UK: Wiley.

Dainty, A.R.J., Cheng, M.-I., Moore, D.R. (2004). A competency-based performance model for construction project managers. *Construction Management and Economics* 22(8), 877–886.

Dansereau, F., Graen, G.B., Haga, W.J. (1975). A vertical dyad linkage approach to leadership within formal organizations. *Organizational Behavior and Human Performance* 13, 46–78.

Dedobbeleer, N., Béland, F. (1998). Is risk perception one of the dimensions of safety climate? In A. Feyer, A. Williamson (eds.), *Occupational Injury: Risk, Prevention and Intervention,* 73–81. London: Taylor & Francis.

DeJoy, D.M., Schaffer, B.S., Wilson, M.G., Vandenberg, R.J., Butts, M.M. (2004). Creating safer workplaces: Assessing the determinants and role of safety climate. *Journal of Safety Research* 35, 81–90.

De Meuse, K.P., Dai, G., Wu, J. (2011). Leadership skills across organizational levels: A closer examination. *Psychologist-Manager Journal* 14, 120–131.

Eid, J., Mearns, K., Larsson, G., Laberg, J.C., Johnsen, B.H. (2012). Leadership, psychological capital and safety research: Conceptual issues and future research questions. *Safety Science* 50, 55–61.

EU-OSHA [European Agency for Safety and Health at Work]. (2010). Mainstreaming OSH into business management. Luxembourg: Office for Official Publications of the European Communities. https://osha.europa.eu/en/publications/reports/mainstreaming_osh_business (accessed December 2014).

Fernández-Muñiz, B., Montes-Peón, J.M., Vázques-Ordás, C.J. (2009). Relation between occupational safety management and firm performance. *Safety Science* 47, 980–991.

Flin, R. (2003). Danger—men at work: Management influence on safety. *Human Factors and Ergonomics in Manufacturing* 13(4), 261–268.

Flin, R., Mearns, K., O'Connor, P., Bryden, R. (2000). Measuring safety climate: Identifying the common features. *Safety Science* 34, 177–192.

Flin, R., Yule, S. (2004). Leadership for safety: Industrial experience. *Quality and Safety in Health Care* 13, 45–51.

Fong, P.S.W., Chan, C. (2004). Learning behaviours of project managers. In *Proceeding of the IRNOP VI,* Turku, Finland, 2004, 200–219.

Fruhen, L.S., Mearns, K.J., Flin, R., Kirwan, B. (2013). Skills, knowledge and senior managers' demonstrations of safety commitment. *Safety Science* 69, 29–36.

Garavan, T.N., McGuire, D. (2001). Competencies and workplace learning: Some reflections on the rhetoric and the reality. *Journal of Workplace Learning* 13(3–4), 144–163.

Gardner, W.L., Avolio, B.J., Luthans, F., May, D.R., Walumba, F. (2005). Can you see the real me? A self-based model of authentic leader and follower development. *Leadership Quarterly* 16, 343–372.

Griffin, M.A., Hu, X. (2013). How leaders differentially motivate safety compliance and safety participation: The role of monitoring, inspiring, and learning. *Safety Science* 60, 196–202.

Griffin, M.A., Neal, A. (2000). Perceptions of safety at work: A framework for linking safety climate to safety performance, knowledge, and motivation. *Journal of Occupational Health Psychology* 5, 347–358.

Guldenmund, F.W. (2000). The nature of safety culture: A review of theory and research. *Safety Science* 34, 215–257.

Hale, A.R., Guldenmund, F.W., van Loenhout, P.L.C.H., Oh, J.I.H. (2010). Evaluating safety management and culture interventions to improve safety: Effective intervention strategies. *Safety Science* 48, 1026–1035.

Hale, A.R., Hovden, J. (1998). Management and culture: The third age of safety. In Feyer, A.-M., Williamson, A. (eds.), *Occupational Injury: Risk, Prevention and Intervention*, 129–166. London: Taylor & Francis.

Hardison, D., Behm, M., Hallowell, M.R., Fonooni, H. (2014). Identifying construction supervisor competencies for effective site safety. *Safety Science* 65, 45–53.

Hoffmeister, K., Gibbons, A.M., Johnson, S.K., Cigularov, K.P., Chen, P.Y., Rosecrance, J.C. (2014). The differential effects of transformational leadership facets on employee safety. *Safety Science* 62, 68–78.

Hofmann, D.A., Morgeson, F., Gerras, S. (2003). Climate as a moderator of the relationship between leader–member exchange and content specific citizenship: Safety climate as an exemplar. *Journal of Applied Psychology* 88(1), 170–178.

Hofmann, D.A., Morgeson, F.P. (1999). Safety-related behavior as a social exchange: The role of perceived organizational support and leader-member exchange. *Journal of Applied Psychology* 84, 286–296.

Hofmann, D.A., Morgeson, F.P. (2004). The role of leadership in safety. In J. Barling, M.R. Frone (eds.), *The Psychology of Workplace Safety*, 159–180. Washington, DC: American Psychological Association.

Hofmann, D.A., Stetzer, A. (1996). A cross-level investigation of factors influencing unsafe behaviours and accidents. *Personnel Psychology* 49, 307–339.

Kapp, E.A. (2012). The influence of supervisor leadership practices and perceived group safety climate on employee safety performance. *Safety Science* 50, 1119–1124.

Kelloway, E.K., Mullen, J., Francis, L. (2006). Divergent effects of transformational and passive leadership on employee safety. *Journal of Occupational Health Psychology* 11(1), 76–86.

Kelloway, E.K., Turner, N., Barling, J., Loughlin, C. (2012). Transformational leadership and employee psychological well-being: The mediating role of employee trust in leadership. *Work and Stress* 26, 39–55.

Königová, M., Urbancová, H., Fejfar, J. (2012). Identification of managerial competencies in knowledge-based organizations. *Journal of Competitiveness* 4(1), 129–142.

Köper, B., Möller, K., Zwetsloot, G. (2009). The occupational safety and health scorecard— a business case example for strategic management. *Scandinavian Journal of Work, Environment and Health* 35(6), 413–420.

Künzle, B., Kolbe, M., Grote, G. (2010). Ensuring patient safety through effective leadership behavior: A literature review. *Safety Science* 48, 1–17.

Leka, S., Jain, A., Widerszal-Bazyl, M., Żołnierczyk-Zreda, D., Zwetsloot, G. (2011). Developing a standard for psychosocial risk management: PAS 1010. *Safety Science* 49, 1047–1057.

Lord, R.G., Hall R.J. (2005). Identity, deep structure and the development of leadership skill. *Leadership Quarterly* 16, 591–615.

Lu, C.-S., Yang, C.-S. (2010). Safety leadership and safety behavior in container terminal operations. *Safety Science* 48, 123–134.

Martínez-Córcoles, M., Gracia, F., Tomás, I., Peiró, J.M. (2011). Leadership and employees' perceived safety behaviours in a nuclear power plant: A structural equation model. *Safety Science* 49, 1118–1129.

Mearns, K.J., Reader, T. (2008). Organizational support and safety outcomes: An un-investigated relationship. *Safety Science* 46, 388–397.

Michael, J., Guo, Z., Wiedenbeckt, J., Ray, C. (2006). Production supervisor impacts on subordinates' safety outcomes: An investigation of leader–member exchange and safety communication. *Journal of Safety Research* 37(5), 469–477.

O'Dea, A., Flin, R. (2001). Site managers and safety leadership in the offshore oil and gas industry. *Safety Science* 37, 39–57.

Pedler, M., Burgoyne, J., Boydell, T. (1986). *A Manager's Guide to Self-Development*. London: McGraw-Hill.

Pfeffer, J. (1998). Seven practices of successful organizations. *California Management Review* 40(2), 96–124.

Pickett, L. (1998). Competencies and managerial effectiveness: Putting competencies to work. *Public Personnel Management* 27(1), 103–115.

Schoonenboom, J., Tattersall, C., Miao, Y., Stefanov, K., Aleksieva-Petrova, A. (2007). A four-stage model for lifelong competence development. In *Proceedings of the 2nd TENCompetence Open Workshop*, Manchester, UK, January 2007, 131–136.

Shannon, H.S., Mayr, J., Haines, T. (1997). Overview of the relationship between organisational and workplace factors and injury rates. *Safety Science* 26(3), 201–217.

Siegrist, J., Starke, D., Chandolab, T., Godinc, I., Marmot, M., Niedhammer, I., Peter, R. (2004). The measurement of effort–reward imbalance at work: European comparisons. *Social Science and Medicine* 58(8), 1483–1499.

Sievänen, M., Nenonen, N., Hämäläinen, P. (2013). The economic impacts of occupational health and safety interventions—a critical analysis based on the nine-box model of profitability. In *Proceedings of the 45th Nordic Ergonomics and Human Factors Society (NES) Conference*, Reykjavik, Iceland, August 2013.

Stinglhamber, F., Vandenberghe, C. (2003). Organizations and supervisors as sources of support and targets of commitment: A longitudinal study. *Journal of Organizational Behavior* 24, 251–270.

Tappura, S., Hämäläinen, P. (2011). Promoting occupational health, safety and well-being by training line managers. In *Proceedings of the 43th Nordic Ergonomics Society (NES) Conference*, Oulu, Finland, September 2011, 295–300. http://www.kotu.oulu.fi/nes2011/docs/Proceedings_NES2011_Oulu.pdf (accessed December 2014).

Tappura, S., Hämäläinen, P. (2012). The occupational health and safety training outline for the managers. In P. Vink (ed.), *Advances in Social and Organizational Factors*, Advances in Human Factors and Ergonomics Series, vol. 9, 356–365. Boca Raton, FL: Taylor & Francis, CRC Press.

Tappura, S., Nenonen, N., Heikkilä, J., Reiman, T., Rasa, P.-L., Ratilainen, H. (2013). Estimating overall costs of occupational accidents in the Finnish industry. In *Proceedings of the 45th Nordic Ergonomics and Human Factors Society (NES) Conference*, Reykjavik, Iceland, August 2013.

Tappura, S., Sievänen, M., Heikkilä, J., Jussila, A., Nenonen, N. (2015). A management accounting perspective on safety. *Safety Science* 71, 151–159.

Törner, M. (2011). The "social-physiology" of safety. An integrative approach to understanding organisational psychological mechanisms behind safety performance. *Safety Science* 49, 1262–1269.

Veltri, A., Pagell, M., Johnston, D., Tompa, E., Robson, L., Amick III, B.C., Hogg-Johnson, S., Macdonald, S. (2013). Undestanding safety in the context of business operations: An exploratory study using case studies. *Safety Science* 55, 119–134.

Viitala, R. (2005). Perceived development needs of managers compared to an integrated management competency model. *Journal of Workplace Learning* 17(7), 436–451.

White, R., Hodgson, P., Crainer, S. (1996). *The Future of Leadership. Riding the Corporate Rapids into the 21st Century.* London: Pitman Publishing.

Williams, J.H. (2002). Improving safety leadership: Using industrial/organizational psychology to enhance safety performance. *Professional Safety* 47(4), 43–47.

Wu, T.-C., Chen, C.-H., Li, C.-C. (2008). A correlation among safety leadership, safety climate and safety performance. *Journal of Loss Prevention in the Process Industries* 21, 307–318.

Yukl, G. (2008). How leaders influence organizational effectiveness. *Leadership Quarterly* 19, 708–722.

Zacharatos, A., Barling, J., Iverson, R.D. (2005). High-performance work systems and occupational safety. *Journal of Applied Psychology* 90, 77–93.

Zohar, D. (1980). Safety climate in industrial organizations: Theoretical and applied implications. *Journal of Applied Psychology* 65, 95–102.

Zohar, D. (2002a). The effects of leadership dimensions, safety climate, and assigned priorities on minor injuries in work groups. *Journal of Organizational Behavior* 23, 75–92.

Zohar, D. (2002b). Modifying supervisory practices to improve subunit safety: A leadership-based intervention model. *Journal of Applied Psychology* 87, 156–163.

Zohar, D. (2010). Thirty years of safety climate research: Reflections and future directions. *Accident Analysis and Prevention* 42(5), 1517–1522.

Zohar, D., Luria, G. (2003). The use of supervisory practices as leverage to improve safety behavior: A cross-level intervention model. *Journal of Safety Research* 34, 567–577.

18 Women with Upper Limb Repetitive Strain Injury (RSI) and Housework

Zixian Yang and Therma Wai Chun Cheung

CONTENTS

18.1 WHAT IS UPPER LIMB RSI?

Repetitive strain injury (RSI) refers to "a soft tissue disorder caused by the overloading of particular muscle groups from repetitive use or maintenance of constrained posture" (Australia National Occupational Health Safety Commission, 1986). This definition attributes the injury to biomechanical exposure that has resulted in either frequent repetitive movements of the limbs or the maintenance of fixed postures for long periods (Hall and Morrow, 1988).

RSI has many alternative names. The term *RSI* is more commonly used in Australian literature (Bammer and Martin, 1992; Ewan et al., 1991; Hall and Morrow, 1988; Hocking, 1987; Hopkins, 1990; Miller and Topliss, 1988; Quintner, 1995) since its first use "in 1982 in a report from the Australian National Health and Medical Research Council" (Tyrer, 1998, p. 175). Alternative terms commonly used in recent literature include musculoskeletal disorders (MSDs) (Dahlberg et al., 2004; Habib et al., 2006; Johnston et al., 2008); work-related musculoskeletal disorders (WRMSDs) (Barr et al., 2004; Hansson ct al., 2009; Naidoo and Haq, 2008), cumulative trauma disorders (CTDs) (Armstrong et al., 2005; Gangopadhyay et al., 2003; Goodman et al., 2012), and occupational overuse syndrome (OOS) (Jaye and Fitzgerald, 2011; Laoopugsin and Laoopugsin, 2012; White et al., 2003).

Upper limb RSI simply refers to the location of the condition in the body. There are also many alternative terms that describe the condition. These include upper extremity musculoskeletal disorders (UEMSDs) (Arvidsson et al., 2003) or work-related neck and upper limb musculoskeletal disorders (WRULDs) (Buckle and Devereux, 2002) or upper extremity disorders (UEDs) (Huisstede et al., 2006). All refer to a similar cluster of conditions in the upper limbs. Despite extensive research focusing on this cluster of conditions, researchers have not agreed upon a universally recognized or accepted term. Adding to this confusion are variations in the diagnostic classification systems used in different literature. Van Eerd et al. (2003) reviewed a total of 27 classification systems used to define upper limb MSDs in workers and found that each of these systems differed.

Two classification systems have been proposed that provide comparatively clearer criteria for inclusion of diagnostic groups, and both have resulted from extensive collaboration by experts in the field. One is the classification system developed by 47 experts of UEMSDs in the Netherlands (Huisstede et al., 2007). This Dutch system includes 23 disorders using three criteria: diagnosable, not caused by acute trauma, and not caused by any systemic disorders.

Another classification system was offered by a group of European experts who worked under the Samarbetsprogram mellan Arbetslivsinstitutet, LO, TCO och Saco (SALTSA) program established by the Swedish National Institute for Working Life (NIWL) (Sluiter et al., 2001). They used three criteria for the inclusion of 12 diagnostic groups. Similar to the Dutch system, NIWL used a well-defined diagnostic criterion as one of its requirements. In contrast to the Dutch classification system, which has a more general exclusion of conditions caused by acute trauma and systemic disorders, the NIWL system includes only specific diagnoses supported by evidence as work related. In addition, the NIWL system includes only diagnoses that are highly prevalent.

Comparing the two classification systems, the NIWL system provides much clearer and more specific criteria for the inclusion of diagnostic groups with respect to the work-relatedness of the conditions than does the Dutch system. Twelve diagnostic groups are considered upper limb WRMSDs (Sluiter et al., 2001) under the NIWL classification system, and were referred to as upper limb RSI within the context of this chapter. They include radiating neck pain, rotator cuff syndrome, medial epicondylitis, lateral epicondylitis, ulna nerve entrapment in the cubital tunnel, radial tunnel syndrome, flexor and extensor tendinitis at the hand and fingers, De Quervain tenosynovitis, carpal tunnel syndrome, ulnar nerve entrapment in Guyon's tunnel, Raynaaud's phenomenon or peripheral neuropathy related to vibrations of the hand and arm, and osteoarthritis of the elbow, wrist, and fingers.

18.2 PREVALENCE OF UPPER LIMB RSI IN WOMEN

RSI is more common in women than in men (World Health Organization, 2006, p. 12). The incidence of upper limb RSI was also higher in women than in men in a major review of 56 articles published from 1981 to 2002 (Treaster and Burr, 2004). Based on 26 studies that used both self-report and physical examination as measures, the review found that with men as the referent, the odds or prevalence ratio of UEMSDs in women ranged from 0.66 to 11.4. Among these 26 studies, 70% of the prevalence rates or odds ratios of MSDs reported on various locations of the upper limbs showed a significantly higher prevalence in women than in men. These ratios were adjusted for confounding factors such as age and physical work, validating the higher prevalence rate of upper limb RSI in women than in men. Since the review, other recent studies have also generated evidence supporting a higher prevalence of upper limb RSI in women than in men (Chiron et al., 2008; Dahlberg et al., 2004; Ha et al., 2009).

Not only is upper limb RSI more prevalent in women, but also its occurrence in elderly women could increase the chance of disability. In a recent retrospective cross-sectional study on 104 computer workers with upper limb RSI, being both female and older were risk factors for disability for this cluster of conditions (van Eijsden-Besseling et al., 2010). Although upper limb RSI is more prevalent in women than in men, and is associated with a risk of disability with its occurrence in older women, research in this area is still limited. According to a report by the World Health Organization (2006, pp. 25–26), there is a gender bias in occupational health research: women are not receiving the research attention they deserve. This brings up a question: Despite the limited information available in literature in this area, what do we know about the reasons for the higher prevalence of upper limb RSI among women?

18.3 WOMEN WITH UPPER LIMB RSI: PAID WORK OR UNPAID HOUSEWORK?

Most studies examined the risk factors of upper limb RSI in paid work. Some studies focus specifically on the physical risk factors related to upper limb RSI for women in paid jobs. The identified common risk factors in these studies include

- Excessive force and repetition with prolonged exposure (Chen et al., 2009)
- Excessive repetition with high-frequency movements of the upper limbs (Arvidsson et al., 2003; Hansson et al., 2000)
- Posture-related risk factors with prolonged exposure, as well as psychosocial factors such as negative affectivity and individual factors such as age (Johnston et al., 2008)

Although most of these studies focus only on paid work carried out by women, women's unpaid duties (housework at home) were often suggested as an explanation of the higher prevalence rate in upper limb RSI. A study on 60 female supermarket employees found that non-work-adverse life events and stress were positively associated with the presence of musculoskeletal UEDs, suggesting that it is necessary to understand the interrelational nature of women's lives in managing their condition (Vroman and MacRae, 2001). A review carried out by Lundberg (2002) also suggested that stress from unpaid duties such as housework and child care could explain the higher prevalence of work-related UEDs in working women.

Child care as a risk factor for upper limb RSI for working women was supported by two studies. In a study by Bjorksten et al. (2001) on 173 Swedish female blue-collar workers, among other physical and psychosocial risk factors, women with a partner and children less than 13 years old had an increased number of musculoskeletal problems of their neck and shoulder. This was supported by the findings of an earlier study by Bergqvist et al. (1995), who reported that the risk of having a musculoskeletal diagnosis of the neck and shoulder was higher for women with children under 16 years of age than those without. These studies demonstrated that home strain in the care of children could play a part in increasing the risk of upper limb RSI for women in the working population.

Although these studies suggested that housework, including duties related to child care, may contribute to women's high prevalence rate in upper limb RSI, these reasons have not been closely examined. This raises two more questions: first, if women are doing more housework than men, and second, if there are specific biomechanical risks associated with housework.

18.4 WOMEN'S PARTICIPATION IN UNPAID HOUSEWORK

Research related to the division of labor in housework emerged in the 1970s when Oakley (1974) published *The Sociology of Housework*, based on a study interviewing 40 young urban housewives in the United States. She revealed a major inequality in the distribution of housework that existed between men and women (Oakley, 1974). Although there is some evidence this inequality may have decreased in recent years, it still exists, and women are responsible for the majority of housework in the home. This claim is examined and supported in the following paragraph.

One study in the 1990s showed that although women did around 3%–7% less housework hours than women in the late 1970s, they were still responsible for 60%–64% of the total housework hours in homes (Ferree, 1991). In fact, since the 1970s, all studies and reviews that investigated gender ideology and household labor agreed that women still carry out the majority of housework at home (Cunningham, 2008; Ferree, 1991; Grote et al., 2002; Hank and Jurges, 2007; Knudsen and Waerness, 2008; Pimentel, 2006). More recently, in a major review of housework distribution in 34 countries using data extracted from the International Social Survey Program, married women performed a mean of 21 hours of housework per week, compared to 7.77 hours carried out by men (Knudsen and Waerness, 2008). The mean division of housework between men and women was 1:3.9 (Knudsen and Waerness, 2008). Evidence from this study confirmed that in many sociocultural environments, women have continued to do more housework than men.

18.5 BIOMECHANICAL RISKS ASSOCIATED WITH UNPAID HOUSEWORK

Although only a limited number of studies could be found on the specific risk factors involved in housework related to upper limb RSI, the results of these studies provide consistent evidence of the association between specific housework tasks and upper limb RSI. A survey of 1000 families in Italy with a response rate of 31.7% found that upper limb disorders are associated with housework tasks such as washing dishes, cleaning carpets, and washing clothes (Rosano et al., 2004).

The use of surveys such as those described above is common in RSI research. Only a limited number of studies were found that involved more objective measures, such as ergonomic site evaluations or biomechanical risk analysis in housework tasks. To date, only two studies employing such a methodology could be found. One study analyzed biomechanical risks in nine housework tasks using video recordings of 12 participants among the 104 participants who participated in the Italian survey (Apostoli et al., 2012; Sala et al., 2007). According to this study, although all tasks were found to impose biomechanical risk when performed for 4 hours a day, higher overloads were detected specifically in ironing, cleaning floors, and cleaning kitchen benchtops (Sala et al., 2007).

In another study with 943 full-time housewives in Hong Kong (Yip and Hung, 2002), the presence of musculoskeletal symptoms reported by 84% of the 794 survey respondents was associated with risk factors such as psychological status, awkward posture, and prolonged duration of housework (Yip and Hung, 2002). These results were reported as consistent with on-site ergonomic evaluations of three patients with tennis elbow (Yip and Hung, 2002). Although the study by Yip and Hung (2002) did not specify the housework tasks being analyzed in their ergonomic site evaluations, it is probably safe to assume that they could differ from the other two studies in Italy due to cultural and climate differences. For example, as carpet is not common flooring in Hong Kong, carpet cleaning most likely would not have been analyzed within their small sample size of three patients. Despite the possible sociocultural differences in the environments in which these studies were conducted, there is support that excessive biomechanical load related to repetitiveness of movements, awkward postures, and duration of housework tasks is the main biomechanical risk factor for upper limb RSI in housework, consistent with those identified in paid work.

18.6   PREVALENCE OF UPPER LIMB RSI IN FEMALE HOMEMAKERS

With women's major participation in housework, and the biomechanical risks identified in housework, a high prevalence of upper limb RSI in female homemakers is expected. A cross-sectional survey using a convenience sample of 216 female homemakers from the Hong Kong Federation Women's Centre, and from a street in one of the oldest urban districts in Hong Kong (age range 20 to >60 years old), found that 63 (29.2%) of the respondents were in paid work; 153 (70.8%) were full-time homemakers (Fong and Law, 2008). Using a standardized Nordic questionnaire in the study, they found that more than 60% of participants had experienced at least one musculoskeletal symptom in the past 12 months (Fong and Law, 2008). Consistent with this finding is another study on 435 women from Nabaa in Lebanon. The women in this study were of a similar age range (from 18 to 62 years old) and employment distribution, with 23.9% in paid jobs and 73.1% full-time homemakers (Habib et al., 2011). This study found that 77% of these women had reported musculoskeletal symptoms in the past 12 months (Habib et al., 2011). These two studies, involving women with a similar percentage distribution in paid work and full-time housework duties, found similar results and agreed on the high prevalence rate of RSI in women who performed housework. However, these studies also included a percentage of women in paid work.

When a similar research methodology was used on a group of full-time homemakers, excluding women in paid work, Habib et al. (2006) found that the prevalence rate dropped significantly. In their study on 1266 married women—all full-time homemakers in three communities on the outskirts of Beirut, Lebanon—they found the prevalence of MSDs was only 19% (Habib et al., 2006). One possible reason for the higher prevalence rate of MSDs among a group of women including both employed women and full-time homemakers could be explained by the higher total workload (paid and unpaid) of women in paid work. A cross-sectional survey of 61 workers in a manufacturing company in the middle of Sweden, which found that a higher proportion of women than men reported shoulder problems, also found that women in their study spent more time on household activities (Dahlberg et al., 2004). This is consistent with the findings of a Korean study on 950 female bank tellers. The study found that daily time spent on housework was one factor associated with the presence of WRMSD symptoms (Yun et al., 2001).

When women spend more time on housework, they will naturally spend less time on leisure activities. This could be associated with the occurrence of RSI. In a study of 737 Australian public service employees, 73% of them women, 83% of the women reported upper body symptoms, compared to 77% of men reporting the same (Strazdins and Bammer, 2004). The same study noted that women reported significantly less time to exercise and relax than men, due to their greater participation in parenthood and domestic work. These are the interlinking factors that explain gender differences in MSDs (Strazdins and Bammer, 2004). These findings are supported by a cross-sectional survey on 651 workers of various occupations in Portugal (Monteiro et al., 2006). This study determined that the occurrence of musculoskeletal diseases was associated with long housework hours and absence of leisure activities (Monteiro et al., 2006).

Although several studies supported the significant role of housework in the etiology of RSI and upper limb RSI among women, more specific studies relating to housework and RSI are necessary. Habib et al. (2010) reviewed 56 articles and compared the findings with their observations of four homes in Beirut, Lebanon. They concluded that "housework activities expose homemakers to known risk factors for MSDs" (Habib et al., 2010, p. 113). However, out of the 56 articles reviewed, they also found that only 12 reported specific findings relating musculoskeletal problems or pain to those who do housework (Habib et al., 2010). This indicates the need for more research.

18.7 INTERNATIONAL STANDARD CLASSIFICATION OF OCCUPATIONS (ISCO-08)

As discussed above, current literature has confirmed the biomechanical risk in housework and its impact on the prevalence of upper limb RSI among women who need to perform housework activities at home. In a retrospective study that we carried out in an occupational therapy outpatient clinic in Singapore to explore the occupational distribution of patients with upper limb RSI who were referred to the clinic, it was found that out of 6715 outpatients referred to the clinic in 2012, 1108 (16.5%) of them were diagnosed as upper limb RSI. Among these patients, the majority of them were females: 827 females (74.6%) with a mean age of 54.32 years (s.d. 12.1 years). The majority of these women (43.5%) were within the age group of 51–60 years old. A summary of the demographic information obtained from the records of these patients is shown in Table 18.1.

TABLE 18.1

Demographic Information of Patients

Hand Dominance	n	%	Injured Side	n	%
Right	817	73.7	Right	481	43.4
Left	33	3.0	Left	386	34.8
Unspecified	258	23.3	Bilateral	241	21.8
Gender			**Dominant Side Injured**		
Female	827	74.6	Yes	551	49.7
Male	281	25.4	No	557	50.3

To find out the occupation distribution of all 6715 patients with upper limb RSI in our study, the ISCO-08, an international standard classification of occupations, was used (International Labour Office). ISCO was first published in 1988 to facilitate international communication on occupational information as ISCO-88 (International Labour Office) and has been used in research to investigate risks of diseases in occupations (Furby et al., 2010; Roquelaure et al., 2012). The latest version, ISCO-08, was updated through a tripartite meeting of experts on labor statistics held in December 2007, and consists of 10 major occupational categories: (1) managers, (2) professionals, (3) technicians and associate professionals, (4) clerical support workers, (5) service and sales workers, (6) skilled agricultural, forestry and fishery workers, (7) crafts and related trades workers, (8) plant and machines operators and assemblers, (9) elementary occupations, and (10) armed forces occupations (International Labour Office). As ISCO-08 only includes job categories that are considered paid work, an 11th category (unemployed) was used in our study for a group of patients who did not engage in any form of paid work. It was then found that a significant proportion of the patients were categorized under the unemployed group (41.4%). A more in-depth analysis performed within the group of unemployed patients found that at least half of them (50.9%) were full-time homemakers, with the rest being either retirees or truly unemployed. Despite the significant role of unpaid housework in the occurrence of upper limb RSI among female homemakers, unpaid housework has not received the attention it deserves in ergonomic research. Most ergonomic research continues to focus on paid work.

18.8 UNPAID HOUSEWORK, ERGONOMIC RESEARCH, AND ISCO-08

Literature has confirmed the biomechanical risks inherent in housework. Our study has also confirmed that female homemakers who need to carry out unpaid housework make up of a major proportion of patients with upper limb RSI referred to an occupational therapy outpatient clinic in Singapore. These findings provide a logical explanation on the high prevalence of upper limb RSI in women as reported in previous studies, confirming the significant role of biomechanical strain in housework in the occurrence of upper limb RSI among women who need to perform housework activities at home. Currently, ISCO-08's 10 categories of paid work do not include this major population group. As such, we recommend that participation in housework should be considered the 11th major category in the ISCO-08, within the context of ergonomic research related to upper limb RSI. It is time for female homemakers to receive the attention they deserve in the field of ergonomic research.

REFERENCES

Apostoli, P., Sala, E., Curti, S., Cooke, R. M., Violante, F. S., and Mattioli, S. (2012). Loads of housework? Biomechanical assessments of the upper limbs in women performing common household tasks. *International Archives of Occupational and Environmental Health*, 85(4), 421–425.

Armstrong, A. J., McMahon, B. T., West, S. L., and Lewis, A. (2005). Workplace discrimination and cumulative trauma disorders: The National EEOC ADA research project. *Work*, 25, 49–56.

Arvidsson, I., Akesson, I., and Hansson, G.-A. (2003). Wrist movements among females in a repetitive, non-forceful work. *Applied Ergonomics*, 34, 309–316.

Australia National Occupational Health Safety Commission. (1986). Repetition strain injury (RSI), a report and model code of practice/National Occupational Health and Safety Commission. Canberra: Australian Government Public Service.

Bammer, G., and Martin, B. (1992). Repetitive strain injury in Australia: Medical knowledge, social movement and de facto partisanship. *Social Problems*, 39(3), 219–237.

Barr, A. E., Barbe, M. F., and Clark, B. D. (2004). Work-related musculoskeletal disorders of the hand and wrist: Epidemiology, pathophysiology, and sensorimotor changes. *Journal of Orthopaedic and Sports Physical Therapy*, 34(10), 610–627.

Bergqvist, U., Wolgast, E., Nilsson, B., and Voss, M. (1995). Musculoskeletal disorders among visual display terminal workers: Individual, ergonomic, and work organizational factors. *Ergonomics*, 38, 763–776.

Bjorksten, M. G., Boquist, B., Talback, M., and Edling, C. (2001). Reported neck and shoulder problems in female industrial workers: The importance of factors at work and at home. *International Journal of Industrial Ergonomics*, 27, 159–170.

Buckle, P. W., and Devereux, J. J. (2002). The nature of work-related neck and upper limb musculoskeletal disorders. *Applied Ergonomics*, 33, 207–217.

Chen, H.-C., Chang, C.-M., Liu, Y.-P., and Chen, C.-Y. (2009). Ergonomic risk factors for the wrists of hairdressers. *Applied Ergonomics*, 41(1), 98–105.

Chiron, E., Roquelaure, Y., Ha, C., Touranchet, A., Chotard, A., Bidron, P., et al. (2008). MSDs and job security of employees aged 50 years and over: A challenge for occupational health and public health [in French]. *Sante Publique (Vandoeuvre-Les-Nancey)*, 20(Suppl. 3), S19–S28.

Cunningham, M. (2008). Influences of gender ideology and housework allocation on women's employment over the life course. *Social Science Research*, 37, 254–267.

Dahlberg, R., Karlqvist, L., Bildt, C., and Nykvist, K. (2004). Do work technique and musculoskeletal symptoms differ between men and women performing the same type of work tasks? *Applied Ergonomics*, 35, 521–529.

Ewan, C., Lowy, E., and Reid, J. (1991). 'Falling out of culture': The effects of repetition strain injury on sufferer's roles and identity. *Sociology of Health and Illness*, 13(2), 168–192.

Ferree, M. M. (1991). The gender division of labor in two-earner marriages. *Journal of Family Issues*, 12(2), 158–180.

Fong, K. N., and Law, C. Y. (2008). Self-perceived musculoskeletal complaints: Relationship to time use in women homemakers in Hong Kong. *Journal of Occupational Rehabilitation*, 18(3), 273–281.

Furby, A., Beauvais, K., Kolev, I., Rivain, J., and Sébille, V. (2010). Rural environment and risk factors of amyotrophic lateral sclerosis: A case–control study. *Journal of Neurology*, 257(5), 792–798.

Gangopadhyay, S., Ray, A., Das, A., Das, T., Ghoshal, G., Banerjee, P., and Bagchi, S. (2003). A study on upper extremity cumulative trauma disorder in different unorganised sectors of West Bengal, India. *Journal of Occupational Health*, 45(6), 351–357.

Goodman, G., Kovach, L., Fisher, A., Elsesser, E., Bobinski, D., and Hansen, J. (2012). Effective interventions for cumulative trauma disorders of the upper extremity in computer users: Practice models based on systematic review. *Work*, 42(1), 153–172.

Grote, N. K., Naylor, K. E., and Clark, M. S. (2002). Perceiving the division of family work to be unfair: Do social comparisons, enjoyment, and competence matter? *Journal of Family Psychology*, 16(4), 510–522.

Ha, C., Roquelaure, Y., Leclerc, A., Touranchet, A., Goldberg, M., and Imbernon, E. (2009). The French Musculoskeletal Disorders Surveillance Program: Pays de la Loire network. *Occupational and Environmental Medicine*, 66(7), 471–479.

Habib, R. R., Fadi A, F., and Messing, K. (2010). Full-time homemakers: Workers who cannot "go home and relax." *International Journal of Occupational Safety and Ergonomics*, 16(1), 113–128.

Habib, R. R., Hamdan, M., Nywayhid, I., Odaymat, F., and Campbell, O. M. R. (2006). Musculoskeletal disorders among full-time homemakers in poor communities. *Women and Health*, 42(2), 1–14.

Habib, R. R., Zein, K. E., and Hojeij, S. (2011). Hard work at home: Musculoskeletal pain among female homemakers. *Ergonomics*, 55(2), 201–211.

Hall, W., and Morrow, L. (1988). 'Repetition strain injury': An Australian epidemic of upper limb pain. *Social Science Medicine*, 27(6), 645–649.

Hank, K., and Jurges, H. (2007). Gender and the division of household labor in older couples—a European perspective. *Journal of Family Issues*, 28(3), 399–421.

Hansson, G.-A., Balogh, I., Ohlsson, K., Granqvist, L., Nordander, C., Arvidsson, I., et al. (2009). Physical workload in various types of work: Part I. Wrist and forearm. *International Journal of Industrial Ergonomics*, 39, 221–233.

Hansson, G.-A., Balogh, I., Ohlsson, K., Palsson, B., Rylander, L., and Skerfving, S. (2000). Impact of physical exposure on neck and upper limb disorders in female workers. *Applied Ergonomics*, 31, 301–310.

Hocking, B. (1987). Epidemiological aspects of "repetition strain injury" in Telecom Australia. *Medical Journal of Australia*, 147(5), 218–222.

Hopkins, A. (1990). The social recognition of repetitive strain injuries: An Austalian/American comparison. *Social Science and Medicine*, 30(3), 365–372.

Huisstede, B., Bierma-Zinstra, S. M. A., Koes, B. W., and Verhaar, J. A. N. (2006). Incidence and prevalence of upper-extremity musculoskeletal disorders. A systematic appraisal of the literature. *BioMed Central Musculoskeletal Disorders*, 7(7).

Huisstede, B., Koes, B. W., Miedema, H. S., Verhaar, J. A. N., and Verhagen, A. P. (2007). Multidisciplinary consensus on the terminology and classification of complaints of the arm, neck and/or shoulder. *Occupational and Environmental Medicine*, 64, 313.

International Labour Office. ISCO-08 structure and preliminary correspondence with ISCO-88. Geneva: International Labour Office. http://www.ilo.org/public/english/bureau/stat/isco/isco08/index.htm (accessed December 2, 2014).

Jaye, C., and Fitzgerald, R. (2011). Embodying occupational overuse syndrome. *Health: An Interdisciplinary Journal for the Social Study of Health, Illness and Medicine*, 15(4), 385–400.

Johnston, V., Souvlis, T., Jimmieson, N. L., and Jull, G. (2008). Associations between individual and workplace risk factors for self-reported neck pain and disability among female office workers. *Applied Ergonomics*, 39, 171–182.

Knudsen, K., and Waerness, K. (2008). National context and spouses' housework in 34 countries. *European Sociological Review*, 24(1), 97–113.

Laoopugsin, N., and Laoopugsin, S. (2012). The study of work behaviours and risks for occupational overuse syndrome. *Hand Surgery*, 17(2), 205–212.

Lundberg, U. (2002). Psychophysiology of work: Stress gender, endocrine response and work-related upper extremity disorders. *American Journal of Industrial Medicine*, 41(5), 383–392.

Miller, M. H., and Topliss, D. J. (1988). Chronic upper limb pain syndrome (repetitive strain injury) in the Australian workforce: A systematic cross sectional rheumatological study of 229 patients. *Journal of Rheumatology*, 15(11), 1705–1712.

Monteiro, M. S., Alexandre, N. M., and Rodrigues, C. M. (2006). Musculoskeletal diseases, work and lifestyle among public workers at a health institution. *Doenças músculo-esqueléticas, trabalho e estilo de vida entre trabalhadores de uma instituição pública de saúde*, 40(1), 20–25.

Naidoo, R. N., and Haq, S. A. (2008). Occupational use syndromes. *Best Practice and Research Clinical Rheumatology*, 22(4), 677–691.

Oakley, A. (1974). *The Sociology of Housework*. New York: Pantheon Books.

Pimentel, E. E. (2006). Gender ideology, household behaviour, and backlash in urban China. *Journal of Family Issues*, 27(3), 341–365.

Quintner, J. (1995). Clinical commentary: The Australian RSI debate: Stereotyping and medicine. *Disability and Rehabilitation*, 17(5), 256–262.

Roquelaure, Y., Petit LeManach, A., Ha, C., Poisnel, C., Bodin, J., Descatha, A., and Imbernon, E. (2012). Working in temporary employment and exposure to musculoskeletal constraints. *Occupational Medicine*, 62(7), 514–518.

Rosano, A., Moccaldi, R., Cioppa, M., Lanzieri, G., Persechino, B., and Spagnolo, A. (2004). Musculoskeletal disorders and housework in Italy [in Italian]. *Annali di Igiene*, 16(3), 497–507.

Sala, E., Mattioli, S., Violante, F. S., and Apostoli, P. (2007). Risk assessment of biomechanical load for the upper limbs in housework [in Italian]. *Medicina del Lavoro*, 98(3), 232–251.

Sluiter, J. K., Rest, K. M., and Frings-Dresen, M. H. W. (2001). Criteria document for evaluating the work-relatedness of upper-extremity musculoskeletal disorders. *Scandinavian Journal of Work, Environment and Health*, 27(Suppl. 1), 1–102.

Strazdins, L., and Bammer, G. (2004). Women, work and musculoskeletal health. *Social Science and Medicine*, 58, 997–1005.

Treaster, D. E., and Burr, D. (2004). Gender differences in prevalence of upper extremity musculoskeletal disorders. *Ergonomics*, 47(5), 495–526.

Tyrer, S. P. (1998). Repetitive strain injury. In P. Manu (ed.), *Functional Somatic Syndromes: Etiology, Diagnosis and Treatment* (pp. 175–201). Melbourne, Australia: Cambridge University Press.

Van Eerd, D., Beaton, D., Cole, D., Lucas, J., Hogg-Johnson, S., and Bombardier, C. (2003). Classification systems for upper-limb musculoskeletal disorders in workers: A review of literature. *Journal of Clinical Epidemiology*, 56, 925–936.

van Eijsden-Besseling, M. D., van den Bergh, K. A., Staal, J. B., de Bie, R. A., and van den Heuvel, W. J. (2010). The course of nonspecific work-related upper limb disorders and the influence of demographic factors, psychologic factors, and physical fitness on clinical status and disability. *Archives of Physical Medicine and Rehabilitation*, 91(6), 862–867.

Vroman, K., and MacRae, N. (2001). Non-work factors associated with musculoskeletal upper extremity disorders in women: Beyond the work environment. *Work*, 17, 3–9.

White, J. W., Hayes, M. G., Jamieson, G. G., and Pilowsky, I. (2003). A search for the pathophysiology of the nonspecific "occupational overuse syndrome" in musicians. *Hand Clinics*, 19(2), 331–341.

World Health Organization. (2006). Gender equality: Work and health. Geneva: World Health Organization. http://www.who.int/gender/documents/Genderworkhealth.pdf.

Yip, V. W. Y., and Hung, L. K. (2002). The musculoskeletal discomforts among Hong Kong housewives. *Medical Section*, 7(2), 3–4.

Yun, M. H., Lee, Y. G., Eoh, H. J., and Lim, S. H. (2001). Results of a survey on the awareness and severity assessment of upper-limb work-related musculoskeletal disorders among female bank tellers in Korea. *International Journal of Industrial Ergonomics*, 27(5), 347–357.

Index